Dimethyl Sulfoxide (DMSO) in Trauma and Disease

Dimethyl Sulfoxide (DMSO) in Trauma and Disease

STANLEY W. JACOB
JACK C. DE LA TORRE

CRC Press
Taylor & Francis Group
Boca Raton London New York

CRC Press is an imprint of the
Taylor & Francis Group, an **informa** business

CRC Press
Taylor & Francis Group
6000 Broken Sound Parkway NW, Suite 300
Boca Raton, FL 33487-2742

First issued in paperback 2017

© 2015 by Taylor & Francis Group, LLC
CRC Press is an imprint of Taylor & Francis Group, an Informa business

No claim to original U.S. Government works

ISBN-13: 978-1-4987-1467-9 (hbk)
ISBN-13: 978-1-138-89462-4 (pbk)

Visit the Taylor & Francis Web site at
http://www.taylorandfrancis.com

and the CRC Press Web site at
http://www.crcpress.com

Contents

Foreword

I have personally known Dr. Stanley W. Jacob and Dr. Jack C. de la Torre over a period of many years. My first encounter with Dr. Jacob occurred when I was a surgical intern at the Boston City Hospital, Boston, Massachusetts. It was my first month as an intern working in the emergency room when Dr. Jacob seemingly appeared out of nowhere. I was treating a patient, and he began to drill me with questions as to what I was doing and for what reasons. At the time, I believed he was a presumptuous medical resident from another service trying to show off his knowledge at my expense and so I told him to "shove off." Dr. Jacob, in his neatly starched white lab coat, looked at me and softly asked, "Do you know who I am?" and then proceeded to inform me that he was the chief resident in the Harvard Surgical Service, a position revered by all who were training in surgery in Boston. I immediately apologized and told him anything he could teach me would be most welcome. Dr. Jacob and I have shared a good laugh over the years recalling this amusing case of mistaken identity.

I was introduced to dimethylsulfoxide, also known as DMSO, by Dr. de la Torre many years ago, who told me of the unusual properties of this drug in central nervous system injuries. I have since used DMSO successfully in my research on spinal cord trauma.

Drs. Jacob and de la Torre have written a book that encompasses the extensive and unique biological actions of DMSO, a drug that was first synthesized in 1866, when it became an important and ubiquitous solvent in most chemistry laboratories. Although the multiple uses of DMSO have been explored and reported over the past half century, this book explains and discusses the diverse areas of research and therapy that have been generated directly and indirectly by the use of DMSO. The book is filled with insights on the physiological and biochemical actions of DMSO in mammalian tissue and presents a highly accurate account of these actions using evidence-based research derived from thousands of studies performed by scientists around the world. One must wonder if many in the medical community are aware of DMSO's potential as a treatment for a number of medical problems.

The clinical possibilities of DMSO were revealed in 1964, when a chemist observed to Dr. Jacob that after DMSO had touched his skin, he rapidly felt a garlic-like taste on his tongue. Although many individuals over the years may have had the same experience when handling DMSO, it was Dr. Jacob who realized that after contact with the chemist's skin, the DMSO must have penetrated his dermis and entered his bloodstream with rapid transportation to his tongue. This observation led to years of study of the cell biology of DMSO and its potential for clinical uses, which has resulted in the publication of thousands of scientific and medical papers.

Dr. de la Torre's reporting on the basic and clinical uses of DMSO for traumatic brain injuries, stroke, and dementia is compelling, especially regarding the favorable effects of the drug for decreasing intracranial pressure, increasing cerebral blood flow, and stabilizing respiration.

The chapters of the book are organized to introduce the reader to the structure and chemical actions of DMSO followed by its basic and clinical pharmacology and medical actions on pain, inflammatory disorders, infection, DNA damage, cancer, stroke, brain trauma, and spinal cord injury. A substantial number of references lend support to the evidence presented.

Drs. Jacob and de la Torre are renowned investigators in their respective fields, surgery and neuroscience, and have devoted large portions of their academic lives to the study of DMSO. These authors have produced the first evidence-based collection of data on this unusual molecule. It is reasonable to expect that this book will become a standard reference on DMSO and that readers will refer to it in the future for research and treatment of a host of clinical conditions that may be improved with the use of this drug. Although the book is aimed at professionals in the health-care system, it should appeal to lay readers who have used DMSO as an off-label medication.

Harry S. Goldsmith, MD, FACS
Clinical Professor of Neurological Surgery
University of California, Davis
Davis, California

Authors

Dr. Stanley W. Jacob, MD, is the discoverer of the medicinal properties of dimethylsulfoxide (DMSO). This discovery has led to more than 20,000 publications in such areas as pain, inflammation, scleroderma, interstitial cystitis, arthritis, resistant tuberculosis, cancer, cryobiology, free radicals, stroke, and neuroprotection. Dr. Jacob received his MD from Ohio State University and completed his surgical training at Beth Israel Hospital in Boston. He was chief resident in surgery at Harvard Surgical Services and instructor in surgery at Harvard School of Medicine before he was appointed to the faculty at Oregon Health Sciences University, where he became director of the Organ Transplant Program and later rose to the rank of Gerlinger professor of surgery. Dr. Jacob received many honors during his career, mainly for his research on DMSO, including the Kemper Foundation Research Scholar, Markle Scholar in Medical Sciences, Governor's Award for Outstanding Northwest Scientist, and Humanitarian of the Year from the National Health Foundation.

Dr. Jacob has chaired or cochaired 16 international research conferences on DMSO held throughout the world. He is the author of a dozen medical textbooks including the classics *Elements of Anatomy and Physiology* (WB Saunders, 1989) and *Structure and Function in Man* (WB Saunders, 1978). His publication list includes more than 170 peer-reviewed articles, dozens of book chapters, and entries in the *Encyclopaedia Britannica*. Dr. Jacob is presently professor emeritus of surgery at the Oregon Health Sciences University. Dr. Stanley Jacob passed away on January 17, 2015. His legacy lives on.

Jack C. de la Torre, MD, PhD, began his work on dimethylsulfoxide (DMSO) in the early 1970s shortly after his appointment as assistant professor in neurosurgery at the University of Chicago. Dr. de la Torre's research revealed DMSO's ability to quickly reduce intracranial pressure, restore cerebral blood flow, and stabilize respiration in nonhuman primates sustaining lethal traumatic brain injuries. For the next eight years, he extended his brain trauma findings and showed that intravenous administration of DMSO was effective in treating primary and secondary damage following experimental ischemic stroke and spinal cord trauma. These neuroprotective properties of DMSO have been broadly confirmed by other investigators in the last three decades.

A host of other ailments amenable to DMSO therapy were reported at the international conference "Biological Actions and Medical Applications of Dimethyl Sulfoxide" in 1983, chaired by Dr. de la Torre and sponsored by the New York Academy of Sciences.

Besides his first faculty appointment at the University of Chicago, Dr. de la Torre has held professorial appointments in neurosurgery and neuroscience at Northwestern University in Chicago, the University of Ottawa, Canada, and Case Western University in Cleveland. He was visiting professor at the University of California, San Diego, University of Valencia, Spain, and the Laboratoire de Physiopathologie Cerebrovasculaire, Paris, France. Dr. de la Torre has published more than 180 peer-reviewed research articles and has written or edited a dozen medical texts on pathology, neurotransmitters, and Alzheimer's disease. Dr. de la Torre's training in neuroanatomy and neuropathology has led him to specialize in disorders of the brain, including stroke, cerebral trauma, and dementia. He continues his research as an adjunct professor of psychology at The University of Texas in Austin.

Introduction

First isolated as a peculiar chemical compound by the Russian chemist Alexander Saytzeff in 1866, dimethylsulfoxide (DMSO) first proved to be a near-perfect solvent for decades until its other remarkable attributes became known to researchers and to millions of people around the planet.

DMSO is one of the most remarkable agents ever to come out of the world of drug development. Its wide range of biological actions involving plants, animals, and humans has led to tens of thousands of articles written in the scientific literature from all parts of the planet since its medicinal discovery in 1964 by Jacob and Hershler.[1] These investigators first pointed out DMSO's clinical applications and, in so doing, opened a Pandora's box of medical and scientific research into the alleviation of disease, pain, and trauma. The earliest and most impressive observation made by Jacob and Hershler[1] was the ability of DMSO to penetrate the skin without damaging it.

A few drops of DMSO spilled on the skin would quickly cause a distinctive garlic taste on the tongue, indicating its immediate transcutaneous penetration into the circulation. This fact alone placed DMSO in the exclusive club where small molecules (under 500 Da) can cross the integument and be considered in the possible realm of topical dermatotherapy. In later years, DMSO was tested intravenously and was shown to exert rapid and safe activity as a neuroprotective agent against various types of trauma in animals and humans.

With a team of collaborators, Jacob's initial article and others that followed described the effects of DMSO on pain, anti-inflammation, vasodilation, reduction of scar tissue, and antimicrobial activity, including its killing action against tuberculosis-resistant bacilli.

These findings have led to a steady cascade of scientific papers on both *in vitro* and *in vivo* systems that continue to the present time. Also, several dozen international conferences have been held where investigators have reported the properties of DMSO on the biochemical, molecular, physiological, pharmacological, and medical systems linked to trauma and disease.

This book examines the major clinical uses of DMSO in humans as supported by basic evidence derived from experiments in animals, including its effects as a powerful pain reliever, and in disorders such as osteoarthritis, interstitial cystitis, gastrointestinal inflammatory changes, scleroderma, respiratory distress, myasthenia gravis, cardiac disease, traumatic brain injury, stroke, Alzheimer's disease, and spinal cord trauma.

In the case of DMSO, how does one categorize an agent with such a wide range of biological and pharmacological activities? Can one small and simple molecule affect these medical conditions in a beneficial manner? And if so, what mechanisms are at play in these processes?

In order to explore the pharmacotherapeutic actions of DMSO on diverse medical conditions, the book offers a critical analysis of the scientific research that examined DMSO's action as a free radical scavenger; an anti-inflammatory, antipain, anticancer, cell membrane stabilizer; and aneuroprotective agent in preventing and treating trauma and disease.

The ability of DMSO to act as a chemical chaperone and to enhance the penetration of an accompanying compound applied topically on the skin has opened the door to the field of drug delivery by percutaneous transport of selective molecules.

The book explores how the chemical structure of DMSO is able to react and deactivate toxic molecules generated by DNA damage, free radical formation, inflammation, infection, and tissue injury. Many toxic and potentially lethal molecules suppressed by DMSO include chemokines, interleukins, tumor necrosis factor, thromboxane, tissue factor, thrombin, and glutamate excitotoxicity.

Many of these toxic molecules appear as a secondary reaction to traumatic injuries and countless of disorders to overwhelm the clinical picture and disease outcome. For this reason, the biological actions and therapeutic usefulness of DMSO have been barely scratched on the surface despite thousands of articles indicating its clinical potential.

We hope this book will serve as a guide to explain the many facets of how DMSO works and to open new doors in the search for better ways to help the ill and the suffering.

This book will be of interest to health-care workers, clinicians, basic scientists, and health-conscious consumers who wish to know more about the biological diversity and medical applications of DMSO.

REFERENCE

1. Jacob SW, Bischel M, Hershler RJ. Dimethyl sulfoxide (DMSO): A new concept in pharmacotherapy. *Curr Ther Res Clin Exp.* 1964; Feb;6:134–135.

1 Chemistry of DMSO

CHEMICAL STRUCTURE AND PROPERTIES OF DMSO

Dimethyl sulfoxide (DMSO) was first synthesized in 1866, and the finding was reported a year later by the Russian organic chemist Alexander Saytzeff.[1] Other common names of DMSO are methyl sulfoxide and methylsulfinylmethane. A sulfoxide is a chemical compound containing a *sulfinyl functional group* attached to two carbon atoms. The DMSO molecule has a pyramidal structure containing sulfur, oxygen, and carbon atoms. The sulfur atom is in the center, while the two methyl groups, the oxygen atom, and a nonbonding electron pair are located at the corners (Figure 1.1). The sulfur atom serves as the positive dipole and activates the hydrogen atoms of the methyl groups making them weakly acidic and very resistant to removal in free radical reactions and allowing removal only by strongly basic reagents such as sodium hydride.[2]

DMSO is produced as a by-product of wood pulping. Lignin, in the kraft pulping liquor wastes, is a polymer of unknown structure and the principal source of DMSO. DMSO synthesis is derived from dimethyl sulfide, a by-product of the manufacture of paper.

Wood contains about 25%–30% lignin and sulfite waste liquors up to 6%.[3] DMSO is commercially made available in the United States by Gaylord Chemical Corporation, which also makes the highly pure grade of DMSO for drug delivery and health-care applications.

The principal impurity present in DMSO is water, and the purified solvent will readily absorb water vapor if exposed to the atmosphere.

Solvents are classified as either polar or nonpolar, and the most common measure of solvent polarity is the dielectric constant.

The sulfur–oxygen bond in DMSO is quite polar so that the liquid has a high dielectric constant. This polarity and geometry lead to considerable organization in the liquid state in which DMSO molecules are associated into chains by dipole attraction.[3] Electrophiles are atoms, molecules, and ions that behave as electron acceptors. Nucleophilic reagents such as DMSO behave as electron donors. The sulfur center in DMSO is nucleophilic toward soft electrophiles, which are molecules that readily accept electrons during a primary reaction step, while the oxygen is nucleophilic toward hard electrophiles, that is, electrons with a positive charge not easily delocalized.[4]

Liquids with a high dielectric constant such as DMSO and water have a higher solubility of substances when immersed in a medium. For that reason, the high dielectric constant of DMSO, which is higher than most organic liquids, makes it a uniquely universal solvent[2] (Table 1.1).

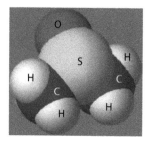

FIGURE 1.1 (**See color insert.**) Structural 3D and 2D formulas of the DMSO molecule. Sulfur (S), oxygen (O), carbon (C), and hydrogen (H).

TABLE 1.1

Physical Properties of DMSO

Formula	M.W.	Boiling Point (°C)	Melting Point (°C)	Density	Solubility H_2O (g/100 g)	Dielectric Constant	Flash Point
C_2H_6OS	78.13	189	18.4	1.092	Miscible	47	95°C

Note: The sulfur–oxygen bond in DMSO is quite polar so that the liquid has a high dielectric constant. This polarity and geometry lead to considerable organization in the liquid state in which DMSO molecules are associated into chains by dipole attraction. See text for details.

  The presence of a liquid structure and association forces is reflected in a rather high freezing point, a low entropy of fusion, a high boiling point, and a high entropy of vaporization[5] (Table 1.1). DMSO is one of the most exceptional and widely used solvents in organic chemistry. It is miscible with water and readily soluble in almost all organic solvents such as alcohols, esters, ketones, chlorinated solvents, and aromatic hydrocarbons. DMSO has been used with thousands of organic compounds to determine the quantitative strength of an acid in solution (pK_a value).[2]

  The versatility of DMSO both as a solvent and as a reagent is partly explained by the dipolar nucleophilic characteristics of the molecule, which are due to the availability of free electron pairs at the oxygen or sulfur terminals. The reason is that these electrons may be shared or transferred to other protons.[6] It has been pointed out by Szmant[6] that although DMSO is not the hydrogen donor that water is, it is superior to water in associations based on the induction of dipoles in aromatic rings and methylmercapto and disulfide bonds. This alteration in the configuration of immobile protein structures induced by DMSO exchanging sites with bound water molecules could explain the penetrant ability of DMSO through the skin of people and animals.[7]

  DMSO is an odorless, colorless, transparent, hygroscopic fluid over a wide range of temperatures with a slightly bitter taste. It has a low level of toxicity, and this issue is discussed in detail in the "Teratology and LD_{50}" section in Chapter 2. DMSO evaporates slowly at normal atmospheric pressure due to its high boiling point (189°C). At its boiling point, DMSO can decompose and has the risk of explosion.

An explosive reaction can also occur when DMSO is combined with halogen compounds, metal nitrides, metal perchlorates, sodium hydride, periodic acid, and fluorinating agents.[8] When DMSO is mixed with water, considerable heat is released. The heat given off by DMSO mixed in water indicates that a stronger interaction between DMSO and water is taking place than that among DMSO molecules themselves.[9]

Since most chemicals used in drug formulation contain thousands of electrons, the behavior of these electrons in chemical reactions is hard to predict, especially as the chemical becomes more complex, because the process governing the motion of one electron depends on the motion of all the others. By contrast, DMSO is a simple molecule with only 90 electrons orbiting around their nuclei and is made up of 3 carbon, 6 hydrogen, 1 sulfur, and 1 oxygen atoms.

DMSO SOLVATION AND CHEMICAL ACTIVITY

As a dipolar aprotic solvent, DMSO differs from protic solvents, such as water and alcohols, because it tends to accept rather than donate protons.[10] In chemistry, a protic solvent is one that has a hydrogen atom bound to an oxygen (as in a hydroxyl group) or a nitrogen (as in an amine group). The molecules of such solvents can donate a H^+ (proton). Thus, protic solvents are those that are capable of hydrogen bonding, which implies that they must have O–H or N–H bonds in their structure. In most cases, aprotic solvents like DMSO in various liquid media do not act as an acid or a base.[7]

The most common examples of protic solvents would be water, alcohols, and carboxylic acids, which all have O–H bonds in their structures. Aprotic solvents, by contrast, are those that do not donate protons. These solvents, including DMSO, generally have high dielectric constants and high polarity.[11] High polarity refers to the polar nature of a molecule that enables it to have some interesting properties, such as being a universal solvent and possessing a high surface tension. Generally, a low-polarity solvent is said to be nonpolar if its dielectric constant is below 15. Every material has a dielectric constant, including solvents. Generally, the dielectric constant of the solvent provides a rough measure of its polarity. The higher the dielectric constant, the more polar the molecule. Water, the most common solvent known, is highly polar and has a dielectric constant of 80 at 20°C, while DMSO has a dielectric constant of 47 at the same temperature and is considered a dipolar solvent.

Being aprotic, DMSO can tolerate relatively strong bases and can dissolve a wide range of solutes while being miscible with many other solvents, a characteristic property of polar compounds.[12]

DMSO is also an amphiphilic molecule. The property of being amphiphilic refers to a chemical compound that possesses both hydrophilic (*water-loving*) and lipophilic (*fat-loving*) properties. Amphiphilic molecules display a special molecular structure that contains both a water-loving polar part (hydrophilic) and a water-hating nonpolar part (hydrophobic). The high polarity of DMSO allows it to dissolve many compounds that other solvents cannot.[5] This polar property turns out to be very important in determining the special structures formed by amphiphilic molecules.

Amphiphiles are a special class of *surface-active* molecules called surfactants. They are called surface active because they have the unique properties of getting adsorbed at various interfaces, for example, oil–water and air–water, while simultaneously altering the properties of the interface. The hydrophilic part of an amphiphilic molecule is a polar molecule called the head, whereas the hydrophobic part, called the tail, is a hydrocarbon chain. When the fluid is water, the terms hydrophilic and hydrophobic parts can be used, respectively, to describe surfactants. Usually, surfactants are classified according to the chemical structure of their hydrophilic groups. In medicine, amphiphilic molecules such as glycolipids or sphingolipids can be found in a biological membrane having a polar water-soluble terminal group attached to a water-insoluble hydrocarbon chain. In such biological membranes, the polar parts of lipid molecules remain in water, but the hydrocarbon chains are localized in the nonpolar phase.[13]

PROTEIN FOLDING

The potential energy from dipole interactions is important for living organisms. The most important effect of dipole interactions on living organisms is their effect on protein folding. Every biological process involved in protein synthesis, from the binding of individual amino acids to the formation of secondary structures, tertiary structures, and even quaternary structures, is dependent on dipole–dipole interactions. The dipole interaction of DMSO may be the basis of its many biological actions on living organisms. For example, the unique ability of DMSO to penetrate living tissues including the skin without causing significant damage is most likely related to its relatively polar nature, its capacity to accept hydrogen bonds, and its relatively small and compact structure. This combination of properties results in the ability of DMSO to associate with water, proteins, carbohydrates, nucleic acid, ionic substances, and other constituents of living systems.[11]

DMSO is metabolized to dimethyl sulfone, which is excreted in the urine and feces. The other DMSO metabolite, dimethyl sulfide, is eliminated through the skin and via the breath, giving off an odor resembling garlic.[14] Dimethyl sulfide is normally present at very low levels in healthy people at concentrations less than 7 nM in blood, about 3 nM in urine, and 0.13–0.65 nM on expired breath.[15]

PERMEABILITY ENHANCEMENT

The stratum corneum represents the primary barrier to molecules penetrating the skin. One approach to overcoming this barrier for the purpose of delivering active molecules into or via the skin is to employ chemical permeability enhancers, such as DMSO. DMSO can also act as a carrier of other substances, a feature that was initially demonstrated when DMSO was topically given together with compounds such as heparin, insulin, sodium salicylate, Evans Blue dye, sulfadiazine, aminophylline, and thioTEPA, and these substances are found to be carried through the intact urinary bladder of the dog.[16] These findings are consistent with the experimental evidence that high concentrations of DMSO fluidize the stratum corneum lipids to enhance permeability.

DMSO also promotes the lipophilic and hydrophilic penetration of many substances through the skin, including antiviral agents, antibiotics, and steroids. This ability of

DMSO has led to a number of commercial preparations that combine these substances with DMSO for topical application.[17] Skin permeation efficacy of many substances combined with DMSO requires a greater than 60% concentration of DMSO, which can cause wheals and erythema of the stratum corneum.[17] Following DMSO topical application, a mildly bitter taste can be felt on the tongue within seconds, which is quickly followed by an unpleasant odor on the breath due to dimethyl sulfide, a breakdown metabolite of DMSO.[18]

DMSO AS A CHEMICAL CHAPERONE

One interesting structural property of DMSO is its action as a molecular or chemical chaperone.

Chemical chaperones were first defined in the studies of mutant cystic fibrosis transmembrane conductance regulator proteins.[19] Chemical chaperones are small molecules that can nonselectively stabilize mutant proteins and have the ability in some cases to rescue protein-folding mutants within cells to assist in their folding. This feature can be very helpful in the so-called conformational or *folding* diseases as discussed later in the text. Chemical chaperones not only aid in the folding of proteins but also provide a type of *quality control system* where they recognize, retain, and target misfolded proteins for their eventual degradation.[19]

Many human diseases are known to arise as a consequence of specific point mutations or deletions within genes that encode essential proteins. In many cases, these mutations and deletions do not totally destroy the biological activity of the specific gene but instead result in only subtle folding abnormalities that can lead to the newly synthesized protein being retained at the endoplasmic reticulum of the cell by the actions of the cellular quality control system.[19]

For example, DMSO acting as a chemical chaperone was used in endoplasmic reticulum canine kidney cells expressing mutants involved in nephrogenic diabetes insipidus and showed improved maturation and plasma membrane rescue.[20] The protective action by DMSO on this model was believed to be due to an increase in the relative hydration that formed around a polypeptide, thereby inducing a tighter packing of the protein and a stabilization of the protein's conformation.[20]

Proteins are the biological workhorses that carry out the vital functions of each cell. Proteins exist as unfolded polypeptides or amino acid chains and derive from a sequence of messenger ribonucleic acid (mRNA) to a linear chain of amino acids. These amino acids interact with each other and, after folding, will create a 3D structure with specific and essential biological functions. If the polypeptide does not fold correctly, it will not function properly and may damage the cell.

The problem with proteins is that, in some circumstances, they can become *misfolded*, that is, folded incompletely or incorrectly. This misfolding can lead to serious illness because their correct molecular structure is degraded, denatured, or lost, and their specific function in the body no longer works. It has been estimated that about half of all human diseases are caused by protein folding defects, making this phenomenon a clearly important medical issue.[21]

Misfolding disorders can result in neurodegenerative disorders. These disorders include Gaucher's disease (misfolded beta glucocerebrosidase), Marfan syndrome (misfolded fibrillin), Fabry's disease (misfolded alpha galactosidase), and retinitis pigmentosa 3 (misfolded rhodopsin). Moreover, some cancers may be associated with misfolded proteins that become dysfunctional. One example is von Hippel Lindau disease, a dominantly inherited hereditary cancer syndrome that leads to a variety of malignant and benign tumors of the brain, spinal cord, eye, kidney, pancreas, and adrenal glands and is caused by a gene mutation that encodes the tumor suppressor protein known as pVHL.

Protein misfolding disorders are not necessarily characterized by the disappearance of a protein but by its deposition in insoluble aggregates within the cell. Diseases caused by protein aggregation include Alzheimer's disease, where excessive deposits of amyloid beta and tau are found in the brain; diabetes type 2, where deposits of amylin accumulate; Parkinson's disease, where brain deposits of α-synuclein are characteristic; and the spongiform encephalopathies such as Creutzfeldt–Jakob disease, described later. Hereditary transthyretin amyloidosis is caused by the aggregation of transthyretin in various tissues, such as the heart where congestive heart failure can follow or the nerves where peripheral neuropathy often occurs.

One of the most crucial biological activities of cells is to apply the correct folding of a newly produced polypeptide in the lumen of the endoplasmic reticulum of the cell into a 3D conformation or complex. Doing so ensures the stability, intracellular trafficking, and targeting of the protein to its final destination and initiation of its specific function. Protein molecules in aqueous buffer are in equilibrium between unfolded and folded states,[22] and when a protective osmolyte is added to the buffer, such as DMSO, there is a shift in this equilibrium toward the correct conformational shape of the protein. However, not all osmolytes are helpful in protein folding, and the most notorious in denaturing proteins is urea, a very simple molecule that is a waste product of nitrogen metabolism found normally in all our cells.

Proteins that have a particularly complicated or unstable conformation sometimes cannot achieve their properly folded or assembled form. In these cases, other specialized chemical chaperones can help them return from their misfolded state to their correct folded form. It is not clear how chemical chaperones are able to nonselectively stabilize the mutant proteins and facilitate their proper folding. One theory is that chemical chaperones may facilitate the proper folding of a polypeptide essentially by unfolding improperly folded structures prior to their final assembly.

Enzymes known as proteases can degrade those proteins that have misfolded incorrectly by cleaving the peptide bonds via a hydrolysis reaction. The process of protein degradation involves forming a complex with the protease, which after cleavage allows the released amino acids to be reused by the cell.

Chemical chaperones are usually osmotically active and include, besides DMSO, glycerol and deuterated water.[23] These chaperones are presently being investigated in conformational disorders where mutations involving improper protein folding occur.

Chemical chaperone research has largely focused on the central nervous system because of the sensitivity of this system to folding diseases. This sensitivity to protein misfolding is due to the inability of the central nervous system to regenerate itself following neuronal death.

Chemical chaperones are usually osmotically active, and this is the case with DMSO, glycerol, and deuterated water.[23] Osmotically active compounds are known to be potentially useful in treating or preventing conformational disorders where protein folding is misguided.[24]

For example, Creutzfeldt–Jakob disease is a form of brain damage that can lead to the rapid development of increased involuntary movements and progressive dementia. The brain tissue develops holes much like a sponge, hence the name spongiform encephalopathy, which, like its medical name, takes on a sponge-like aspect.[24] This encephalopathic outcome is due to an infectious protein called prion. Prions are misfolded proteins that replicate by converting their properly folded counterparts into acting as a template to guide the chronic misfolding of more proteins into a prion form.[25] It is theorized that a conformational change from the normal cellular prion protein (PrP^C) is the decisive step in the formation of the abnormal scrapie PrP^{Sc} isoform.[26]

The prion found in Creutzfeldt–Jakob disease can become quickly lethal because it induces refolding of native proteins into the diseased state.[24] DMSO and glycerol have been found to revert the mutated form of the scrapie form prion (PrP^{Sc}), the protein responsible for the onset of prion disease,[27] raising the hope of curing or managing Creutzfeldt–Jakob disease and other protein misfolding disorders.

Although the mechanism by which PrP^C is converted into PrP^{Sc} remains obscure, DMSO was used as a chemical chaperone and a protein-stabilizing solvent to explore the possibility of retarding the formation of PrP^{Sc} in scrapie-infected mouse neuroblastoma cells.[28] It was found that DMSO proved to be effective in reducing the extent of PrP^C conversion into its detergent insoluble form in ScN2a cells, a model of scrapie-infected neuroblastoma cells.[28]

Confirmation of these findings were reported by Shaked and his group.[29] They showed that DMSO acted as a chemical chaperone to partially inhibit the aggregation of either PrP^{Sc} or that of its protease-resistant core, designated PrP 27–30, the major scrapie prion protein.

From this perspective, DMSO might be useful in conformational disorders where protein misfolding may play an important role in the pathologic process. Such conformational protein disorders, as mentioned earlier, ostensibly include neurodegenerative disorders such as Alzheimer's disease and Parkinson's disease, which share some features with the prion diseases.

DMSO has been used as a chemical chaperone in the autosomal neurodegenerative disorder Machado–Joseph disease, a type of spinocerebellar ataxia affecting muscle control and coordination of the upper and lower extremities.[30] The disease is caused by a gene that contains lengthy irregular repetitions of the code *CAG*, producing the truncated form of a mutated protein called ataxin-3.

Ataxin-3 causes aggregation and cell death *in vitro* and *in vivo*. In one *in vitro* study by Yoshida et al.,[31] large intracellular aggregate bodies were remarkably decreased when exposed to 0.5% of DMSO. At this concentration, DMSO also significantly suppressed neuron and non neuronal cell death frequency.[31]

These studies suggest that the use of DMSO in conformational disorders is an area of molecular research that could provide fundamental understanding of how and why proteins misfold and clues as to their eventual treatment.

Due to the pleiotropic nature of DMSO, it is reasonable to assume that the structural configuration of this molecule may have the ability to target a diverse spectrum of pathogenic events that emerge from trauma and disease either acutely or chronically. This brings up the tantalizing possibility that further research into the structure of DMSO may uncover other more powerful and efficient molecules with similar molecular and chemical properties as DMSO.

DMSO AS AN ELECTROLYTE

One of the more curious chemical uses found for DMSO in the last few years is as an electrolyte in rechargeable batteries. Rechargeable batteries are widely used in modern society for portable consumer devices, automobile starters, electric tools, light vehicles such as electric bicycles, motorized wheelchairs, golf carts, standby generators, and large-scale electricity storage in smart or intelligent grids. The performance of rechargeable batteries depends essentially on the thermodynamics and kinetics of the electrochemical reactions involved in the components. The components are made up of an anode, cathode, electrolyte, and separator of the cells. Extensive efforts have been dedicated in the last decade to develop advanced batteries with large capacity, high energy and power density, high safety, long cycle life, fast response, and low cost.[32]

Prototype lithium–oxygen (Li^+–O_2) batteries typically are made of a metal Li^+ anode, a porous O_2^- diffusion cathode, and an organic electrolyte. DMSO has been proposed as an electrolyte for Li^+–O_2 batteries.[33] Operation depends on the reduction of O_2 at the cathode to O_2^{2-}, which combines with Li^+ from the electrolyte to form Li_2O_2 on discharge. The reverse reaction occurs when the battery is charging.

An ideal electrolyte is considered to be one with low volatility and high oxygen solubility and to be inert to superoxide radicals. It was reported that an electrochemical reaction in Li_2O_2 batteries using a DMSO-based electrolyte was obtained and showed resistance of the battery after charge recovery was deemed very good when compared with that produced by fresh batteries.[34] The authors of this study concluded that a superior good rate performance, reflected by a high discharge capacity and low charge potential with an optimized cathode, was observed when DMSO was used in rechargeable Li_2O_2 batteries as compared with other electrolytes.

These findings may open new and exciting possibilities to promote the development of rechargeable Li–O_2 batteries, especially those used to power hybrid and electric automobiles. However, Lithium-ion batteries for electric automobiles may not be practical due to the inadequate charging energy that is generated. Nevertheless, this shortcoming may be bridged by using rechargeable, nonaqueous Li–air batteries that use carbon, which is lighter and cheaper and reacts with oxygen from ambient air to produce an electric current. In fact, a lithium–air battery can safely hold 10 times more energy than the best lithium-ion batteries in the market today. New research is now being carried out where the carbon-based material used for the cathode is replaced with another material, containing inert gold nanoparticles, and with DMSO as the electrolyte that replaces the polycarbonates or polyethers that once served as the electrolytes. DMSO is preferred for this reaction because it is less prone to react with the cathode, is 10 times faster than previous Li–air batteries containing carbon cathodes, and expends much less power loss when the battery is charging or discharging.[34]

Due to its high dissolution properties and permeability, DMSO has an extensive application as an electronic component cleaner, as a mold release solvent for motor parts molding, and as a plant cleaning agent.

DMSO can ligate metal atoms by bonding in one of two possible ways: through the oxygen atom or through the sulfur atom.

CRYOPROTECTION

Preservation of biological materials for an extended period of time can be achieved by the process known as cryopreservation. This technique involves freezing human or animal tissues to subzero temperatures. This frozen state effectively stops chemical reactions from occurring and causing any damage to the materials, thus allowing them to remain viable for medical use later on. The two main advantages of tissue cryoprotection are storing and transporting the frozen material to when and where it is needed.

One problem involved in cryoprotection is that the material to be frozen can contain a considerable amount of water, which can create ice crystals when frozen causing lethal damage to the tissue.

Because unprotected freezing of tissue is lethal, cryopreservants are used at very low temperatures to preserve structurally intact living cells and tissues. Cryoprotectants can reduce the amount of ice formed at any given temperature, but to be biologically acceptable, they must be able to penetrate into the cells and have low toxicity. Many compounds have such properties, including DMSO, glycerol, ethanediol, propylene glycol, and formamide, but often require high concentrations of the cryoprotective agent. DMSO is often the preferred cryoprotectant because it has been shown to cause the least harm to the tissue being frozen.[35]

DMSO is a commonly used cryoprotectant also due to its high water solubility, rapid penetrability, and diminished osmotic damage at high concentrations normally used in cryopreservation protocols.[35]

Preservation of cells and tissues at low temperatures requires the presence of a cryoprotectant that displays these qualities and display a low toxicity to the cell plasma membrane. DMSO possesses all three qualities.[36] The unusual ability of DMSO to dramatically depress the freezing point of water while maintaining viability with many cell and tissue types has led to the use of this agent in cryopreservation media for stem cells derived from human umbilical cord blood and reproductive tissue.

The role of hematopoietic stem cell transplants in the treatment of hematologic and nonhematologic malignancies is now widely used clinically. Hemopoietic stem cell preservation and transplantation is an important part of the treatment of cancer patients who have undergone high-dose chemotherapy and irradiation for cancers such as lymphoma and leukemia. DMSO is a common cryopreservative for these cells, which are derived from bone marrow, peripheral circulatory system, or umbilical cord blood.[37] DMSO is chosen as the standard cryoprotectant because it best prevents freezing damage to living cells.

Many reports have appeared in the literature concerning adverse reactions in DMSO-preserved stem cells following transplantation in recipients. These reactions include nausea, vomiting, abdominal cramps, chills, cardiac arrhythmias, neurological symptoms, and respiratory arrest.[38,39] These adverse reactions by stem cell transplants preserved in DMSO may be caused by the amount of infusate used, amount of cells being infused, and the rate of infusion.[36]

These adverse reactions are transient for the most part and can be minimized when the infusate is optimized to the medical condition of the individual patient.[40] Also, when DMSO is used with normal saline and albumin at a concentration of 10%, the incidence of these toxic reactions is diminished.[39] Severe adverse reactions after stem cell transplants are infrequent although, when they occur, patients may require clinical treatment. One trick that has shown antiemetic properties following stem cell transplants using DMSO as cryoprotectant is the use of orange-flavored ice lollies. The relief of nausea and vomiting may be due to the vasoconstrictive action of ice on the stomach and the masking of DMSO's garlic odor by the orange flavor of the ice.[41]

A robust and slow-freezing cocktail containing DMSO and hydroxyethyl starch in saline was recently developed to freeze human pluripotent stem cells with a high recovery rate (80%) of the cells after thawing.[42] The method has not been tested clinically on patients yet but promises to provide storage of human pluripotent stem cells easily, effectively, and economically in the future.

The potential of gonadal cryobiology in the conservation of fertility is now fully realized. Freezing semen to safeguard male fertility is an option available to men undergoing sterilizing chemotherapy.

DMSO is the cryoprotectant most commonly used in many hospitals as a standard procedure for autologous peripheral blood progenitor cell transplantation for the treatment of chemosensitive malignancies.

DMSO IN THE SEA AND THE ATMOSPHERE

A salty sea smell can usually be felt when one is near the ocean but the smell is not from the salt alone. The smell comes from wind-driven white caps and breaking waves, which is composed not only of salt but also of gases that diffuse across the air–sea interface, many of which are synthesized and emitted by microalgae and degraded by bacteria and phytoplankton.[43] One of these gases found to occur in ocean surface waters is dimethyl sulfide, a DMSO metabolite with a sulfur-based compound that has a strong characteristic odor. While dimethyl sulfide may sound like a noxious pollutant, it is a naturally produced biogenic gas essential for the earth's biogeochemical cycles.

Sulfur is also an element of the earth and is essential to life. It is the eighth most prevalent element in the human body, not as sulfur itself but always in combination with other elements, most often, in complex molecules and, less often, in simple molecules such as DMSO.

DMSO can be found in the atmosphere.[44] Water vapor is transferred to the atmosphere as a gaseous dimethyl sulfide, which is then oxidized to tropospheric sulfate aerosols, and these particle aerosols condense on cloud particles, forming the water droplets that make up clouds.[45] Clouds as well as marine life can affect the earth's radiation balance and markedly influence its temperature and climate.

Recent findings show that the amount of dimethyl sulfide entering the atmosphere could decrease in the future due to atmospheric oxidation of this DMSO metabolite by greenhouse gases.[46,47] Such oxidation of dimethyl sulfide can lead to climate changes by increasing cloud albedo, a condition where clouds reflect more solar radiation contributing to the greenhouse effect. The greenhouse effect is a process by which thermal radiation from the planet's surface is absorbed by greenhouse gases, such as CO_2 in the atmosphere, and reradiated in all directions, including back to the earth's surface where the average temperature is increased.[47] As CO_2 levels rise in the atmosphere due to human activities, the ocean pH is acidified, an effect that leads to lower plankton production of dimethyl sulfide concentrations, with the consequence of less dimethyl sulfide in the atmosphere and more global warming.[48]

Dimethyl sulfide makes up 95% of the natural marine flux of sulfur gases to the atmosphere, and scientists estimate that the flux of marine dimethyl sulfide supplies about 50% of the global biogenic source of sulfur to the atmosphere.[46] Dimethyl sulfide is oxidized in the marine atmosphere to other sulfur-containing compounds, such as DMSO and its second metabolite dimethyl sulfone, as well as sulfur dioxide, methanesulfonic acid, and sulfuric acid.[49]

There is evidence that DMSO and its second metabolite dimethyl sulfone may be as significantly present as dimethyl sulfide in marine rain and marine air masses. Measures in central equatorial Pacific rain were reported to contain 1–10 µg/L of both DMSO and dimethyl sulfone, which are near the concentrations found for dimethyl sulfide in marine sulfur transport.[48,49] This observation suggests that a partially reversible loop in the sulfur transport cycle of marine rain and air gases may be operating between DMSO and its metabolites and could play a critical role in climate, temperature, and marine life.

REFERENCES

1. Saytzeff A. Ueber die Einwirkung von Saltpetersäure auf Schwefelmethyl und Schwefeläthyl (On the effect of nitric acid on methyl sulfide and ethyl sulfide). *Ann Chem Pharm.* 1867;144:148–156.
2. MacGregor WS. The chemical and physical properties of DMSO. *Ann NY Acad Sci.* 1967;141(1):3–12.
3. Stecker PG. (ed.). *Lignin*, Vol. 8. Merck Index, Rahway, NJ, 1968; p. 619.
4. Epstein WW, Sweat FW. Dimethyl sulfoxide oxidations. *Chem Rev.* 1967;67:247–260.
5. Yu ZW, Quinn PJ. Dimethyl sulphoxide: A review of its applications in cell biology. *Biosci Rep.* 1994;14(6):259–281.
6. Szmant HH. Chemistry of DMSO. In *Dimethyl Sulfoxide, Basic Concepts*, eds. S.W. Jacob, E.E. Rosenbaum, D.C. Wood. Marcel Dekker, Inc., New York, 1971; pp. 1–94.
7. David NA. The pharmacology of dimethyl sulfoxide. *Annu Rev Pharmacol.* 1972;12:353–374.
8. *Römpp Chemie-Lexikon*, Bde, Thieme, Stuttgard, 2006; p. 114.
9. Martin D, Hauthal G. *Dimethyl Sulfoxide* (Trans by Halberstadt ES). Van Nostrand Reinhold, Akademie-Verlag, Berlin, 1971.

10. Kharasch N, Thyagarajan BS. Structural basis for biological activities of dimethyl sulfoxide. *Ann NY Acad Sci.* 1983;411:391–402.

11. Szmant HH. Physical properties of dimethyl sulfoxide and its function in biological systems. *Ann NY Acad Sci.* 1975;243:20–23.

12. Windholz M. (ed.). *The Merck Index*, 10th edn. Merck, Rahway, NJ, 1983.

13. Przestalski S, Sarapuk J, Kleszczyńska H, Gabrielska J, Hladyszowski J, Trela Z, Kuczera J. Influence of amphiphilic compounds on membranes. *Acta Biochim Pol.* 2000;47(3):627–638.

14. Jacob SW, Rosenbaum EE, Wood DC. *Dimethyl Sulfoxide*, Vol. 1. Marcel Dekker, New York, 1971.

15. Tangerman, A. Measurement and biological significance of the volatile sulfur compounds hydrogen sulfide, methanethiol and dimethyl sulfide in various biological matrices. *J Chromatogr B.* 2009;877(28):3366–3377.

16. Jacob SW, Bischel M, Herschler RJ. Dimethyl sulfoxide (DMSO): A new concept in pharmacotherapy. *Curr Ther Res Clin Exp.* 1964;6:134–135.

17. Williams AC, Barry BW. Penetration enhancers. *Adv Drug Deliv Rev.* 2004;56(5):603–618.

18. Jacob SW, Wood DC. Dimethyl sulfoxide (DMSO). Toxicology, pharmacology, and clinical experience. *Am J Surg.* 1967;114(3):414–426.

19. Welch WJ, Brown CR. Influence of molecular and chemical chaperones on protein folding. *Cell Stress Chaperones.* June 1996;1(2):109–115 (Review: Erratum in: *Cell Stress Chaperones.* September 1996;1(3):207.)

20. Robben JH, Sze M, Knoers NV, Deen PM. Rescue of vasopressin V2 receptor mutants by chemical chaperones: Specificity and mechanism. *Mol Biol Cell.* January 2006;17(1):379–386.

21. Burg MB, Ferraris JD. Intracellular organic osmolytes: Function and regulation. *J Biol Chem.* 2008;283(12):7309–7313.

22. Street TO, Bolen DW, Rose GD. A molecular mechanism for osmolyte-induced protein stability. *Proc Natl Acad Sci USA.* 2006;103(38):13997–14002.

23. Papp E, Csermely P. Chemical chaperones: Mechanisms of action and potential use. *Handb Exp Pharmacol.* 2006;172:405–416.

24. Mackay GA, Knight RSG, Ironside JW. The molecular epidemiology of variant CJD. *Int J Mol Epidemiol Genet.* 2011;2(3):217–227.

25. Aguzzi A. Unraveling prion strains with cell biology and organic chemistry. *Proc Natl Acad Sci.* 2008;105(1):11–12.

26. Stahl N, Baldwin MA, Teplow DB, Hood L, Gibson BW, Burlingame AL, Prusiner SB. Structural analysis of the scrapie prion protein using mass spectrometry and amino acid sequencing. *Biochemistry.* 1993;32:1991–2002.

27. Gu Y, Singh N. Doxycycline and protein folding agents rescue the abnormal phenotype of familial CJD H187R in a cell model. *Brain Res Mol Brain Res.* 2004;123:37–44.

28. Tatzelt J, Prusiner SB, Welch WJ. Chemical chaperones interfere with the formation of scrapie prion protein. *EMBO J.* 1996;15(23):6363–6373.

29. Shaked GM, Fridlander G, Meiner Z, Taraboulos A, Gabizon R. Protease-resistant and detergent-insoluble prion protein is not necessarily associated with prion infectivity. *J Biol Chem.* 1999;274(25):17981–17986.

30. Nakano KK, Dawson DM, Spence A. Machado disease. A hereditary ataxia in Portuguese emigrants to Massachusetts. *Neurology.* 1972;22(1):49–55.

31. Yoshida H, Yoshizawa T, Shibasaki F, Shoji S, Kanazawa I. Chemical chaperones reduce aggregate formation and cell death caused by the truncated Machado-Joseph disease gene product with an expanded polyglutamine stretch. *Neurobiol Dis.* 2002;10(2):88–99.

32. Cheng F, Liang J, Tao Z, Chen J. Functional materials for rechargeable batteries. *Adv Mater.* 2011;23(15):1695–1715.

33. Xu D, Wang ZL, Xu JJ, Zhang LL, Zhang XB. Novel DMSO-based electrolyte for high performance rechargeable Li-O$_2$ batteries. *Chem Commun (Camb).* 2012;48(55):6948–6950.
34. Peng Z, Freunberger SA, Chen Y, Bruce PG. A reversible and higher-rate Li-O$_2$ battery. *Science.* 2012;337(6094):563–566.
35. McGann LE, Walterson ML. Cryoprotection by dimethyl sulfoxide and dimethyl sulfone. *Cryobiology.* 1987;24(1):11–16.
36. Cox MA, Kastrup J, Hrubiško M. Historical perspectives and the future of adverse reactions associated with haemopoietic stem cells cryopreserved with dimethyl sulfoxide. *Cell Tissue Bank.* 2012;13(2):203–215.
37. Benekli M, Anderson B, Wentling D, Bernstein S, Czuczman M, McCarthy P. Severe respiratory depression after dimethyl sulphoxide-containing autologous stem cell infusion in a patient with AL amyloidosis. *Bone Marrow Transplant.* 2000;25(12):1299–1301.
38. Petropoulou AD, Bellochine R, Norol F, Marie JP, Rio B. Coronary artery spasm after infusion of cryopreserved cord blood cells. *Bone Marrow Transplant.* 2007;40(4):397–398.
39. Martino M, Morabito F, Messina G, Irrera G, Pucci G, Iacopino P. Fractionated infusions of cryopreserved stem cells may prevent DMSO-induced major cardiac complications in graft recipients. *Haematologica.* 1996;81(1):59–61.
40. Fahmy MD, Almansoori KA, Laouar L, Prasad V, McGann LE, Elliott JA, Jomha NM. Dose-injury relationships for cryoprotective agent injury to human chondrocytes. *Cryobiology.* 2014;68(1):50–56.
41. Gonella S, Berchialla P, Bruno B, Di Giulio P. Are orange lollies effective in preventing nausea and vomiting related to dimethyl sulfoxide? A multicenter randomized trial. *Support Care Canc.* 2014;22:2417–2424.
42. Imaizumi K, Nishishita N, Muramatsu M, Yamamoto T, Takenaka C, Kawamata S, Kobayashi K, Nishikawa S, Akuta T. A simple and highly effective method for slow-freezing human pluripotent stem cells using dimethyl sulfoxide, hydroxyethyl starch and ethylene glycol. *PLOS ONE.* 2014;9(2):e88696.
43. Simpson D, Winiwarter W, Börjesson G et al. Inventorying emissions from nature in Europe. *J Geophys Res.* 1999;104:8113–8152.
44. Berresheim H, Tanner DJ, Eisele FL. Real-time measurement of dimethyl sulfoxide in ambient air. *Anal Chem.* 1993;65(1):84–86.
45. Avgoustidi V, Nightingale P, Joint I, Steinke M, Turner S, Hopkins F, Liss PS. Decreased marine dimethyl sulfide production under elevated CO$_2$ levels in mesocosm and in vitro studies. *Environ Chem.* 2012;9:399–404.
46. Bates TS, Lamb BK, Guenther A, Dignon J, Stoiber RE. Sulfur emissions to the atmosphere from natural sources. *J Atoms Chem.* 1992;14:315–337.
47. Schneider SH. The greenhouse effect: Science and policy. *Science.* 1989;243:771–781.
48. Lucas DD, Prinn RG. Parametric sensitivity and uncertainty analysis of dimethyl-sulfide oxidation in the clear-sky remote marine boundary layer. *Atmos Chem Phys.* 2005;5(6):1505–1525.
49. Harvey GR. Dimethylsulfoxide and dimethylsulfone in the marine atmosphere. *Geophys Res Lett.* 1986;13:49–53.

2 DMSO in Basic Pharmacology

ABSORPTION, FATE, AND EXCRETION OF DMSO

Pharmacokinetics is the branch of pharmacology that deals with what the body does to administered drugs. The opposite of pharmacokinetics is pharmacodynamics, which deals with what the drug does to the body. Pharmacokinetics of a drug can be examined by administering the drug to animals or humans and obtaining blood, urine, and feces specimens for the quantitative analysis of the drug's adsorption, metabolism, and excretion patterns. The rate of appearance and elimination of the drug in the bloodstream as well as the drug's transformation into other compounds as it passes through the liver and excretion of the drug's metabolites can then be recorded.

The pharmacokinetics of dimethyl sulfoxide (DMSO) following its topical application has been well studied. Findings show that DMSO has a wide distribution in tissue and body fluids. DMSO and its metabolite dimethyl sulfone are excreted in the urine and feces, while its second metabolite, dimethyl sulfide, is eliminated through the breath and skin and is responsible for the characteristic *garlic* odor that is noticed after DMSO application. Dimethyl sulfone can persist in serum 2 weeks after a single intravesical instillation. No residual accumulation of DMSO occurs after its prolonged use.[1]

Human volunteers were given DMSO orally and dermally at a dose of 1 g/kg as a 70% aqueous solution. DMSO was quickly absorbed when administered dermally. Peak serum levels occurred after 4–8 h. Orally administered DMSO was rapidly absorbed, reaching a peak serum level after 4 h. Unchanged DMSO and its metabolite dimethyl sulfone appeared in the serum after about 48 h and persisted for as long as 2 weeks. Urinary excretion of DMSO after dermal and oral administration was approximately 13% and 30%–68% of the dose, respectively. Renal clearance of DMSO was about 14 mL/min.[2]

Topical application of radioactive ^{35}S-DMSO in humans can be observed in blood after 5 min and in bones 1 h later.[3]

Radioactive ^{35}S-DMSO was orally administered to rats. Animals were killed at specific times, and the tissues were assayed for total radioactivity.[4] All tissues showed radioactivity after 30 min. Plasma, kidney, spleen, lung, heart, and testes seemed to have higher levels than liver, fat, small intestine, brain, skeletal muscle, and red cells. Concentrations in the testes, skeletal muscle, heart, and brain increased after 30 min but remained virtually constant in other tissues. Levels of DMSO had declined to minimal values in all tissues after 24 h.[4] The ratio of dimethyl sulfone to DMSO 4 h after oral administration of ^{35}S-DMSO was found to be virtually constant in liver, testes, kidney, spleen, small intestine, heart, and plasma.[4] The lowest levels of ^{35}S-DMSO were found in the lens of the eye. The half-life of ^{35}S-DMSO was prolonged in hard tissues, and virtually no radioactive DMSO is found in any tissue 1 week after the last dose of a single or repeated administration.[4]

Rhesus monkeys were given daily oral doses of 3 g DMSO/kg for 14 days, and its major metabolite, dimethyl sulfone, was measured in serum, urine, and feces by gas–liquid chromatography.[5] DMSO was quickly absorbed and reached a steady-state blood level after 24 h and then was cleared from blood within 72 h after treatment ended. Serum DMSO declined in a linear fashion as described by second-order kinetics.[5] Dimethyl sulfone half-life in serum was calculated to be 38 h and appeared in blood within 2 h, reaching a steady-state concentration after 4 days of treatment. Following DMSO administration, dimethyl sulfone was cleared from serum after 120 h. Urinary excretion of unmetabolized DMSO and dimethyl sulfone accounted for about 60% and 16%, respectively, of the total ingested dose.[5] The absorption of DMSO by monkeys appears similar to that for humans, but its conversion to dimethyl sulfone and its urinary elimination are more rapid in monkeys.[5]

PENETRATION

There are four principal variables that influence the penetration of a solute through any given membrane: (1) the diffusion coefficient through the membrane, (2) the concentration of the agent in the vehicle, (3) the partition coefficient between the membrane and the vehicle, and (4) the thickness of the membrane barrier.[6] Penetration agents are designed to affect one or more of these variables without causing permanent structural or chemical modification of the physiological barrier. Alteration of membrane thickness is less practical for drug delivery (it is difficult to conceive of nontoxic agents that could reversibly decrease the thickness of the stratum corneum), so most penetration agents, including DMSO, attempt to reversibly alter variables 1–3 given earlier.

There is some evidence to suggest that DMSO can increase diffusion through the stratum corneum by disruption of the barrier function.[7] This probably occurs through aprotic interactions with intercellular lipids and may also include reversible distortion of lipid head groups that produce a more permeable packing arrangement.[8,9] DMSO may play a role in partitioning as well by forming solvent microenvironments within the tissue that can effectively extract solute from vehicle. Finally, DMSO can have a profound solubilizing effect on less soluble agents in a variety of vehicles, increasing penetration simply by delivering a higher concentration to the membrane barrier.[10]

CHEMICAL CHAPERONES

The actions of DMSO as a chemical chaperone are discussed in Chapter 1. Chemical chaperones such as DMSO, glycerol, trimethylamine n-oxide, and other small molecules have been found to stabilize and help in the folding of proteins.[11]

Chemical chaperones differ from molecular chaperones, which are small molecules found in the endoplasmic reticulum. These chemical chaperones also differ from pharmacological chaperones, which are small synthetic molecules that can induce *correct* folding by binding to the hydrophobic domains of a given protein in an adenosine triphosphate (ATP)-dependent fashion.[12] The older the cell, the more difficult it is to cope with protein misfolding.[13] Whether DMSO can also act as a pharmacological chaperone has not been determined. Although DMSO has not been used extensively in misfolded protein diseases, it may offer some hope as a treatment due to its osmolytic activity in preventing aggregate formation in *neurotoxic gain* disorders such as Alzheimer's disease (deposition of amyloid-β peptide in brain tissue and neurofibrillary tangles within neurons), Parkinson's disease (deposition of aggregated α-synuclein in Lewy bodies), scrapie or mad cow disease (pathological prion protein production), and Huntington's disease (excessive number of CAG repeats in the huntingtin gene).

It follows that if DMSO is successful in treating neurotoxic gain disorders, it could also be useful in *loss-of-function* diseases such as nephrogenic diabetes mellitus, Gaucher's disease, Fabry's disease, and cystic fibrosis.[14] Loss-of-function disorders are generally caused by missense point mutations. Examples of point mutation disorders are sickle-cell disease and amyotrophic lateral sclerosis. Loss-of-function mutations can alter the normal folding proteins and lead to their degradation, a process that could be blocked by DMSO and by the cell's ability to try and stay healthy.[14] Toxic gain disorders involve proteins that fail to fold into their normal configuration, and in their misfolded state, they can become toxic in some way and quickly aggregate, leading to the formation of inclusion bodies that eventually result in cell death.[15]

DMSO AS A PAIN MEDICATION

Many ancient cultures believed that pain was a test of faith or a punishment by the gods for human folly. The remedy for pain was to appease the wrath of gods with magic, rituals, or sacrifice of animals and humans.

Ancient philosophers considered pain to be an emotion. Aristotle in his *Nicomachean Ethics* saw pain not as an internal noxious sensation but as spirits entering the injured body to affect the soul. This explanation that pain was an external force that entered the body and could be treated by prayer endured for centuries. This thinking was partly replaced by the seventeenth-century discoveries of a mechanistic theory to explain pain. One of the earliest expositions of pain mechanics were the descriptive 1644 writings of the French philosopher and mathematician René Descartes, who proposed a more pragmatic approach to pain. Descartes correctly believed that pain did not come from the outside but rather within the body where it traveled through nerves to the brain.

Pain is the most common reason for people to seek medical help and is a major presenting complaint in a multitude of medical conditions and disorders. It has been common knowledge for thousands of years that acute or chronic pain can significantly interfere with a person's quality of life and normal daily function.

Woolf[16] has described three types of pain. First, there is *nociceptive* pain, which acts as an early warning, protective system that can feel a harmful stimulus, such as an object that is too hot or cold, or too sharp, and react to minimize the pain.

The second type of pain is *inflammatory* and occurs when there is tissue injury, for example, with an inflamed joint, a local or systemic infection, or following surgery.

This type of pain is adaptive in the sense that the individual limits the movement of the body part in pain in order to shorten the duration or the degree of the pain and accelerate its eventual recovery.

The third type of pain is *pathological. Pathological pain* is not adaptive or protective *as* in types 1 and 2 but results from tissue damage affecting the nervous system.[16] This type of pain can occur in the absence of inflammation or noxious stimuli and may persist for a long duration when stimuli are excessively intense or prolonged. The quality of pathological pain can be severe or moderate and may interfere with sleep or daily function. At the same time, pain can downgrade the health and psychological well-being of the individual.[17]

Pain killers, particularly opioid drugs, are the leading cause of unintentional overdose deaths in teens and adults, and according to the Centers for Disease Control and Prevention (CDC), such deaths have reached epidemic proportions in the United States. The number of unintentional drug poisoning deaths exceeds either motor vehicle accidents or suicides, two of the leading causes of death in the United States.[18] One of the reasons for easily overdosing and dying from narcotics is their powerful action as central nervous system (CNS) depressants, especially when the narcotic is combined with alcohol or another pain medication.[19]

For these reasons, an analgesic medication with a high margin of safety, which lacks CNS depressant activity and is void of the danger of unintentional overdose and death, has been an elusive Golden Fleece quest by the pharmaceutical industry and by academic and government-sponsored research.

These pharmacological properties in an analgesic agent and avoidance of death by overdose appeared to have been solved in 1964 when Jacob et al.[20] first reported on the local analgesic potential of DMSO in humans. The use of DMSO as a pain medication in humans is discussed in detail in Chapter 3.

The basic biological actions of DMSO as a pain medication using animals will be reviewed here. Although there is extensive literature on the pain mechanisms exerted by DMSO in animal models, its physiological and pharmacological properties are incompletely understood.

It should be pointed out that there is considerable difficulty in using methods that can accurately characterize pain sensibility more so in animals than humans with respect to its quality, intensity, and duration, and this fact makes it challenging to discover treatments that truly improve the outlook of patients with pain. A case in point is the thousands of papers written on brain endorphins in the past three decades where great hope for controlling pain was overshadowed by the fact that not one effective treatment emerged from this vast field of research.

In experimental pain research, there are two common techniques used in rodent models to evaluate potential analgesic agents, the tail flick test and the hot plate test. The tail flick test measures a rodent's pain response to a light beam focused on the animal's tail and the time for the tail to flick that is recorded as the pain threshold.[21] In the hot plate test, the rodent is placed on a warm plate, and temperature is increased until a thermal nociceptive response is observed, indicated by the animal rapidly licking or flicking its hind paws.[22] In these tests, a baseline latency response without the analgesic agent is first obtained and compared to the latency following administration of the agent to be tested.

Following clinical reports on the action of DMSO to relieve pain in humans,[20,23–28] animal research was begun to answer multiple questions regarding whether DMSO was an analgesic agent and, if so, how it worked.

One early study to address some of these questions was reported by Haigler and Spring.[29]

Using the tail flick and hot plate tests, these investigators applied two criteria to determine (1) if DMSO had analgesic properties and, if so, (2) whether the analgesic effect observed with DMSO was related to opiate receptors.

It was found that animals given DMSO IV and IP at high doses (5.5 g/kg), and administered topically on the tail, produced a profound analgesia on both the hot plate and tail flick tests.[29]

The analgesic effect obtained with DMSO was apparently not due to a general anesthetic effect because the rats were not ataxic and did not lose their righting reflex. By contrast, morphine given 10 mg/kg intraperitoneally produced an analgesic effect as frequently as DMSO on the hot plate, but the duration of morphine analgesia was typically less than 2 h both on the hot plate and on tail flick. The analgesic time course for DMSO was longer than that for morphine and, in several rats, lasted over 24 h. Additionally, the analgesic effect by morphine, but not by DMSO, was blocked with the opioid antagonist naloxone indicating that the analgesic action on pain receptors by DMSO and morphine differed significantly.[29]

If the analgesic action of DMSO was not achieved by blocking morphine receptors, how was pain relieved in animals or humans?

The answer remained unsettled until 1993, when Evans et al.[30] reported their findings in a study where 2%, 9%, and 15% DMSO was applied directly to exposed cat sural nerves. These unmyelinated sural nerves mediate somatic pain from C-fibers that terminate on the skin, muscle, and joint capsules.[31] It was found that C-fiber conduction velocity was slowed by 5% DMSO and was concentration dependent. At 9% concentration, DMSO completely blocked C-fiber conduction, and at 15% concentration, the onset of nerve block was virtually immediate[30] (Figure 2.1). The study by the Evans et al.[30] was an extension of previous findings by Sawada and Sato,[32] who had reported that the mechanism involved in the nerve conduction block induced by DMSO appeared to be due to potassium channel blockade. These investigators observed that DMSO caused a rapid membrane depolarization and decrease in membrane conductance of *Aplysia* neurons, a finding consistent with the blockade of potassium leak channels.[32] These *leak* potassium channels function to set the negative membrane potential of neurons. They differ from voltage-gated channels because they are always open to the passage of sodium and potassium ions across the membrane whereas voltage-gated channels open and close in response to specific changes in the membrane potential.

How DMSO specifically affects potassium channel depolarization has not been made clear, but Davis et al.[33] found that 50% DMSO was necessary to block muscle twitch in a frog sciatic nerve-gastrocnemius muscle preparation. Interestingly, Becker et al.[34] found that 75% DMSO blocked both myelinated A-delta and unmyelinated C-nerve fibers (Figure 2.1). In a study of small fiber afterdischarge in spinal cord, medulla, and tegmentum following stimulation of the superficial radial and sural nerves, Shealey[35]

DMSO action in pain relief

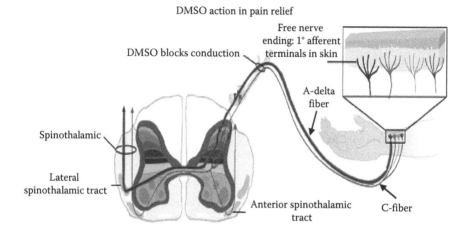

FIGURE 2.1   (**See color insert.**) DMSO topical application can block both myelinated A-delta and unmyelinated C-pain fibers.[30,34] See text for details.

also found blocking of pain impulses by exposing the peripheral nerve to 5%–10% DMSO, a result that they concluded might be due to blockade of C-fibers.

The effects of DMSO and a DMSO-containing saline solution were used on thermal- and chemical-induced nociception and inflammation, and the locomotor activity after treatment was evaluated in CD1 mice.[36] This mouse type is a multi-purpose research animal model that can be used in such fields as toxicology to test safety and efficacy of drug products, and on *in vivo* nerve preparations.

Centrally or orally administered DMSO was shown to display anti-nociception to thermal (hot plate and tail flick test) and chemical (formalin test) stimuli.[36] Additionally, oral and central, but not local, administrations of DMSO were seen to improve locomotor activity in the CD1 mice.[36]

Thus, the collective evidence on the analgesic effects of DMSO in reducing pain sensation differs from that of complete anesthesia since it appears to involve blockade or partial blockade of A-delta fibers that mediate acute sharp pain as well as C-fibers that mediate somatic signals involving temperature, sensual touch, and muscle and joint pain. This evidence implies that DMSO could be useful in musculoskeletal aching pain and sharp arthritic pain impulses, which are conveyed by lightly myelinated A-delta fibers, and somatic pain mediated by C-fibers. The merit of this conclusion will be further examined in Chapter 3. The effects of DMSO on these experimental pain-related conditions offer the potential by DMSO to reduce or abolish pain by mechanisms ostensibly distinct from the opioids and the dangers posed by narcotic prescriptive agents.

Another possible mechanism exerted by DMSO on pain receptors may be its effect on the excitotoxic expression of voltage-gated sodium channels and glutamate, NMDA (*N*-methyl-D-aspartate), and AMPA (α-amino-3-hydroxy-5-methylisoxazole-4-propionate)[37] as suggested by Hoang et al.[38]

Sodium channels are located in all sensory neurons, and some sodium channels that are sensitive to tetrodotoxin are found only on nociceptive-related sensory neurons implicated in pathological pain states.[38] Lidocaine, for example, is a widely used local anesthetic that functions through the blockade of sodium channels, a mechanism distinct from the opioids.[39]

Since DMSO is reported to have an effect on blocking sodium and calcium ion entry into cells[40,41] and several research groups have shown that cancer-related neuropathic pain[42,43] has an underlying mechanism involving membrane hyperexcitability due to overexpression of sodium channels and glutamate buildup stemming from the upregulation of NMDA/AMPA pathways, it has been suggested that DMSO could be used as a primary or adjuvant analgesic agent in controlling pain in advanced cancer patients.[38]

ANTI-INFLAMMATORY

Inflammation is a localized and complex protective reaction by the organism to remove damaging stimuli by removing offending factors and mobilizing molecules responsible for the healing process. The damaging stimuli can involve, but are not limited to, adhesion molecules, neutrophils, complement, IgG, irritants, dying cells, or pathogens. Inflammation is implicated in the pathogeneses of arthritis, cancer, and stroke, as well as in neurodegenerative and cardiovascular diseases. The acute phase of inflammation is characterized by the rapid influx of blood granulocytes, typically neutrophils, followed swiftly by monocytes that mature into inflammatory macrophages that can subsequently proliferate and cause tissue damage. When the inflammation occurs at or just under the skin, the characteristic signs of inflammation are pain (dolor), heat (calor), redness (rubor), and swelling (tumor), responses that can lead to scarring and loss of organ function.

A classical and popular model of inducing experimental inflammation and hyperalgesia from edema formation is the carrageenan-induced inflammation in the rat paw. This model has been used extensively in the development of nonsteroidal anti-inflammatory drugs and selective cyclooxygenase-2 (COX-2) inhibitors.[44]

Evidence indicates that the COX-2-mediated increase in prostaglandin E_2 (PGE_2) and prostaglandin I_2 (PGI_2) production contributes to the severity of the inflammatory and pain responses in this rat paw model.[45]

Carrageenan-induced inflammation in the rat paw was used by Formanek and Kovak[46] and later by Görög and Kovács[47] to demonstrate the anti-inflammatory action of DMSO applied to the edematous rat paw. These investigators showed a significant reduction in paw and limb edema after topical DMSO application, but no biochemical explanation was available to explain how DMSO exerted its action. The anti-inflammatory actions of DMSO generated suspicion that its therapeutic benefit could be explained by its activity on prostaglandin pathways since it was well known that prostaglandins play a key role in the generation of the inflammatory response.

PROSTAGLANDINS

It remained for Panganamala and his research team[48] in 1976 to show that DMSO had an effect on prostaglandin biosynthesis. Using a liver microsomal preparation, DMSO was seen to stimulate PGE_1 and inhibit $PGF_2\alpha$ by blocking arachidonic-acid-induced platelet aggregation, a reaction involving endoperoxide metabolites, PGE_2, thromboxane A_2 (TXA_2), and PGI_2.[48] These authors concluded that anti-inflammatory properties shown by DMSO in these studies were likely due to its hydroxyl radical–scavenging ability and inhibition of prostaglandin synthesis.[48]

Prostaglandins and TXA_2 are collectively called prostanoids, which form when arachidonic acid, a 20-carbon unsaturated fatty acid, is released from the plasma membrane by phospholipases and metabolized by the sequential actions of prostaglandin G/H synthase or cyclooxygenase (COX).[40] Prostaglandin production remains very low in uninflamed tissues, but when acute inflammation occurs, it increases immediately prior to the recruitment of leukocytes and the infiltration of immune cells.

The presence of TXA_2 in blood will cause strong vasoconstriction and platelet aggregation, while the formation of $PGF_2\alpha$, or PGE_2, will also negatively affect the vascular and platelet systems, but the reaction will not be as severe as with TXA_2.[49]

There is presently considerable evidence that the resulting inhibition of $PGF_2\alpha$, PGE_2, and TXA_2 synthesis has the effect of reducing inflammation, as well as promoting antipyretic, antithrombotic, and analgesic effects. The effects of PGE_1 and PGE_2 on platelet function are believed to be mediated through prostanoid receptors that are linked to the control of intracellular levels of cAMP.

The outer cell membranes of bovine corpora lutea are reported to contain discrete and specific receptors for PGE_1 and $PGF_2\alpha$.[50] Preincubation of membranes for 1 h at 4°C with increasing concentrations of DMSO resulted in a progressive inhibition of [^{3}H]$PGF_2\alpha$ binding but had no effect on [^{3}H]PGE_1 binding.[51] It should be noted that PGE_1 and PGE_2 are known to modify platelet function. PGE_1 has been shown in many studies to inhibit platelet aggregation in a concentration-dependent manner.[52,53] In contrast, the effects of PGE_2 are less consistent and have been reported to either stimulate or inhibit platelet aggregation depending on the concentration of PGE_2 and the conditions used.[54]

The action by DMSO on PGI_2, PGE_1, and $PGF_2\alpha$ led de la Torre[55] to theorize that DMSO may play an important clinical role in neutralizing tissue injury by acting on prostaglandin pathways (Figure 2.2). Assuming the tissue injury is of sufficient magnitude, one of the initial molecular reactions to occur in the body is an immediate reduction of oxygen delivery to the tissue. As this happens, mitochondrial oxidative phosphorylation is negatively affected and results in a reduction of ATP, the major source of cellular energy. As ATP concentration decreases, sequestered calcium is released to activate mitochondrial and microsomal phospholipases.

The phospholipases chemically attack long-chain phospholipids, thus uncoupling esterified free fatty acids. At the same time, inhibition of mitochondrial oxidation results in hypoxia and further buildup of free fatty acids. Polyunsaturated free fatty acids provide the substrate for conversion to prostaglandins. PGI_2 (prostacyclin) and, to a lesser extent, PGE_1 act as vasodilators and platelet deaggregators. The antiplatelet aggregation activity by PGI_2, and PGE_1, results from its increase in cAMP levels in platelets (Figure 2.2).[56,57]

Therapeutic use of PGE_1 has been shown to have anti-ischemic and anti-inflammatory effects on patients with cardiovascular disease, such as in peripheral arterial occlusive disease.[58,59] Also, it has recently been proposed that naturally occurring PGE_1 may have a modulatory role in the development of atherosclerosis.[60]

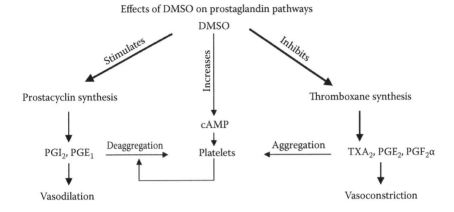

Effects of DMSO on prostaglandin pathways

FIGURE 2.2 Presumed actions of DMSO based on experimental evidence derived from experiments on the prostaglandin–thromboxane (PG–TX) and platelet systems. DMSO stimulates prostacyclin synthase to form prostacyclin (PGI_2), a powerful platelet deaggregator and vasodilator. DMSO has been shown to antagonize the synthesis or release of TXA_2, $PGF_2\alpha$, and PGE_2 to counteract their effect of platelet aggregation and vasoconstriction in model systems. The antiplatelet aggregation activity by PGI_2 and PGE_1 is reported to result from a stimulating increase in platelet cAMP levels by DMSO. See text for details.

By contrast, studies in mice have revealed that PGE_2 is released from inflamed vessels and atherosclerotic plaques,[61] and evidence indicates that PGE_2 is involved in platelet thrombus formation,[62] making this prostaglandin a possible therapeutic target for stroke or cardiac thrombosis prevention.

One of the first observations that noted DMSO's anti-inflammatory properties used ultraviolet light (PUVA) to produce an inflammatory reaction that killed or reduced proliferation of epidermal melanocytes whose role is to protect the skin against ultraviolet radiation. In a mice skin model exposed to prolonged PUVA known to induce a strong inflammatory reaction, DMSO application to the skin increased the numerical density of pigmented cells, thus allowing better recovery of the inflammatory response.[63]

Epithelial cell injury under hyperinflammatory conditions is critical in the development of septic acute lung injury (ALI). A model of ALI using the human alveolar epithelial cell line A549 was created to test the cytoprotective effects of anti-inflammatory agents, including DMSO.[64] It was found that 0.02% DMSO acted as a powerful antioxidant, attenuating a cocktail of cytotoxic mediators made up of IL-1β/TNF-α/IFN-γ exerted from the A549 cancer cell line.[64]

The effects of DMSO on inflammatory cytokines have been reported in intestinal cells exposed to a variety of inflammatory cytokines *in vitro*.[65] DMSO at 0.5% was shown to significantly decrease mRNA levels of inflammatory proteins, including IL-6, IL-1α, IL-1β, and dose dependently reduced COX-2-derived PGE_2.[65] In addition, DMSO decreased the production of IL-6 and macrophage chemoattractant protein-1 (MCP-1) secretions, both in a dose-dependent manner in the intestinal cells.[65] MCP-1 is a major chemokine responsible for inducing macrophage migration and appears to play several roles in tumor growth and metastasis.

Plasma cholesterol and bile acid metabolism were significantly reduced by 2% DMSO treatment in the drinking water for 9 days, a finding consistent with the reduced CH-7 alpha-hydroxylase activity, the rate-limiting enzyme of bile acid biosynthesis in the treated rat group.[66] Although the exact mechanism for this action by DMSO on cholesterol and bile production is unclear, it did not affect either microsomal cholesterol content or hepatic glutathione content.[66]

These actions by DMSO on toxic cytokines could be a major clue as to its action as a pain reliever and powerful anti-inflammatory agent. In addition, the complex interactions of DMSO in prostaglandin biosynthesis and production have important implications in cerebral and cardiac physiology.

CARDIAC DISEASE

The human heart receives about 1/20th of cardiac output under basal conditions, and when this blood flow delivery is reduced or compromised, cardiac damage or death can occur.

Several intriguing studies have examined the effects of DMSO on experimental myocardial infarction and cardiac hemodynamics. One early study reported that DMSO provided significant protection of rat heart exposed to reoxygenation-induced cell injury.[67] This model of anoxic injury to cardiac cells involves perfusing the heart with an anoxic perfusate followed by reoxygenation with an oxygen perfusate, a technique that induces a rapid release of oxygen–creatine kinase (CK), causing reduced coronary flow and severe cardiac contraction with the formation of contraction bands. Prior to reoxygenation, 10% DMSO was given by mixing it with the anoxic perfusate, an action that resulted in increased coronary flow rate, a marked reduction of the oxygen–CK release, and a significant decrease in the number of contraction bands from 43% in control hearts to 23% using DMSO.[67] This study indicated that DMSO has the potential to protect cardiac tissue from a hypoxic insult, an ability that could be useful in cardiac ischemia or heart failure.

One of the actions of DMSO in the ischemic myocardium is to aid in the diffusion of oxygen into the thick myocardial tissue, thus aiding the myocyte protection from anoxia or ischemia.[68]

The effects of DMSO infusion on a canine model of myocardial ischemia and systemic hemodynamics were reported by Levett et al.[69] Hourly measurements of cardiac output, pulmonary wedge pressure, pulmonary artery pressure, heart rate, mean arterial pressure, cerebral blood flow, intracranial pressure, and blood gases were monitored before and after the ligation of the left descending coronary artery. After 15 min of coronary artery ligation, control saline or 500 mg/kg of a 50% DMSO solution was injected as an intravenous (IV) bolus.[69] By the third hour after ligation, cardiac output increased significantly, and both systemic and cerebrovascular resistances decreased in the DMSO-treated animals as compared to controls. CBF also increased after DMSO but not in controls, and there were no significant differences in heart rate, mean arterial pressure, pulmonary artery wedge pressure, or cerebral or pulmonary resistances in the control as compared to the DMSO group.[69] This study concluded that DMSO infusion in a canine model of myocardial ischemia has protective activity and merits further evaluation by extending the time course after infarction to a more realistic clinical setting.[69]

The study by Levett et al.[69] recently received confirmation using a rat model of acute myocardial infarction caused by left anterior descending coronary artery ligation.[70] Before myocardial infarction was induced, 500 µL/kg of DMSO was given intraperitoneally daily for 3 days with the last dose given 15 min before ligation.[70] After 90 min of occlusion, postmortem examination showed markedly less cardiac damage reflected by a reduction of mean necrotic volume in the infarcted left ventricular wall area and by significantly lowered levels of the cardiac injury markers troponin I and myoglobin, as compared to controls. The mechanism by DMSO in protecting cardiac tissue from ischemic injury remains unsettled but appears to be similar to its protection of focal ischemic brain tissue following occlusion of the middle cerebral artery in rats.[71]

A review of the experimental cardiac effects of DMSO may provide some clues.[72] For example, DMSO can produce either positive or negative inotropic effects on cardiac tissue depending on the dose used and animal species and specific cardiac tissue studied.[73] Myocardial tissue preparations using 70 mM DMSO concentration or less will increase positive inotropic contractile response, while using more than 70 mM concentration, the inotropic effect will be species dependent.[74] Moreover, none of the contractile effects of DMSO is antagonized by blocking beta-adrenergic receptors or by blocking other classical myocardial receptors at doses that do not intrinsically alter cardiac contractility.[75] In isolated atrial preparations, DMSO decreases contractile rate without appreciably altering cardiac rhythm. However, when the DMSO concentration is above 140 mM, a concentration not likely to be used *in vivo*, both rate and rhythm are decreased possibly by DMSO's effect on acetylcholinergic response to slow down conduction via the atrioventricular node.[74] However, these effects on the myocardium are reversible when DMSO exposure stops.

A reduction of systemic vascular resistance and an increase in cardiac output were found after slow DMSO infusion at a dose of 2 g/kg 50% solution in canine myocardium.[76] These values were temporary and returned to normal 10 min after DMSO infusion had been completed although cardiac output remained high for several hours when compared to saline control infusions.[76] These findings were considered to result from a possible transient expansion of plasma volume that increased cardiac preload and cardiac output and could be relevant in treating hypovolemic patients with clinical brain injury or cardiac ischemia.[76]

The case for using DMSO as a primary agent in the clinical treatment of myocardial infarction and coronary artery ischemia or hypoxia can be argued by noting some of the specific properties reported for this drug. These cardiac conditions diminish blood flow to the heart either by coronary vessel narrowing due to platelet aggregation on vessel walls or by atherosclerotic epicardial plaque-related stenosis. Present treatments for these conditions include coronary vasodilators such as ranolazine and nitrate preparations, platelet deaggregators such as aspirin or clopidogrel bisulfate, coronary artery balloon angioplasty or stenting, and coronary artery bypass surgery. Other pharmacological treatments such as calcium channel blockers, beta blockers, angiotensin-converting enzyme inhibitors, cholesterol-lowering drugs, and angiotensin receptor blockers may be useful in long-term therapy, but they do not reverse coronary artery ischemia or myocardial infarction acutely. DMSO addresses an important aspect related to the pathology seen in coronary artery ischemia.

First, DMSO, as we have reviewed earlier, is a powerful platelet deaggregator. A report by Pace et al.[77] showed that when DMSO was used as a solubilizing agent to test the potential of cardiac anti-aggregatory drugs in experimental coronary artery stenosis, it was discovered that the active agent was DMSO, not the drugs that were tested. When used alone, DMSO was observed to reverse the reduction of coronary blood flow induced by a critical stenosis on the canine circumflex coronary artery without changing other hemodynamic measures in the animals tested.[77] These findings suggested that DMSO acted a platelet deaggregator in reversing the coronary artery stenosis that allowed platelet aggregation and lowered coronary blood flow. The findings by Pace et al.[77] were confirmed by Dujovny et al.,[78] who reported scanning electron microscopic examination of rat carotid artery–induced occlusion induced by arteriotomy and suture closure where the formation of heavy platelet deposits, fibrin, and red blood cell clumps at the arteriotomy site collected. The administration of DMSO 2 g/kg slowly infused intravenously resulted in minimal signs of fibrin clumping and platelet deposition at the carotid artery suture line.[78] The platelets appeared spherical with no pseudopod formation, while the adjacent endothelium appeared intact in contrast to control non-DMSO-treated animals who showed heavy platelet deposits and fibrin and erythrocyte clumping.[78]

We have discussed the antiplatelet action of DMSO from the studies of Panganamala et al.,[48] who showed that DMSO inhibited arachidonic-acid-induced platelet aggregation possibly by inhibiting prostaglandin biosynthesis, particularly PGE_2. Holtz and Davis[79] and Harrison et al.[80] have suggested that DMSO is a sulfhydryl inhibitor and that this action prevents platelet-to-platelet bonding, whereas Wieser et al.[81] showed that DMSO could increase cyclic AMP by inhibiting phosphodiesterase in a non platelet system (Figure 2.2). DMSO was found to inhibit the cardiac hypertrophy, anemia, and depression in a model of copper deficiency of the heart. This finding suggests that the hydroxyl radical–scavenging ability by DMSO may contribute to the protection of cardiovascular defects caused by dietary copper deficiency.[82]

The second property DMSO possesses that may be useful in coronary artery pathology is its hydroxyl radical–scavenging action and its ability to function as an antioxidant and inhibitor of superoxide anion formation.[83–86]

Oxygen-derived free radicals, such as the superoxide anion, hydrogen peroxide, and the hydroxyl radical, are implicated as mediators in myocardial injury and ischemia during reoxygenation of ischemic tissue.[87–89] The naturally occurring antioxidant enzymes, superoxide dismutase and catalase, are reported to prevent the formation of the cytotoxic hydroxyl radical during physiological conditions but may not be able to cope with the free radical generation that follows ischemia and reperfusion.[90]

The third property by DMSO that may be useful in cardiac ischemia serves as a sodium channel blocker. Heart muscle homeostasis depends on the optimal flux of Na^+ and Ca^+ for its proper rhythmic contractions, and when an equilibrium of myocyte influx and efflux of these two ion species fails, it results in rhythm and

DMSO cardiac actions

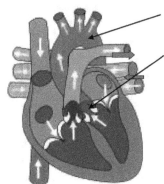

Prevents VSM migration and proliferation

Restores cardiac output and cerebral blood flow
after LAD occlusion

Blocks tissue factor expression

Inhibits platelet aggregation after LAD occlusion

Blocks abnormal Na$^+$ and Ca$^+$ cell entry

FIGURE 2.3 Cardiac actions reported following administration of DMSO. DMSO prevents the migration and proliferation of vascular smooth muscle cells (VSM) following cardiac ischemia, a common occurrence seen after heart attacks and stroke. Restoration of cardiac output is reported following occlusion of the lateral anterior descending (LAD) coronary artery, which mimics a heart attack. Tissue factor, a powerful cardiac thrombogen, is suppressed by DMSO. DMSO may have salutary effects on injured cardiac tissue by blocking abnormal Na$^+$ and Ca$^+$ influx into cardiac myocytes, preventing myocardial ischemia. See text for details.

contractile dysfunction (Figure 2.3). Drugs that prevent abnormal sodium influx into heart tissue provide an effective protection against Na$^+$ and Ca$^+$ overload. It has been reported that DMSO has an effect on blocking Na$^+$ and Ca$^+$ entry into cells.[48,91] Since substantial Na$^+$ and Ca$^+$ entry into myocytes typically occurs after cardiac arrhythmias and myocardial infarction, DMSO administration may prevent this inward cellular ion flux while preserving K$^+$ outflux from cardiac tissue. The mechanisms exerted by DMSO on Na$^+$ and Ca$^+$ channels need to be further investigated in mammalian models since the results of such studies could produce an extremely useful and relatively safe agent for a variety of cardiac disorders affected by changes involving these cations.

The fourth property shown by DMSO as a potential cardiac agent is its protective activity against tissue factor (TF) expression in human endothelial cells in response to tumor necrosis factor-α (TNF-α) or thrombin exposure[91] (Figure 2.3). There is general agreement that TF is a key protein in the activation of coagulation and thrombus formation.[92] This activation of thrombus formation and coagulation is a cause of acute coronary syndromes and myocardial infarction.[93]

High levels of TNF-α may result in acute coronary disease where it can induce increasing levels of TF in coronary vessels.[92]

DMSO was found to suppress, in a concentration-dependent manner, TF expression and activity in response to TNF-α or thrombin exposure in human endothelial cells, monocytes, and vascular smooth muscle cells (VSM) *in vitro*.[91] Moreover, it was additionally observed that DMSO prevented proliferation and migration of VSM from human aorta, an outcome that could have important clinical application in treating coronary thrombosis and myocardial infarction.[91]

In vivo, DMSO treatment significantly suppressed TF activity and prevented thrombotic occlusion in the mouse carotid artery model subjected to a photochemical injury[91] (Figure 2.3).

In summary, DMSO combines four important properties that could be beneficial in cardiac disease: *platelet deaggregation, sodium channel blocker, TF antagonist,* and *free radical scavenger* and *antioxidant.*

There is no drug therapy at the present time that combines these four properties shown by DMSO, thus giving this agent a novel therapeutic advantage as a potential treatment of coronary syndromes. As an antiplatelet agent, DMSO could have beneficial effects in coronary artery disease by inhibiting blood clot formation and thus prevent vascular ischemic events involving atherosclerosis in the periphery as well as in the heart and brain vasculature. As a sodium channel blocker, DMSO may show usefulness in the treatment of cardiac arrhythmias.[94,95]

DMSO's antagonism of TF could prevent or treat clot formation in coronary syndromes as well as other disorders associated with TF elevation, such as hypertension, diabetes, and dyslipidemia.[93] Since these cardiac conditions involve tissue injury to cardiac cells, the formation of oxygen and hydroxyl radical formation is to be expected. DMSO, as a powerful free radical scavenger, may be pivotal as a primary or adjuvant therapeutic agent in controlling cytotoxic damage from radical formation.

ULCERATIVE COLITIS

Current use of DMSO in inflammatory bowel disease is limited, and few studies are available in experimental intestinal disorders.

Ulcerative colitis is a nonspecific inflammatory disease of unknown origin, arising in the colonic mucosa and clinically characterized by recurrent attacks of bloody diarrhea interspersed with asymptomatic intervals. Pathologically, the disease begins by a diffuse inflammation of the rectosigmoid area and may extend proximally to the entire colon. Crypt abscesses, mucosal ulceration, and epithelial necrosis eventually develop. There is no cure for ulcerative colitis, and treatment usually aims at the inflammatory aspect of the disease and the use of corticosteroids as immune system suppressants and with aminosalicylate preparations as the main anti-inflammatory drugs.

Free radicals such as DMSO have been tried for ulcerative colitis on the theory that oxygen-derived free radicals may be involved in the mechanism of this disorder and that removing these free radicals may result in the treatment of attacks and in protecting the colon against the recurrence of attacks. In a mouse model of colitis induced by dextran sodium sulfate, 500 μL of DMSO was reported to suppress the aggravation of the inflammatory response when given daily via the rectum to the mice for 7 days.[96] Inflammatory suppression by DMSO of experimental colitis may be due to its inhibition of superoxide radical formation generated by inflammatory cells.[97] The damaging process could involve reactive oxygen metabolites generated intracellularly by either NADPH oxidase in the neutrophils or by xanthine oxidase, which is present in gut mucosa.[98] This pathway seems to be a major source of oxygen free radicals in ischemic/reperfusion injury of the gut.[99]

SKIN PENETRANT

DMSO has been reported to enhance drug penetration, promote membrane transport, stabilize membrane integrity, improve percutaneous penetration, and accelerate drug absorption and drug effects.[100] The ability of DMSO to quickly and safely cross the dermal barrier when topically applied to the skin has been used in clinical medicine using DMSO alone or as a percutaneous carrier of other drug molecules.[101,102]

Penetration of the skin by an active agent is dependent on the barrier function of the stratum corneum.[103] Four variables influence the penetration of an active agent through any given membrane: (1) the relative diffusion coefficient through the membrane, (2) the concentration of the agent, (3) the partition coefficient between the membrane and the agent, and (4) altering the thickness of the membrane barrier.[104] Agents that penetrate the stratum corneum may affect one or more of these variables and, except for the fourth variable, can do so without causing permanent structural or chemical modification of the dermal barrier. Alteration of membrane thickness is less practical for drug delivery because of the damage it would cause to the stratum corneum, so most penetration agents, including DMSO, do so by relative diffusion coefficient through the membrane or by variables 1–3.[105]

There is some evidence to suggest that DMSO can increase diffusion through the stratum corneum by the disruption of the barrier function.[106] This probably occurs through aprotic interactions with intercellular lipids and may also include reversible distortion of lipid head groups that produce a more permeable packing arrangement. DMSO may also play a role in partitioning as well by forming solvent microenvironments within the tissue that can effectively extract solute from vehicle.[107]

WOUND HEALING

Matrix metalloproteinases, especially metalloproteinase-9 (MMP-9), degrade various proteins of the extracellular matrix, including collagen type IV, the major component of basement membranes that also separate the epidermis from the dermis. Limited activity of MMP-9 is essential, while excessive activity is deleterious to the healing process.

The antioxidant effects of DMSO have been reported to be useful as a topical agent in the treatment of chronic wounds. The effects of DMSO on MMP-9 were studied in disordered wound healing using TNF-α to induce MMP-9 from human keratinocytes.[108] Results showed that DMSO inhibited the production of both MMP-9 levels and MMP-9 mRNA expression when TNF-α-stimulated keratinocyte cells in a concentration-dependent manner. Inhibition of MMP-9 levels was statistically significant at DMSO concentrations of 0.75% and higher.[108] These findings suggest that DMSO may reduce the harmful effects of MMP-9 through downregulation at the transcription level and may be useful to prevent TNF-α-induced proteolytic activity in cutaneous inflammatory reactions.

A systematic review looked at the efficacy of topical DMSO on wound healing of decubitus ulcers and its use as a topical anti-inflammatory drug.[109] The collective evidence reported indicates that the use of DMSO in these conditions was beneficial, both for wound healing and for analgesia, while the toxicity was deemed to be low. The most frequent outcome measures were reduction of erythema and rapid healing of ulcers, along with decreased signs of inflammation, such as rubor, dolor, calor, and tumor. DMSO was found to be dramatically effective for healing severe skin necrosis caused by accidental extravasation of the anticancer drug mitomycin C during IV administration.[110] This involved using 90% DMSO combined with a 10% alpha-tocopheryl acetate applied prophylactically on the skin after extravasation into the tissue of antineoplastic agents but before ulceration.[110] This same combination of DMSO–alpha-tocopherol has been found to universally prevent severe ulceration and tissue breakdown in patients sustaining extravasation by a cytotoxic drugs, in most cases preventing skin ulceration induced by anthracyclines and mitomycin.[111]

BURNS AND SCAR TISSUE

Partial-thickness burn wounds on rat abdominal skin were treated using 1% metronidazole or 1% norfloxacin using a petrolatum base with 0.25% w/w DMSO. These formulations with DMSO as a base appeared more effective for wound contraction of the partial-thickness burn wound and against aerobic and anaerobic bacteria than a marketed formulation composed of 1% silver sulfadiazine cream.[112] Histopathology of the burn wound tissue confirmed the effectiveness of these preparations in treating the partial-thickness burn wound.[112]

As an antimicrobial applied topically to thermal burn wounds in rats, 1% w/vol DMSO mixed with 1% silver sulfadiazine showed better response than when 1% silver sulfadiazine was used alone possibly by potentiating the absorption of silver sulfadiazine across the mucosal membrane in the treated animals.[113]

Antimicrobial photodynamic therapy (PDT) is able to photoinactivate microorganisms produced by a light-activated photosensitizer (PS). PDT mediated by meso-mono-phenyl-tri(N-methyl-4-pyridyl)-porphyrin (PTMPP) to treat burn wounds in mice was used with 25% of 500 μM DMSO in mice with third-degree burn wounds 1 day after being infected with bioluminescent *Staphylococcus aureus*. When a PDT–DMSO–PTMPP mixture was used,[114] a substantial fluence-dependent reduction of the luminescence signal was obtained, indicating that more than 98% of the bacteria were eradicated. By contrast, PDT alone or light alone had no antimicrobial effect, and 500 μM PTMPP in PBS had only a small light dose-dependent effect on reducing the luminescence signal and did not appear to have significant antimicrobial activity against *S. aureus*.[114] The effect of DMSO in this preparation was thought to have been to disaggregate the PS molecules and to increase the antimicrobial PDT efficiency by the light-activated PS.[114]

Chemical burns have been treated with DMSO. Subcutaneous injections of 20% calcium gluconate in a solution of 25% DMSO significantly slowed the progress of hydrochloric acid burn in rat hindquarters during the first 24 h, markedly reduced damage, and enhanced tissue recovery for the remainder of the observation period.[115] Subcutaneous injections of calcium gluconate–DMSO were found to be more effective in this chemical burn model than topical application of this preparation.[115]

The anti-inflammatory and antioxidant effects of DMSO were tested in rats exposed to alkali-acid burns of the esophagus. The resulting corrosive esophagitis was treated with 10 mL/kg of a 10% DMSO solution given intraperitoneally 15 min following the corrosive esophagitis.[115a] DMSO treatment was continued every 12 h for 4 days, and the esophagi were removed postmortem after 7 days. Histopathological examination revealed that the alkali burn group treated with DMSO showed significantly reduced inflammation than the control alkali burn group. When alkali and acid burn groups treated with DMSO were compared, there was more inflammation in the DMSO-treated acid burn group, but histological analysis still showed significantly less damage than the acid control group. DMSO treatment also resulted in a significant decrease in the immunoreactivity of the oxidative stress marker nuclear factor-kappa B (NF-κB) in both the acid and alkali groups.[115a] This study suggested that DMSO is a promising agent in the treatment of experimental corrosive esophagitis possibly by suppressing NF-κB from rapidly activating transcription factors that express genes critical in the acute phase inflammatory response.

Corneal alkali injury was induced in rabbits and treated with daily topical applications of 20% DMSO. Corneal opacification at day 3 and corneal ulceration were significantly lower in the treatment group than in the control group, but this effect was not sustained during the rest of the study period.[116]

Not many studies have been reported on the effects of DMSO in scar formation. A study by Erk et al.[117] reported that incubation of human scar with 99% DMSO showed disruption of collagen fibers and dramatic loss of periodic cross-striations when examined by electron microscopy. Scar tissue incubated with 50% DMSO showed only moderate loss of the periodic striations in the collagen fibers and no fiber fraying.[117] These findings have intriguing implications to limit or abolish undesirable scar tissue in the clinical setting.

DMSO treatment administered intraperitoneally to rats for 35 days decreased experimentally induced intestinal adhesions by 80% over controls as compared to saline, cortisone acetate, or a combination of cortisone and DMSO administered separately.[118]

When administered next to a wound incision but not at the incision site itself in rabbits, 70% DMSO appeared to increase the development of wound tensile strength over controls.[119] Although 70% topical DMSO administered peripherally will also reduce local pain from the wound, effective pain analgesia with intrawound administration of very minute amounts of DMSO may be as effective or more so than peripheral wound administration. This intrawound application approach will be reviewed in detail later (see Section "Intrawound Administration of DMSO").

In 1968, McFarlane et al.[120] reported that DMSO could prevent necrosis in experimental skin flaps by increasing blood perfusion to the ischemic region. This observation was followed by many articles since then, which have confirmed this finding.[121] The mechanism for the protective effect on skin flaps by DMSO remains to be clarified, but one possibility may be the stimulation by DMSO of histamine, a strong vasodilator that could counter the vasoconstriction seen in dying skin flaps.[122,123] Another possibility is DMSO's action as an antioxidant and free radical scavenger that would likely worsen the skin flap necrosis.

Survival of skin flaps in surgery by preventing tissue necrosis is an important aspect of patient recovery. In one study, experimental skin flaps were elevated in rat abdomen, and the epigastric vein was occluded by a microvascular clamp for 8 h.[124] DMSO 1.5 g/kg was injected intraperitoneally only once at reperfusion or administered every day for 5 days. DMSO did not increase percent flap survival when given as a single dose at reperfusion compared with saline solution, but when given for 5 days at reperfusion and postoperatively, DMSO significantly increased flap survival.[124]

In plastic surgery, a direct relationship between heavy smoking and flap necrosis is well known. As an antioxidant and vasodilator, DMSO was given orally at 2 mL/kg to rats with random skin flaps that had been exposed to nicotine.[125] DMSO was shown to reduce the serum and skin levels of malondialdehyde, an indicator of oxidative stress in cells and tissues and a marker of lipid peroxidation, and to lessen the necrotic vasculopathy observed in controls following nicotine exposure.[125]

The topic of scarification after physical injury of CNS tissue and the use of DMSO in scar formation will be further examined in Chapter 8.

RESPIRATORY STIMULATION

The effects of DMSO and two analeptics, doxapram and ethamivan, were tested for their activity on respiratory minute volume (RMV) and respiratory rate in monkeys and rats.[126] These tests were inspired by the dramatic respiratory response of rhesus monkeys given IV DMSO following experimental traumatic brain injury, which produced a short period of apnea due to severe brain swelling[127] (see Chapter 8 for details). In a subsequent experiment, 10 rhesus monkeys were lightly anesthetized with sodium pentobarbital (a respiratory depressant), intubated, and given an IV injection of 0.09% saline; DMSO, 2 g/kg in a 50% solution; doxapram, 0.8 mg/kg; or ethamivan, 5 mg/kg. Doxapram is a stimulant of central respiratory centers used clinically at low IV doses in patients with respiratory depression due to ingestion of excessive doses of narcotics.

Doxapram is also indicated in preterm infants with prolonged periods of apnea that can lead to hypoxemia and neurological damage.[128] Doxapram is also used in pets presenting with acute respiratory problems associated with perioperative respiratory depression and neonatal hypoxia. Although not presently used anymore, ethamivan is a respiratory stimulant drug related to nikethamide and was mainly used in the treatment of barbiturate overdose and chronic obstructive pulmonary disease.[129]

Adverse effects of these and other respiratory stimulants include hypertension, skeletal muscle hyperactivity, tachycardia, and generalized CNS excitation, including seizures.

One minute after DMSO injection in the monkey experiment, RMV and RR showed a sharp rise in RMV of 295% together with a rise in RR of 49% of the preinjection levels. The stimulating effects of DMSO on RMV and RR remained high for 25 min and reached preinjection levels after 50 min. Some animals treated with DMSO urinated copiously during this time.[127] Injections of doxapram also

TABLE 2.1
Survival Rates of Hypoxemic Rats Exposed to 3 min of Pure Nitrogen in a Closed Chamber Pretreated with DMSO, Doxapram, and Ethamivan

Effects of DMSO on Experimental Hypoxemia in Rats

Agent	Survivors/Total	Survival, %	P Value
DMSO	18/20	90	<0.0025
Saline	3/20	15	
Doxapram	18/20	90	<0.0025
Saline	2/20	10	
Ethamivan	6/20	30	>0.4
Saline	3/20	15	

Note: See text for details.

increased RMV to 300%, while RR rose 37% of baseline levels. Ethamivan injection stimulated RMV and RR only slightly, and the effect lasted only 2 min. These changes by ethamivan were accompanied by severe convulsions, arousal, and nystagmus.[126] No changes were observed on RMV or RR when physiological saline was injected.

The results of respiratory stimulation by DMSO led to a second experiment to examine the effects of DMSO on experimental hypoxemia in rats. Rats were treated with intraperitoneal (IP) injections of DMSO 2 g/kg, doxapram 25 mg/kg, and ethamivan 10 mg/kg and were then exposed precisely for 210 s to pure nitrogen in a closed chamber. Each treated rat was paired with a saline control while in the nitrogen chamber. The results of this study are shown in Table 2.1. Survivors of nitrogen exposure experiencing acute hypoxia treated with DMSO were 18 of 20, with doxapram 18 of 20, and with ethamivan 6 of 20. Saline-treated rats' average for survival was 3 of 20 rats.[126]

The effect of DMSO on hypoxemic rats revealed that this treatment was superior in keeping animals alive when compared to the well-known respiratory stimulant ethamivan and had similar effects as doxapram. When nitrogen exposure was increased to 215 s, DMSO-treated rats showed more resistance and better survival rate to hypoxic death than did doxapram-treated rats. The results of these studies on rhesus monkeys and rats undergoing either narcotic-induced respiratory depression or nitrogen-induced hypoxia indicate that DMSO administration may reverse respiratory distress in humans. This aspect is discussed further in the text.

ACUTE RESPIRATORY DISTRESS SYNDROME

These experiments indicated that DMSO has the potential to quickly stimulate respiratory centers in rats and nonhuman primates. These data were the basis for a study in humans presenting with acute respiratory distress syndrome (ARDS). ARDS is characterized by interstitial edema, low paO_2, rising $paCO_2$, and complete opacification of both lung fields on chest x-ray. A 10% DMSO solution was given intravenously to three patients nearing death presenting with ARDS who were refractory to classical therapy. All three patients showed marked improvement 1 h after the application of DMSO with respect to oxygen saturation, increased paO_2, and lowered $paCO_2$. One patient showed a dramatic improvement of the chest x-ray, which resolved over a period of 1 week. It was concluded by the author of the study that DMSO use in ARDS merits further investigation.[130]

A model of ARDS using formylated tripeptide formyl-norleucyl-leucyl-phenylalanine (FNLP) was instilled intratracheally into hamsters to induce acute edematous lung injury. Intratracheal FNLP is time dependent and dose dependent and increases neutrophils to create neutrophil alveolitis and leak of intravenously injected albumin into the extravascular lung space (lung leak).[131]

It was observed that treatment with 0.20% DMSO significantly decreased neutrophil alveolitis and lung leak in hamsters following FNLP intratracheally, possibly by inhibiting neutrophil chemotaxis.[131]

DMSO is reported to increase hypoxic tolerance of brain slices under *hypoxic* or *ischemic* conditions. After the induction of hypoxia or ischemia in the CA1 region of hippocampal slices of adult guinea pigs, evoked potentials recording showed that during superfusion with artificial CSF containing DMSO 14.1 mmol/L the latency of anoxic terminal negativity was markedly improved when compared to the control brain slices, demonstrating a neuroprotective effect of DMSO on ischemia/hypoxia.[132]

EXPERIMENTAL BLUNT CHEST TRAUMA

Blunt chest trauma is an injury to the chest caused by an impact from a blunt object. Blunt chest trauma puts multiple adjoining structures at risk of injury. This type of injury can involve cardiac contusion, cardiovascular injury, and pulmonary lacerations. Traffic accidents are the most frequent cause of blunt chest trauma and cardiac contusions due to acceleration–deceleration and usually involving a direct blow to the chest. Many other causes can result in blunt chest trauma, including a violent fall, explosion, aggression, and various types of high-risk sports.

The majority of blunt chest injuries can be classified into three types: minor chest wall injuries, major but stable chest wall injuries, and flail chest injuries. The first two chest injuries can involve fracture of the bony chest wall or no fracture. Flail chest is a life-threatening medical condition where the rib cage breaks under extreme stress, damaging ribs in multiple places where they become detached from the rest of the chest wall moving independently of the rib cage.

To study traumatic pulmonary contusion, DMSO was compared to the powerful anti-inflammatory agent dexamethasone using a designed thoracic trauma model in rats.[132a] DMSO (1.2 g/kg) and dexamethasone (10 mg/kg) were given intraperitoneally 15 min prior to injury. The effects of blunt trauma to lung tissue was examined biochemically and histopathologically. A number of pathologic signs were observed in all animal groups following blunt trauma, including alveolar edema, congestion, and pulmonary parenchymal hemorrhage.[132a]

The levels of oxidative stress predictors malondialdehyde and nitric oxide were lower with both dexamethasone- or DMSO-treated animals when compared with untreated blunt chest trauma animals. This finding indicated the well-known antioxidant properties of DMSO and dexamethasone in inflammatory conditions. However, further analysis revealed that only DMSO-treated animals prevented further injury by decreasing neutrophil infiltration and endothelial cell injury of blood vessels near the injury site following the lung contusion.[132a]

As we have reviewed in Chapter 8, the production of excessive free radicals and the release of local inflammatory mediators including cytokines such as interleukin 6, tumor necrosis factor-α, and nitric oxide are generally the culprits responsible for the deadly pathologic cascades seen following traumatic brain injury. DMSO is a powerful free radical scavenger and effective anti-inflammatory agent that is able to quell the cytokine *storm* seen after the blunt chest trauma. In addition, these cytokines can lead to complications affecting immunosuppression, leading to multi organ damage.[72]

AUTOPHAGY

Autophagy is a dynamic process of subcellular component degradation, which has sparked a great deal of interest in the last decade since it is now recognized to be involved in various developmental processes affecting certain disorders, including neurodegeneration and cancer.[133]

Autophagy is a basic catabolic mechanism in eukaryotes that involves cell degradation of unnecessary or dysfunctional cellular components in the cytoplasm and sequesters these into double-membrane vesicles where they are delivered to the lysosome/vacuole for breakdown and eventual recycling.[134] The breakdown of cellular components can ensure cellular survival during nutrient starvation by maintaining cellular energy levels.[134] Normally, when autophagy is regulated, synthesis, degradation, and recycling of these cellular components are maintained. Targeted cytoplasmic components are isolated from the rest of the cell within the autophagosomes, which are then fused with lysosomes and degraded or recycled. The relevance of autophagy in disease is seen as an adaptive response to survival, but in certain circumstances, autophagy has also been described as a promoter of cell death and morbidity.[134]

Three different forms of autophagy have been described. These are macroautophagy, microautophagy, and chaperone-mediated autophagy (CMA).[135]

Autophagy can function as a cytoprotective mechanism, but it also has the capacity to cause cell death. Autophagy is a major mechanism for degrading long-lived cytosolic proteins and the only known pathway for degrading organelles.[136] Autophagy is activated by many forms of stress, including nutrient and energy starvation, oxidative stress, mitochondrial dysfunction, endoplasmic reticulum stress, and infections. Autophagy recycles amino acids and fatty acids to produce energy while removing damaged cell components, thereby playing an essential role in cell survival and intracellular surveillance mechanism, which is indispensable for maintaining cell health. When this process is abnormal, autophagy leads to cell death.[137]

CMA is a selective type of autophagy by which specific cytosolic proteins are sent to lysosomes for degradation.

The chemical chaperones in the cytosol normally unfold substrate proteins prior to their translocation across the lysosomal membrane, but when this process is aberrant, DMSO administration, acting a chemical chaperone, may induce CMA and exert a protective effect toward the cell's survival (see also Chapter 1, "DMSO as a Chemical Chaperone").

For example, DMSO appears to have a beneficial role in inducing autophagy as a chemical chaperone by reducing free fatty acid–induced hepatic fat accumulation in a dose-dependent manner.[138] Treatment with 0.01% DMSO for 16 h significantly reduced palmitate-induced triglyceride accumulation in rat hepatocytes.[138] This effect by DMSO was postulated to result from the downregulation of the ATF4/AKT1 pathway, a stress-induced transcription factor that is overexpressed in many cancers and is an important mediator of the unfolded protein response.[138,139]

DIURESIS

DMSO is a powerful diuretic when used intravenously. A fivefold increase in urine excretion as compared to saline controls is seen after 1 h of 40% DMSO IV given at 2.5 g/kg in dogs with spinal cord trauma.[127] In rats, DMSO given topically five times daily increased urine volume 10-fold.[140] The brisk diuretic response to 40% DMSO infusion has been seen in patients treated for high intracranial pressure secondary to head injury.[141] Diuresis after high doses of DMSO is accompanied by electrolyte changes involving sodium and potassium excretion,[142] but these changes can be minimized by reducing excessive DMSO fluid overload and by carefully monitoring electrolytes after high and repeated dosing with DMSO.[143]

These collective findings suggest that DMSO merits clinical evaluation as a potent and primary treatment when diuresis is indicated in water retention for such conditions as cirrhosis, acute or chronic renal failure, nephrotic syndrome, and congestive heart failure.

CHOLINESTERASE INHIBITION

DMSO is reported to inhibit the isolated and innervated guinea pig atrium following incubation in a bath mimicking the action of neostigmine by lowering the vagal threshold to electrical stimulation.[144] It was found that this action by DMSO at doses of 0.01–1.0 mol/L incubated for 30 min in the guinea pig atrium bath was able to inhibit erythrocyte cholinesterase, the enzyme that breaks down the neurotransmitter acetylcholine.[144]

The cholinesterase inhibitory activity by DMSO could reasonably target the lowering of high blood pressure, bronchoconstriction, and intraocular pressure. Despite their annoying gastrointestinal effects, cholinesterase inhibitors are also used in treating the symptoms of Alzheimer's disease,[138] a disorder where DMSO dose adjustment could avoid the gastrointestinal side effects seen with stronger cholinesterase drugs. This aspect of DMSO therapy will be more fully discussed in Chapter 8.

DMSO was studied at the neuromuscular junction of the frog cutaneous pectoris using electrophysiological techniques and found to elevate the amplitude of extracellularly recorded miniature end-plate potentials as well as the time constant of their monoexponential decay. These findings indicated that DMSO partially blocked cholinesterase activity and the presynaptic action of acetylcholine release,[145] an action that has been confirmed on a variety of invertebrate muscles.[146]

SOLVENT ACTION

DMSO is a solvent for many aromatic and unsaturated hydrocarbons as well as inorganic salts and nitrogen-containing compounds. This ability to act as a universal solvent may be due to DMSO's high dielectric constant due to the polarity of the sulfur–oxygen bond.

DMSO has been shown to carry physostigmine and phenylbutazone through the skin of the rat. Moreover, there is evidence that DMSO enhances the percutaneous absorption of many other compounds, including steroids, vasoconstrictors, antiperspirants, and various dyes.[147–149] Additionally, DMSO has been mixed with skin antiseptics such as hexachlorophene, the anthelmintic thiabendazole, and both estrogens and corticoids, which, when applied topically in DMSO, exerted their characteristic biologic effect in immature female rats.[150,151]

There are thousands of compounds used in pharmacology research that have been reported to be solubilized in varying concentrations of DMSO. Balakin et al.[152] have compiled a comprehensive reference database of proprietary data on compound solubility and shown that, as of 2004, 55,277 organic compounds possess good DMSO solubility and 10,223 organic compounds possess poor DMSO solubility.

One caveat when using DMSO as a solvent in testing the activity of a drug as a potential treatment is the singular action exerted by DMSO, which, in some cases, may show more activity than the drug being tested.[76] This error, of course, can be easily avoided by testing DMSO alone as a control during the testing procedure, but it does not avoid the error that the inclusion of DMSO in the drug preparation may synergistically enhance and improve the experimental drug's effect by increasing its penetration into the lesion site or by prolonging its kinetic distribution.

VETERINARY USES

DMSO has been approved for two veterinary products in the United States. Domoso (90% DMSO) is indicated for acute swelling in dogs and horses, which can be applied as a gel or liquid preparation. Synotic is 60% DMSO and 0.01% fluocinolone acetonide, indicated for the relief of pruritus and inflammation associated with acute and chronic otitis in the dog or horse. The majority of veterinary reports published on the use of DMSO in animals concern horses and dogs.

Many of the injuries seen in horses have been successfully treated with DMSO especially when edema and inflammation are involved in the lower extremities. Some conditions amenable to DMSO treatment include tendonitis, periosteal involvement of metacarpal bones, osteoarthritis, bursitis, open wound infections, and traumatic injuries involving tissue swelling and hemorrhage. The usual treatment dose for these equine conditions is 1 g/kg body weight intravenously of DMSO in a 40% solution with a maximum duration of treatment of 5 days.[153]

DMSO has been used for open wounds in horses where it appeared to stimulate healthy granulation of the wound within the first few days.[154] The granulation tissue after a 2-week course of DMSO three times/day was reduced to normal within 30 days.[154]

Endotoxemia is a severe and ubiquitous disease in horses receiving IV lipopolysaccharide. IV DMSO at 1 g/kg body weight improved the effect of lipopolysaccharide on fever.[155]

In a double-blinded, crossover, paired study with a 1-week washout period, midcarpal joints in horses with acute inflammation were topically treated with DMSO gel 15 g at 90%, which was seen to penetrate in the synovial fluid by the use of gas chromatography in sufficient quantities to decrease joint inflammation.[156]

Fungal infection in the eyes of horses has been treated with DMSO in combination with antifungal agents. Horses with keratomycoses were given DMSO–itraconazole 0.25 mL of a 1% itraconazole with 30% DMSO in a petrolatum-based ointment every 4 h for 16–53 days, and it was seen that 8 of 10 eyes treated with this mixture resolved completely.[157]

DMSO has been used in small animal practice mostly involving dogs. The conditions that are reported to be helped by DMSO include traumatic edema, sprains, tendonitis, intervertebral disk disease, mastitis, and soft tissue inflammatory problems.[158]

Uveitis was induced in dogs by intracameral injection of canine lens protein, and after DMSO application, intraocular pressure and fibrin production were decreased after treatment possibly from the antagonism of DMSO to lipoxygenase formation.[159]

One interesting report concerns the anecdotal use of DMSO for 2 years in a dog diagnosed with hypoalbuminemia, proteinuria, and renal amyloidosis. Two years after the initiation of DMSO treatment, the 24 h urinary protein excretion returned to normal, while serum albumin concentrations increased to within normal range.[160] These findings would need to be replicated in a larger study but may serve as a clue of the ability of DMSO to act as a chemical chaperone in protracted disorders such as renal amyloidosis, a progressive and fatal disease that is difficult to treat or cure.

TERATOLOGY AND LD$_{50}$

The intraperitoneal administration of 5.5 g/kg of DMSO as a single dose to pregnant hamsters induced developmental malformations of their embryos.[161]

Both DMSO and diethyl sulfoxide are teratogenic when injected into the chick embryo, the classification of malformations being dependent upon the stage of embryonic development at the time of treatment. The same drugs when administered by various techniques to mice, rats, and rabbits in which fertility had been established did not cause any embryonic malformations.[162]

Teratogenic effects produced in laboratory animals were not observed when therapeutic doses of DMSO were used, and LD$_{50}$ could be reached only by going far beyond the therapeutic doses.[162] The LD$_{50}$ of DMSO in animals varies depending on the species and the route of administration (Table 2.2).

However, the toxicity of DMSO in pregnant or lactating women has not been determined and is consequently not recommended in his population group.

TABLE 2.2
LD$_{50}$ for Various Animal Species in g/kg Using Different Routes of Administration

LD$_{50}$ of DMSO in Animals

Species	Route of Administration	g/kg
Mouse	SQ	13.9–20.5
Mouse	IV	3.8–10.7
Mouse	PO	15–22
Mouse	IP	20.0
Rat	IV	5.2–5.3
Rat	PO	16.0–28.3
Rat	IP	6.5–13.6
Dog	IV	2.5
Guinea pig	IP	6.5
Chicken	PO	12.5

Note: SQ, subcutaneous; IV, intravenous; PO, oral; IP, intraperitoneal.

OCULAR EFFECTS

Following high oral or topical administration of DMSO, certain eye changes in the dog have been reported.[163] The lens changes were first observed in dogs receiving 5 g DMSO/kg after 9 weeks of administration.[163] At lower dose levels, the change was observed later. The lens alteration consisted mainly of a change in the refractive index of the lens.[163] The lens changes were characterized by a decrease in the normal relucency of the lens cortex, causing the normal central zone of the lens to act as a biconvex lens. The functional effect of this refractive change would be a tendency toward myopia.[163]

In swine, dermal application of 4.5 g 90% DMSO/kg twice daily caused similar lens changes by 90 days of treatment. The eye changes were slowly reversible after DMSO was discontinued but with a definite species difference, the dog being the slowest to exhibit improvement.[164]

Ocular effects in dogs started after 5–10 weeks of dosing of DMSO at 9 mL/kg and involved nuclear lens changes with changes of the refractive index plus transitory equatorial opacities during the fifth month and changes in the vitreous humor.[165] Similar changes were observed when DMSO was given more slowly at 3 mL/kg, with alterations to the vitreous being first seen after 9–10 months at this dose level. Progressive refractive index changes occurred when DMSO was given for 6 months at 1 mL/kg but none of the treated animals showed nuclear lens opacity. When DMSO dosing was stopped, refractive nuclear changes were no longer seen.[165]

Daily IV doses of 3 g/kg DMSO in a 40% solution given to rhesus monkeys for nine consecutive days did not lead to any lens changes.[166] The monkeys were monitored before and after treatment for 4 months for changes in blood chemistry, hematology, urine, ocular, neurological, and cardiovascular systems. No significant or long-lasting changes were recorded in any of the parameters studied when these data were compared to monkeys receiving IV daily physiological saline solution.[166] These negative findings of lens changes were confirmed in monkeys receiving 11 g/kg oral DMSO/day for 1 year.[167] The unique lens changes reported in some animal species have not been reported in 213 human volunteers receiving 1 g/kg of 80% DMSO applied dermally per day for 14 days and followed up for 90 days.[167]

INTRAWOUND ADMINISTRATION OF DMSO

Managing pain is very important as we have reviewed earlier, but even more so when pain is due to tissue wounds. Tissue wounds can lead to infection, tissue scarring, and tightening of the skin, which can make joint mobility difficult and painful. These effects can often be minimized with good pain control medication. Thus, pain needs to be treated as soon as possible to avoid further tissue damage.

Most of the pain medications for penetrating wounds of the skin have a systemic effect, and few are effective when given locally. DMSO has never been used following direct intrawound administration. The advantages of intrawound administration are that a local rather than systemic effect on pain can be achieved, thus reducing side effects of the medication.

One study has reported that damage to the plantar surface of a hind paw in rats received intrawound administration of DMSO.[168] Before the closure of the wound, DMSO was administered using a sterile pipette at the minimal dose of 10 µL in a 100% solution and applied deep into the damaged plantar tissue. The application of intrawound DMSO administration resulted in significant control of pain when a thermal hyperalgesia stimulus was applied to the affected paws of the treated rats but not to control rats receiving a placebo. The authors concluded that DMSO appeared effective in the treatment of acute pain and should be further tested in tissue injuries resulting in wound damage and burns.[168]

REFERENCES

1. Hucker HB, Miller JK, Hochberg A, Brobyn RD, Riordan FH, Calesnick B. Studies on the absorption, excretion and metabolism of dimethylsulfoxide (DMSO) in man. *J Pharmacol Exp Ther*. 1967;155(2):309–317.
2. Egorin MJ, Rosen DM, Sridhara R, Sensenbrenner L, Cottler-Fox M. Plasma concentrations and pharmacokinetics of dimethylsulfoxide and its metabolites in patients undergoing peripheral-blood stem-cell transplants. *J Clin Oncol*. 1998;16(2):610–615.
3. Kolb KH, Jaenicke G, Kramer M, Schultze PE. Absorption, distribution and elimination of labeled dimethyl sulfoxide in man and animals. *Ann NY Acad Sci*. 1967;141:85–95.
4. Denko CW, Goodman RM, Miller R, Donovan T. Distribution of dimethyl sulfoxide-35S in the rat. *Ann NY Acad Sci*. 1967;141(1):77–84.
5. Layman DL, Jacob SW. The absorption, metabolism and excretion of dimethyl sulfoxide by rhesus monkeys. *Life Sci*. 1985;37(25):2431–2437.
6. Notman R, Anwar J. Breaching the skin barrier—Insights from molecular simulation of model membranes. *Adv Drug Deliv Rev*. 2013;65:237–250.
7. Yoshiike T, Aikawa Y, Sindhvananda J, Suto H, Nishimura K, Kawamoto T, Ogawa H. Skin barrier defect in atopic dermatitis: Increased permeability of the stratum corneum using dimethyl sulfoxide and theophylline. *J Dermatol Sci*. 1993;5(2):92–96.

8. Capriotti K, Capriotti JA. Dimethyl sulfoxide: History, chemistry, and clinical utility in dermatology. *J Clin Aesthet Dermatol.* 2012;5(9):24–26.

9. Gurtovenko AA, Anwar J. Modulating the structure and properties of cell membranes: The molecular mechanism of action of dimethyl sulfoxide. *J Phys Chem B.* September 6, 2007;111(35):10453–10460.

10. Wang H, Zhong CY, Wu JF, Huang YB, Liu CB. Enhancement of TAT cell membrane penetration efficiency by dimethyl sulphoxide. *J Control Release.* April 2, 2010;143(1):64–70.

11. Welch WJ. Role of quality control pathways in human diseases involving protein misfolding. *Semin Cell Develop Biol.* 2004;15:31–38.

12. Sawkar AR, Cheng WC, Beutler E, Wong CH, Balch WE, Kelly JW. Chemical chaperones increase the cellular activity of N370S beta-glucosidase: A therapeutic strategy for Gaucher disease. *Proc Natl Acad Sci USA.* 2002;99(24):15428–15433.

13. Grune T, Jung T, Merker K, Davies KJ. Decreased proteolysis caused by protein aggregates, inclusion bodies, plaques, lipofuscin, ceroid, and 'aggresomes' during oxidative stress, aging, and disease. *Int J Biochem Cell Biol.* 2004;36(12):2519–2530.

14. Gregersen N, Bross P, Vang S, Christensen JH. Protein misfolding and human disease. *Annu Rev Genomics Hum Genet.* 2006;7:103–124.

15. Arakawa T, Ejima D, Kita Y, Tsumoto K. Small molecule pharmacological chaperones: From thermodynamic stabilization to pharmaceutical drugs. *Biochim Biophys Acta.* 2006;1764(11):1677–1687.

16. Woolf CJ. What is this thing called pain? *J Clin Invest.* 2010;120(11):3742–3744.

17. Chapman CR, Stillman M. Pathological pain. In *Handbook of Perception: Pain and Touch,* ed. L. Krueger. Academic Press, New York, 1996; pp. 315–340.

18. Paulozzi LJ, Weisler RH, Patkar AA. A national epidemic of unintentional prescription opioid overdose deaths: How physicians can help control it. *J Clin Psych.* 2011;72(5):589–592.

19. Dunn KM, Saunders KW, Rutter CM et al. Opioid prescriptions for chronic pain and overdose. *Ann Intern Med.* 2010;152:85–92.

20. Jacob SW, Bischel M, Herschler RJ. Dimethyl sulfoxide (DMSO): A new concept in pharmacotherapy. *Curr Ther Res.* 1964;6:134–135.

21. D'Amour FE, Smith DL. A method for determining loss of pain sensation. *J Pharmacol Exp Ther.* 1941;72:74–78.

22. Eddy NB, Leimbach D. Synthetic analgesics. II. Dithienylbutenyl- and dithienylbutyl-amines. *J Pharmacol Exp Ther.* 1953;107:385–393.

23. Matsumoto J. Clinical trials of dimethyl sulfoxide in rheumatoid arthritis patients in Japan. *Ann NY Acad Sci.* 1967;141(1):560–568.

24. Blumenthal LS, Fuchs M. The clinical use of dimethyl sulfoxide on various headaches, musculoskeletal, and other general medical disorders. *Ann NY Acad Sci.* 1967:141(1):572–578.

25. Arno IC, Wapner PM, Brownstein IE. Experiences with DMSO in relief of postpartum episiotomy pain. *Ann NY Acad Sci.* 1967;141(1):403–405.

26. Vuopala U, Vesterinen E, Kaipainen WJ. The analgetic action of dimethyl sulfoxide (DMSO) ointment in arthrosis. A double blind study. *Acta Rheumatol Scand.* 1971;17(1):57–60.

27. Demos CH, Beckkloff GL, Donin MN, Oliver PM. Dimethyl sulfoxide in musculoskeletal disorders. *Ann NY Acad Sci.* 1967;141:517–523.

28. Percy EC, Carson JD. The use of DMSO in tennis elbow and rotator cuff tendonitis: A double-blind study. *Med Sci Sports Exerc.* 1981;13(4):215–219.

29. Haigler HJ, Spring DD. Comparison of the analgesic effects of dimethyl sulfoxide and morphine. *Ann NY Acad Sci.* 1983;411:19–27.

30. Evans MS, Reid KH, Sharp JB Jr. Dimethylsulfoxide (DMSO) blocks conduction in peripheral nerve C fibers: A possible mechanism of analgesia. *Neurosci Lett.* 1993;150(2):145–148.
31. Douglas WW, Ritchie JM. Mammalian nonmyelinated nerve fibers. *Physiol Rev.* April 1962;42:297–334.
32. Sawada M, Sato M. The effect of dimethyl sulfoxide on the neuronal excitability and cholinergic transmission in *Aplysia* ganglion cells. *Ann NY Acad Sci.* 1975;243:337–357.
33. Davis HL, Davis NL, Clemons AL. Procoagulant and nerve-blocking effects of DMSO. *Ann NY Acad Sci.* 1967;141(1):310–325.
34. Becker DP, Young HF, Nuisen FE, Jane JA. Physiological effects of dimethyl sulfoxide on peripheral nerves: Possible role in pain relief. *Exp Neurol.* 1969;24:272–276.
35. Shealey CN. The physiological substrate or pain. *Headache.* 1966;6:101–108.
36. Colucci M, Maione F, Bonito MC, Piscopo A, Di Giannuario A, Pieretti S. New insights of dimethyl sulphoxide effects (DMSO) on experimental in vivo models of nociception and inflammation. *Pharmacol Res.* 2008;57(6):419–425.
37. Lu C, Mattson MP. Dimethyl sulfoxide suppresses NMDA- and AMPA-induced ion currents and calcium influx and protects against excitotoxic death in hippocampal neurons. *Exp Neurol.* 2001;170(1):180–185.
38. Hoang BX, Levine SA, Shaw DG, Tran DM, Tran HQ, Nguyen PM, Tran HD, Hoang C, Pham PT. Dimethyl sulfoxide as an excitatory modulator and its possible role in cancer pain management. *Inflamm Allergy Drug Targets.* 2010;9(4):306–312.
39. Woosley RL, Funck-Brentano C. Overview of the clinical pharmacology of antiarrhythmic drugs. *Am J Cardiol.* January 15, 1988;61(2):61A–69A.
40. Ogura T, Shuba LM, McDonald TF. Action potentials, ionic currents and cell water in guinea pig ventricular preparations exposed to dimethyl sulfoxide. *J Pharmacol Exp Ther.* 1995;273(3):1273–1286.
41. Hulsmann S, Greiner C, Kohling R. Dimethyl sulfoxide increases latency of anoxic terminal negativity in hippocampal slices of guinea pig in vitro. *Neurosci Lett.* 1999,261:1–4.
42. Campbell JN, Meyer RA. Mechanisms of neuropathic pain. *Neuron.* 2006;52(1):77–92.
43. Zimmermann M. Pathobiology of neuropathic pain. *Eur J Pharmacol.* 2001;429(1–3):23–37.
44. Daher JB, Tonussi CR. A spinal mechanism for the peripheral anti-inflammatory action of indomethacin. *Brain Res.* 2003;962(1–2):207–212.
45. Morris CJ. Carrageenan-induced paw edema in the rat and mouse. *Methods Mol Biol.* 2003;225:115–121.
46. Formanek K, Kovak W. Die wirkung von DMSO auf experimentell erzeugte rattenpfotenodeme. In *DMSO Symposium*, Vienna, Austria. Saladruk, Berlin, Germany, 1966; pp. 18–26.
47. Görög P, Kovács IB. Effect of dimethyl sulfoxide (DMSO) on various experimental inflammations. *Curr Ther Res Clin Exp.* 1968;10(9):486–492.
48. Panganamala RV, Sharma HM, Heikkila RE, Geer JC, Cornwell DG. Role of hydroxyl radical scavengers dimethyl sulfoxide, alcohols and methional in the inhibition of prostaglandin biosynthesis. *Prostaglandins.* 1976;11(4):599–607.
49. Spagnuolo C, Sautebin L, Galli G, Racagni G, Galli C, Mazzari S, Finesso M. PGF2 alpha, thromboxane B2 and HETE levels in gerbil brain cortex after ligation of common carotid arteries and decapitation. *Prostaglandins.* 1979;18(1):53–61.
50. Powell WS, Hammarström S, Samuelsson B, Miller WL, Sun FF, Fried J, Lin CH, Jarabak J. Interactions between prostaglandin analogues and a receptor in bovine corpora lutea. Correlation of dissociation constants with luteolytic potencies in hamsters. *Eur J Biochem.* 1975;59(1):271–276.
51. Rao CV. Differential effects of detergents and dimethylsulfoxide on membrane prostaglandin E1 and F2alpha receptors. *Life Sci.* 1977;20(12):2013–2022.

52. Whittle BJ, Moncada S, Vane JR. Comparison of the effects of prostacyclin (PGI2), prostaglandin E1 and D2 on platelet aggregation in different species. *Prostaglandins.* 1978;3:373–378.

53. Matthews JS, Jones RL. Potentiation of aggregation and inhibition of adenylate cyclase in human platelets by prostaglandin E analogues. *Br J Pharmacol.* 1993;108:363–369.

54. Gray SJ, Heptinstall S. Interactions between prostaglandin E2 and inhibitors of platelet aggregation which act through cyclic AMP. *Eur J Pharmacol.* 1991;194:63–70.

55. de la Torre JC. Role of dimethyl sulfoxide in prostaglandin-thromboxane and platelet systems after cerebral ischemia. *Ann NY Acad Sci.* 1983;411:293–308.

56. Gorman RR, Bunting S, Miller OV. Modulation of human platelet adenylate cyclase by prostacyclin (PGX). *Prostaglandins.* 1977;13(3):377–388.

57. Moncada S, Needleman P, Bunting S, Vane JR. Hamberg M, Samuelsson B. Identification of an enzyme in platelet microsomes which generates thromboxane A, from prostaglandin endoperoxides. *Nature.* 1976;261:558–560.

58. Palumbo B, Oguogho A, Fitcha P, Sinzinger H. Prostaglandin E1 therapy reduces circulating adhesion molecules (ICAM-1, E-selectin, VCAM-1) in peripheral vascular disease. *Vasa.* 2000;29:179–185.

59. Creutzig A, Lehmacher W, Elze M. Meta-analysis of randomised controlled prostaglandin E1 studies in peripheral arterial occlusive diseases stages III and IV. *Vasa.* 2004;33:137–144.

60. Takai S, Jin D, Kawashima H et al. Anti-atherosclerotic effects of dihomo-gamma-linolenic acid in ApoE-deficient mice. *J Atheroscler Thromb.* 2009;16:480–489.

61. Gross S, Tilly P, Hentsch D, Vonesch JL, Fabre JE. Vascular wall-produced prostaglandin E2 exacerbates arterial thrombosis and atherothrombosis through platelet EP3 receptors. *J Exp Med.* 2007;204:311–320.

62. Fabre JE, Nguyen M, Athirakul K et al. Activation of the murine EP3 receptor for PGE2 inhibits cAMP production and promotes platelet aggregation. *J Clin Invest.* 2001;107:603–610.

63. Nordlund JJ, Ackles AE, Traynor FF. The proliferative and toxic effects of ultraviolet light and inflammation on epidermal pigment cells. *J Invest Dermatol.* 1981;77(4):361–368.

64. Muroya M, Chang K, Uchida K, Bougaki M, Yamada Y. Analysis of cytotoxicity induced by proinflammatory cytokines in the human alveolar epithelial cell line A549. *Biosci Trends.* 2012;6(2):70–80.

65. Hollebeeck S, Raas T, Piront N, Schneider YJ, Toussaint O, Larondelle Y, During A. Dimethyl sulfoxide (DMSO) attenuates the inflammatory response in the in vitro intestinal Caco-2 cell model. *Toxicol Lett.* 2011;206:268–275.

66. Hassan AS. The effect of dimethyl sulfoxide on cholesterol and bile acid metabolism in rats. *Proc Soc Exp Biol Med.* November 1987;186(2):205–210.

67. Ganote CE, Sims M, Safavi S. Effects of dimethylsulfoxide (DMSO) on the oxygen paradox in perfused rat hearts. *Am J Pathol.* 1982;109:270–276.

68. Finney JW, Urschel HC, Balla GA, Race GJ, Jay BE, Pingree HP, Dorman HL, Mallams JT. Protection of the ischemic heart with DMSO alone or DMSO with hydrogen peroxide. *Ann NY Acad Sci.* 1967;141(1):231–241.

69. Levett JM, Johns LM, Grina NM, Mullan BF, Kramer JF, Mullan JF. Effects of dimethyl sulfoxide on systemic and cerebral hemodynamic variables in the ischemic canine myocardium. *Crit Care Med.* 1987;15(7):656–660.

70. Parisi A, Alfieri A, Mazzella M, Mazzella A, Scognamiglio M, Scognamiglio G, Mascolo N, Cicala C. Protective effect of dimethyl sulfoxide on acute myocardial infarction in rats. *J Cardiovasc Pharmacol.* 2010;55(1):106–109.

71. Shimizu S, Simon RP, Graham SH. Dimethylsulfoxide (DMSO) treatment reduces infarction volume after permanent focal cerebral ischemia in rats. *Neurosci Lett.* 1997;239:125–127.

72. Jacob SW, de la Torre JC. Pharmacology of dimethyl sulfoxide in cardiac and CNS damage. *Pharmacol Rep.* 2009;61(2):225–235.

73. Shlafer M, Karow AM. Pharmacological effects of dimethyl sulfoxide in the mammalian myocardium. *Ann NY Acad Sci.* 1975;243:110–121.

74. Shlafer M. Cardiac pharmacology of dimethyl sulfoxide and its postulated relevance to organ preservation in ischemic or hypoxic states. *Ann NY Acad Sci.* 1983;411:170–179.

75. Spilker B. Pharmacological studies on dimethyl sulfoxide. *Arch Int Pharmacodyn Therap.* 1972;200:153–167.

76. Hameroff SR, Otto CW, Kanel J, Weinstein PR, Blitt CD. Acute cardiovascular effects of dimethyl sulfoxide. *Ann NY Acad Sci.* 1983;411:94–99.

77. Pace DG, Kovacs JL, Klevans LR. Dimethyl sulfoxide inhibits platelet aggregation in partially obstructed canine coronary vessels. *Ann NY Acad Sci.* 1983;411:352–356.

78. Dujovny M, Rozario R, Kossovsky N, Diaz FG, Segal R. Antiplatelet effect of dimethyl sulfoxide, barbiturates, and methyl prednisolone. *Ann NY Acad Sci.* 1983;411:234–244.

79. Holtz GC, Davis RB. Inhibition of human platelet aggregation by dimethylsulfoxide, dimethylacetamide, and sodium glycerophosphate. *Proc Soc Exp Biol Med.* 1972;141(1):244–248.

80. Harrison MJ, Emmons PR, Mitchell JR. The effect of sulphydryl and enzyme inhibitors on platelet aggregation in vitro. *Thromb Diath Haemorrh.* 1966;16(1):122–133.

81. Wieser PB, Zeiger MA, Fain JN. Effects of dimethylsulfoxide on cyclic AMP accumulation, lipolysis and glucose metabolism of fat cells. *Biochem Pharmacol.* 1977;26(8):775–778.

82. Saari JT. Chronic treatment with dimethyl sulfoxide protects against cardiovascular defects of copper deficiency. *Proc Soc Exp Biol Med.* 1989;190(1):121–124.

83. Rosenblum W. Dimethyl sulfoxide effects on platelet aggregation and vascular reactivity in pial microcirculation. *Ann NY Acad Sci.* 1983;411:110–119.

84. Repine JE, Pfenninger OW, Talmage DW, Berger EM, Pettijohn DE. Dimethyl sulfoxide prevents DNA nicking mediated by ionizing radiation or iron/hydrogen peroxide-generated hydroxyl radical. *Proc Natl Acad Sci USA.* 1981;78(2):1001–1003.

85. Santos NC, Figueira-Coelho J, Martins-Silva J, Saldanha C. Multidisciplinary utilization of dimethyl sulfoxide: Pharmacological, cellular, and molecular aspects. *Biochem Pharmacol.* 2003;65(7):1035–1041.

86. Beilke MA, Collins-Lech C, Sohnle PG. Effects of dimethyl sulfoxide on the oxidative function of human neutrophils. *J Lab Clin Med.* 1987;110(1):91–96.

87. Magovern GJ Jr, Bolling SF, Casale AS, Bulkley BH, Gardner TJ. The mechanism of mannitol in reducing ischemic injury: Hyperosmolarity or hydroxyl scavenger? *Circulation.* 1984;70(3 Pt 2):I91–I95.

88. Stewart JR, Blackwell WH, Crute SL, Loughlin V, Greenfield LJ, Hess ML. Inhibition of surgically induced ischemia/reperfusion injury by oxygen free radical scavengers. *J Thorac Cardiovasc Surg.* 1983;86(2):262–272.

89. Arroyo CM, Kramer JH, Dickens BF, Weglicki WB. Identification of free radicals in myocardial ischemia/reperfusion by spin trapping with nitrone DMPO. *FEBS Lett.* 1987;221(1):101–104.

90. Chambers DJ, Braimbridge MV, Hearse DJ. Free radicals and cardioplegia. Free radical scavengers improve postischemic function of rat myocardium. *Eur J Cardiothorac Surg.* 1987;1(1):37–45.

91. Camici GG, Steffel J, Akhmedov A et al. Dimethyl sulfoxide inhibits tissue factor expression, thrombus formation, and vascular smooth muscle cell activation: A potential treatment strategy for drug-eluting stents. *Circulation.* 2006;114(14):1512–1521.

92. Mackman N. Role of tissue factor in hemostasis, thrombosis, and vascular development. *Arterioscler Thromb Vasc Biol.* 2004;24(6):1015–1022.
93. Steffel J, Luscher TF, Tanner FC. Tissue factor in cardiovascular diseases: Molecular mechanisms and clinical implications. *Circulation.* 2006;113:722–731.
94. Yang T, Atack TC, Stroud DM, Zhang W, Hall L, Roden DM. Blocking Scn10a channels in heart reduces late sodium current and is antiarrhythmic. *Circ Res.* 2012;111(3):322–332.
95. Burashnikov A, Belardinelli L, Antzelevitch C. Atrial-selective sodium channel block strategy to suppress atrial fibrillation: Ranolazine versus propafenone. *J Pharmacol Exp Ther.* 2012;340(1):161–168.
96. Yasukawa K, Miyakawa R, Yao T, Tsuneyoshi M, Utsumi H. Non-invasive monitoring of redox status in mice with dextran sodium sulphate-induced colitis. *Free Radic Res.* 2009;43(5):505–513.
97. Salim AS. Role of oxygen-derived free radical scavengers in the management of recurrent attacks of ulcerative colitis: A new approach. *J Lab Clin Med.* 1992;119(6):710–717.
98. Auscher C, Amory N, Van der Kemp P, Delbarre A. Xanthine oxidase activity in human intestines. Histochemical and radiochemical study. *Adv Exp Med Biol.* 1979;122:197–201.
99. Parks D, Buckley G, Granger N. Role of oxygen-derived free radicals in digestive tract disease. *Surgery.* 1983;94:415–422.
100. Wood DC, Wood J. Pharmacologic and biochemical considerations of dimethyl sulfoxide. *Ann NY Acad Sci.* January 27, 1975;243:7–19.
101. Brayton CF. Dimethyl sulfoxide (DMSO): A review. *Cornell Vet.* January 1986;76(1):61–90.
102. Perlman F, Wolfe HF. Dimethylsulfoxide as a penetrant carrier of allergens through intact human skin. *J Allergy.* 1966;38(5):299–307.
103. Boman A, Wahlberg JE. Percutaneous absorption of 3 organic solvents in the guinea pig (I). Effect of physical and chemical injuries to the skin. *Contact Dermatitis.* 1989;21(1):36–45.
104. Hui X, Lamel S, Qiao P, Maibach HI. Isolated human and animal stratum corneum as a partial model for the 15 steps of percutaneous absorption: Emphasizing decontamination, part II. *J Appl Toxicol.* 2013;33:157–172.
105. Frosch PJ, Duncan S, Kligman AM. Cutaneous biometrics. I. The response of human skin to dimethyl sulphoxide. *Br J Dermatol.* 1980;103(3):263–274.
106. Horita A, Weber LJ. Skin penetration property of drugs dissolved in dimethyl sulfoxide (DMSO) and other vehicles. *Life Sci.* 1964;3:1389–1395.
107. Skog E, Wahlberg JE. Effect of dimethyl sulfoxide on skin. A macroscopic and microscopic investigation on human skin. *Acta Derm Venereol.* 1967;47(6):426–434.
108. Majtan J, Majtan V. Dimethyl sulfoxide attenuates TNF-α-induced production of MMP-9 in human keratinocytes. *Toxicol Environ Health A.* 2011;74(20):1319–1322.
109. Duimel-Peeters IG, Halfens RJ, Snoeckx LH et al. A systematic review of the efficacy of topical skin application of dimethyl sulfoxide on wound healing and as an anti-inflammatory drug. *Wounds.* 2003;15(11):361–370.
110. Alberts D, Dorr RT. Case report: Topical DMSO for mitomycin-C induced skin ulceration. *Oncol Nurs Forum.* 1991;18(4):693–695.
111. Ludwig CU, Stoll HR, Obrist R, Obrecht JP. Prevention of cytotoxic drug induced skin ulcers with dimethyl sulfoxide (DMSO) and alpha-tocopherole. *Eur J Canc Clin Oncol.* 1987;23(3):327–329.
112. Pandey S, Basheer M, Roy S, Udupa N. Development and evaluation of transdermal formulations containing metronidazole and norfloxacin for the treatment of burn wound. *Indian J Exp Biol.* 1999;37(5):450–454.
113. Raskin DJ, Sullivan KH, Rappaport NH. The role of topical dimethyl sulfoxide in burn wound infection: Evaluation in the rat. *Ann NY Acad Sci.* 1983;411:105–109.

114. Lambrechts SA, Demidova TN, Aalders MC, Hasan T, Hamblin MR. Photodynamic therapy for *Staphylococcus aureus* infected burn wounds in mice. *Photochem Photobiol Sci.* 2005;4(7):503–509.

115. Seyb ST, Noordhoek L, Botens S, Mani MM. A study to determine the efficacy of treatments for hydrofluoric acid burns. *J Burn Care Rehabil.* 1995;16(3 Pt 1):253–257.

115a. Kilincaslan H, Ozbey H, Olgac V. The effects of dimethyl sulfoxide on the acute phase of experimental acid and alkali corrosive esophageal burns. *Eur Rev Med Pharmacol Sci.* October 2013;17(19):2571–2577.

116. Skrypuch OW, Tokarewicz AC, Willis NR. Effects of dimethyl sulfoxide on a model of corneal alkali injury. *Can J Ophthalmol.* 1987;22(1):17–20.

117. Erk Y, Raskin DJ, Mace M Jr, Spira M. Dimethyl sulfoxide alteration of collagen. *Ann NY Acad Sci.* 1983;411:364–368.

118. Mayer JH, Anido H, Almond CH, Seaber A. Dimethyl sulfoxide in prevention of intestinal adhesions. *Arch Surg.* 1965;91:920–923.

119. Huu N, Albert HM. Effect of DMSO on wound healing tensile strength measurements in rabbits. *Am Surg.* 1966;32:421–424.

120. McFarlane RM, Laird JJ, Lamon R, Finlayson AJ, Johnson R. Evaluation of dextran and DMSO to prevent necrosis in experimental pedicle flaps. *Plast Reconstr Surg.* 1968;41(1):64–70.

121. Young VL, Boswell CB, Centeno RF, Watson ME. DMSO: Applications in plastic surgery. *Aesthet Surg J.* 2005;25(2):201–209.

122. Kligman AM. Topical pharmacology and toxicology of dimethyl sulfoxide (DMSO). Part 1. *JAMA.* 1965;193:551–554.

123. Kligman AM. Topical pharmacology and toxicology of dimethyl sulfoxide (DMSO). Part 2. *JAMA.* 1965;193:796–804.

124. Carpenter RJ, Angel MF, Morgan RF. Dimethyl sulfoxide increases the survival of primarily ischemic island skin flaps. *Otolaryngol Head Neck Surg.* February 1994;110(2):228–231.

125. Leite MT, Gomes HC, Percário S, Russo CR, Ferreira LM. Dimethyl sulfoxide as a block to the deleterious effect of nicotine in a random skin flap in the rat. *Plast Reconstr Surg.* 2007;120(7):1819–1822.

126. de la Torre JC, Rowed DW. DMSO: A new respiratory stimulant? *J Clin Pharmacol.* 1974;14(7):345–353.

127. de la Torre JC, Kawanaga HM, Johnson CM, Goode DJ, Kajihara K, Mullan S. Dimethyl sulfoxide in central nervous system trauma. *Ann NY Acad Sci.* 1975;243:362–389.

128. Poets CF, Darraj S, Bohnhorst B. Effect of doxapram on episodes of apnoea, bradycardia and hypoxaemia in preterm infants. *Biol Neonate.* 1999;76(4):207–213.

129. Wheeldon PJ, Perry AW. The use of ethamivan in the treatment of barbiturate poisoning. *Can Med Assoc J.* 1963;89:20–22.

130. Klein HA. Dimethyl sulfoxide in adult respiratory distress syndrome. *Ann NY Acad Sci.* 1983;411:389–390.

131. Leff JA, Oppegard MA, McCarty EC, Wilke CP, Shanley PF, Day CE, Ahmed NK, Patton LM, Repine JE. Dimethyl sulfoxide decreases lung neutrophil sequestration and lung leak. *J Lab Clin Med.* 1992;120(2):282–289.

132. Greiner C, Schmidinger A, Hülsmann S, Moskopp D, Wölfer J, Köhling R, Speckmann EJ, Wassmann H. Acute protective effect of nimodipine and dimethyl sulfoxide against hypoxic and ischemic damage in brain slices. *Brain Res.* 2000;887(2):316–322.

132a. Boybeyi O, Bakar B, Aslan MK, Atasoy P, Kisa U, Soyer T. Evaluation of dimethyl sulfoxide and dexamethasone on pulmonary contusion in experimental blunt thoracic trauma. *Thorac Cardiovasc Surg.* August 12, 2013; Epub ahead of print.

133. Huang J, Klionsky DJ. Autophagy and human disease. *Cell Cycle*. 2007;6(15):1837–1849.
134. Bergamini E, Cavallini G, Donati A, Gori Z. The role of autophagy in aging: Its essential part in the anti-aging mechanism of caloric restriction. *Ann NY Acad Sci*. 2007;1114:69–78 [Epub 2007].
135. Mijaljica D, Prescott M, Devenish RJ. The intriguing life of autophagosomes. *Int J Mol Sci*. 2012;13(3):3618–3635.
136. Dong Y, Undyala VV, Gottlieb RA, Mentzer RM Jr, Przyklenk K. Autophagy: Definition, molecular machinery, and potential role in myocardial ischemia-reperfusion injury. *J Cardiovasc Pharmacol Ther*. 2010;15(3):220–230.
137. Shen HM, Codogno P. Autophagy is a survival force via suppression of necrotic cell death. *Exp Cell Res*. 2012;318(11):1304–1308.
138. Sonali N, Tripathi M, Sagar R, Velpandian T, Subbiah V. Clinical effectiveness of rivastigmine monotherapy and combination therapy in Alzheimer's patients. *CNS Neurosci Ther*. 2013;19:91–97.
139. Horiguchi M, Koyanagi S, Okamoto A, Suzuki SO, Matsunaga N, Ohdo S. Stress-regulated transcription factor ATF4 promotes neoplastic transformation by suppressing expression of the INK4a/ARF cell senescence factors. *Canc Res*. 2012;72(2):395–401.
140. Formanek K, Suckert R. Diuretische Wirkung von DMSO. In *DMSO Symposium*, Vienna, Austria. Saladruck, Berlin, Germany, 1966; pp. 21–24.
141. Waller FT, Tanabe CT, Paxton HD. Treatment of elevated intracranial pressure with dimethyl sulfoxide. *Ann NY Acad Sci*. 1983;411:286–292.
142. Marshall LF, Camp PE, Bowers SA. Dimethyl sulfoxide for the treatment of intracranial hypertension: A preliminary trial. *Neurosurgery*. 1984;14(6):659–663.
143. Karaca M, Bilgin UY, Akar M, de la Torre JC. Dimethyl sulphoxide lowers ICP after closed head trauma. *Eur J Clin Pharmacol*. 1991;40(1):113–114.
144. Sams WM Jr, Carroll NV. Cholinesterase inhibitory property of dimethyl sulphoxide. *Nature*. 1966;212(5060):405.
145. Cherki-Vakil R, Meiri H. Postsynaptic effects of DMSO at the frog neuromuscular junction. *Brain Res*. 1991;566(1–2):329–332.
146. Plummer JM, Greenberg MJ, Lehman HK, Watts JA. Competitive inhibition by dimethylsulfoxide of molluscan and vertebrate acetylcholinesterase. *Biochem Pharmacol*. 1983;32(1):151–158.
147. Stoughton RB, Fritsch W. Influence of dimethyl sulfoxide (DMSO) on human percutaneous absorption. *Arch Derm*. 1964;90:512–517.
148. Stoughton RB. Hexachlorophene deposition in human stratum corneum, enhancement by dimethylacetamide and dimethyl sulfoxide and methylene ether. *Arch Derm*. 1966;94:646–648.
149. Katz R, Hood RW. Topical thiabendazole for creeping eruption. *Arch Derm*. 1966;94:643–645.
150. Smith QT, Allison DJ. Uterotrophic effect of topically applied estradiol-17ß benzoate. *Acta Derm Venereol*. 1967;47:435–439.
151. Tjan ID, Gunberg DL. Percutaneous absorption of two steroids dissolved in dimethyl sulfoxide in the immature female rat. *NY Acad Sci*. 1967;141:406–413.
152. Balakin KV, Ivanenkov YA, Skorenko AV, Nikolsky YV, Savchuk NP, Ivashchenko AA. In silico estimation of DMSO solubility of organic compounds for bioscreening. *J Biomol Screen*. 2004;9(1):22–31.
153. Douwes RA, van der Kolk JH. Dimethylsulfoxide (DMSO) in horses: A literature review. *Tijdschr Diergeneeskd*. 1998;123(3):74–80.
154. Levesque F. Effects of DMSO on open wounds in horses. *Ann NY Acad Sci*. 1967;141(1):490–492.

155. Kelmer G, Doherty TJ, Elliott S, Saxton A, Fry MM, Andrews FM. Evaluation of dimethyl sulphoxide effects on initial response to endotoxin in the horse. *Equine Vet J.* 2008;40(4):358–363.

156. Smith G, Bertone AL, Kaeding C, Simmons EJ, Apostoles S. Anti-inflammatory effects of topically applied dimethyl sulfoxide gel on endotoxin-induced synovitis in horses. *Am J Vet Res.* 1998;59(9):1149–1152.

157. Ball MA, Rebhun WC, Gaarder JE, Patten V. Evaluation of itraconazole-dimethyl sulfoxide ointment for treatment of keratomycosis in nine horses. *J Am Vet Med Assoc.* 1997;211(2):199–203.

158. Knowles RP. Clinical experience with DMSO in small animal practice. *Ann NY Acad Sci.* 1967;141(1):478–483.

159. Dziezyc J, Millichamp NJ, Rohde BH, Baker JS, Chiou GC. Effects of lipoxygenase inhibitors in a model of lens-induced uveitis in dogs. *Am J Vet Res.* 1989;50(11):1877–1882.

160. Spyridakis L, Brown S, Barsanti J, Hardie EM, Carlton B. Amyloidosis in a dog: Treatment with dimethylsulfoxide. *J Am Vet Med Assoc.* 1986;189(6):690–691.

161. Marin-Padilla M. Mesodermal alterations induced by dimethyl sulfoxide. *Proc Soc Exp Biol Med.* 1966;122(3):717–720.

162. Caujolle F, Caujolle D, Cros S, Calvet M, Tollon Y. Teratogenic power of dimethyl sulfoxide and diethyl sulfoxide. *C R Acad Sci Paris.* 1965;260:327–330.

163. Rubin LF, Mattis PA. Dimethyl sulfoxide lens changes in dogs during oral administration. *Science.* 1966;153:83–84.

164. Leake CD. Dimethyl sulfoxide. *Science.* 1966;152:1646–1649.

165. Noel PR, Barnett KC, Davies RE et al. The toxicity of dimethyl sulphoxide (DMSO) for the dog, pig, rat and rabbit. *Toxicology.* 1975;3(2):143–169.

166. de la Torre JC, Surgeon JW, Ernest T, Wollmann R. Subacute toxicity of intravenous dimethyl sulfoxide in rhesus monkeys. *J Toxicol Environ Health.* 1981;7(1):49–57.

167. Brobyn RD. The human toxicology of dimethyl sulfoxide. *Ann NY Acad Sci.* 1975;243:497–506.

168. Gautam M, Prasoon P, Kumar R, Singh A, Shrimal P, Ray SB. Direct intrawound administration of dimethylsulphoxide relieves acute pain in rats. *Int Wound J.* April 21, 2014; Epub ahead of print.

3 DMSO Clinical Pharmacology

DMSO IN DISEASE

Although dimethyl sulfoxide (DMSO) was first synthesized by the Russian scientist Alexander Zaytsev in 1866, it took another century for its pharmacotherapeutic properties to be discovered by Jacob et al.,[1] beginning in 1964. Discovering the pharmacotherapeutic activities of DMSO is a lesson in how good science works, involving a keen observation, the hypothesis that was formulated, testing, and the conclusion.

Stanley Jacob, a surgeon and, at the time, the head of the Organ Transplant Program at Oregon Health Sciences University, had observed that besides the cryoprotective properties exhibited by DMSO in the preservation of frozen organs, its topical application seemed to quickly penetrate the skin without damaging it, resulting in a garlicky taste in the mouth within seconds after its application. Intrigued by this suspected cause-and-effect reaction, Jacob and his colleagues[1] began a research journey to find out what else this simple molecule could do and, in so doing, set off an explosion of publications worldwide that would culminate in more than 45,000 scientific articles, which continue growing to this day.

The use of DMSO as a pharmacotherapeutic agent for clinical maladies discussed in this chapter has, in some cases, not been clinically exploited to its full extent despite consistent evidence of its safety and efficacy in a host of medical conditions. It is hoped that this review can aid the reader to better understand the unusual properties of this simple molecule and help dispel both unfounded negative notions and positive expectations that have emerged during the history of DMSO and which are either anecdotal in content or unsupported by scientific evidence.

Some of the useful features that characterize DMSO use are as follows:

1. It can be administered intravenously, orally, topically, or intravesically.[2,3]
2. When combined with other selected compounds, it can enhance their tissue penetration, enhance their pharmacological activity synergistically, and reduce their toxicity by lowering their dose.[4]
3. It is relatively nontoxic at a wide range of concentrations and routes of administration.[5]

PAIN

Pain is essentially an unpleasant perceptual process that arises in response to inflamed or injured tissues.[6] Pain can be perceived from endogenous stressors, including damage and/or inflammation within both neuronal and non neuronal tissues, or by exogenous stress that can be produced by psychosocial factors. Physical pain sensation and suffering can be measured, though not always in a precise manner, to provide useful parameters for clinical assessment. Dangerous tissue-damaging pain often develops from skeletal muscle and from joints and less often from viscera and almost never from liver or lung even when these organs are severely damaged or diseased.[7] The likely explanation for this curious absence of pain sensation in those organs is that they lack a network of sensory nerves for the transmission of nociceptive signals to the brain.[8] Similarly, viscera, such as gut and urinary tract, can elicit pain following ischemic tissue damage or inflammation (e.g., Crohn's disease or cystitis), but this type of pain is not as common as musculoskeletal or joint pain.[9]

Peripheral nerve stimulation in humans has shown the involvement of two afferent fiber groups subserving pain.[10] Due to their higher conduction velocity, fine myelinated fibers of the A-δ (gamma) pain group when activated evoke quick, shallow pain interpreted as sharp or pricking,[11] whereas activity in slow nonmyelinated C-fibers responds to stimuli that have stronger intensities and account for the slow, but deeper and widespread pain over an unspecific area, often characterized as a burning sensation.[12] In response to single shocks, the A-δ pain is more severe than C-fiber pain. But C-fiber pain is more severe when the stimulation is repetitive, which is the basis of chronic pain.[13] Strong C-fiber pain is thought to underlie the suffering felt in neuropathic pain[14,15] (see Chapter 2).

Due to the high subjectivity involved in pain sensation in humans, the assessment of pain medications and even nonpharmacological interventions can be difficult to measure with respect to effectiveness of easing pain sensation since the perception of pain can change over time and even during the course of a day.[16]

For example, chronic neuropathic pain has a complex pathophysiology and is difficult to treat.[17] The clinical measures that evaluate pain alleviation involve patient-reported outcomes where the patient is the most important evaluator of the analgesic being tested.[18] This can present a major problem involving bias for the clinician and the patient as well as to regulatory agencies that must sift through the merit of such evidence before approving a prescriptive analgesic for the market, a market whose revenue classifies it as a billion-dollar industry.

The second problem related to pain medications in clinical studies' assessments is that analgesics are unusually assessed by a single pain rating.[19] This approach has yielded conflicting and confusing results that put into question the true pharmacological action of an analgesic with that of a placebo effect or nocebo effect. The placebo effect describes the improvement seen when patients, unknowingly, are given a sham or inactive ingredient as a pain-relieving treatment that the patients believe will alleviate their pain, as it often does.[20] This is a very real physiological effect that remains a scientific mystery but nevertheless a valid response. The nocebo effect is the opposite: patients believe that even a potent analgesic will make their pain feel worse, and often it does.[21]

Due to this lack of scientific objectivity in measuring analgesic properties of efficacy, information derived from the impact of treatments on (1) the quality of pain symptoms, (2) improved anatomic or physiological functioning, and (3) overall well-being of the patient can add to the validity and confidence that a particular analgesic for a particular pain complaint shows evidence or not that it is a useful pain reliever. We will call the criteria of using these three guidelines to measure analgesic efficacy, the *QFW values*, for quality of easing pain (Q), restoring or significantly improving anatomic–physiological function (F), and general well-being (W) subjectively expressed by the patient following treatment.

Thus, two subjective and one objective values complementary to each other to assess the effectiveness of DMSO on pain add a level of accuracy and confidence into what was previously considered a purely subjective assessment of a drug's action to alleviate pain.

Oral nonsteroidal anti-inflammatory drugs (NSAIDs) and cyclooxygenase-2-selective (COX-2) inhibitors are frequently recommended for the management of osteoarthritis pain. However, serious gastrointestinal and cardiovascular systemic adverse events are associated with oral NSAIDs and COX-2 inhibitors, thus limiting their usefulness.

For these reasons, many clinicians treating patients with an assortment of painful disorders turned to other possible remedies with a better safety margin.

Because pain is a primitive and conspicuously unpleasant sensation, it is generally regarded as a warning signal of danger. This unpleasant or aversive sensation originates from some specific region of the body. Since pain is highly subjective in nature, it is often difficult to treat clinically despite the fact that it is the most common complaint that patients describe when seeking medical help. Specific pain sensations have been described as burning, pricking, stinging, aching, and soreness.[11]

When the clinical potential of DMSO was first discovered,[1] its topical application for the relief of musculoskeletal pain was consistently shown.

DMSO and aspirin work similarly by blocking the production of certain prostaglandins by controlling the on–off switch in cells that regulate pain and inflammation, among other things. That is likely the reason why aspirin stops mild inflammation and pain. However, DMSO goes a step farther than aspirin in that it not only blocks the prostaglandins that can induce pain and inflammation, such as prostaglandin E_2 (PGE_2),[22] but also stops or slows down conduction of pain fibers when it is administered topically or internally.[23] PGE_2 is of particular interest because it is involved in all processes leading to the classic signs of inflammation: redness, swelling, and pain.[24] Also, unlike aspirin, DMSO is not considered toxic to the stomach or gut where aspirin can cause peptic ulcers or gastrointestinal tract irritation at therapeutic doses.[25]

DMSO has been found to block conduction of C-fibers when applied directly to a sensory nerve, even in low concentrations (5%–7% v/v).[23] Higher concentrations completely blocked C-fiber conduction, with a minimum blocking concentration of 9%.[23] When the application of DMSO continues on a nerve, besides blocking large C-fibers, it also blocks small pain fibers, such as A-δ (delta) and sensory receptors, such as those associated with noxious stimuli response.[26] This action makes the repeated application of DMSO often show dramatic effects on musculoskeletal pain disorders 4–7 days after the initial application showed little to no response.[27]

One early study in the use of DMSO for musculoskeletal pain was on six patients presenting with acute calcified supraspinatus bursitis, a painful condition due to deposition of calcium hydroxyapatite within tendons, usually of the rotator cuff.[28] All six patients had been refractory to conventional treatments whose duration of pain had varied from 2 days to 4 weeks. Following one or several topical DMSO applications at a concentration of 70%, dramatic symptomatic relief occurred in all six patients.[28] This relief consisted in pain alleviation, improvement in the range of motion and mobility of the affected joint, and patient satisfaction of the treatment results indicated a significant reduction in disability.[28]

Using the QFW values discussed earlier as the assessing criteria for the effectiveness of DMSO, there is confidence in the conclusion that DMSO treatment improved the quality of symptoms, anatomic functioning of extremities and joints with increased mobility, and overall well-being of the patient. Local skin irritation and a characteristic *clam-like* odor to the breath were the only side effects observed in the study by Lockie and Norcross[28]. In that early study by Lockie and Norcross,[28] over 100 patients were treated with topical DMSO generally with similar results as those treated with calcified supraspinatus bursitis for such conditions as degenerative arthritis, rheumatoid arthritis, periarthritis of varying degrees of acuteness and severity, and various musculoskeletal complaints following muscle trauma sustained by professional football players.[28] In that study, it was observed that patients unable to tolerate 70% DMSO, a 10% addition of glycerin avoided or lessened the skin irritation when it occurred.[28] The Lockie and Norcross[28] study became the gold standard for evaluating DMSO in other musculoskeletal and joint pain resulting in limited mobility of body extremities.

Several additional early studies on the activity of DMSO in musculoskeletal pain disorders were reported. Blumenthal and Fuchs[29] reported on 122 patients ranging in age from 17 to 89 years with musculoskeletal sprains or injuries to the chest, back, pelvis, and extremities, who were treated for several days or months with topical DMSO using undisclosed doses. These authors reported that 74% of patients given DMSO showed excellent–good response to treatment especially if the pain was acute rather than chronic, while 24% of the patients showed no improvement of symptoms.[29] This study failed to provide QFC values and other important information involving treatment times, doses, and manner of assessment and can therefore be considered more anecdotal in nature.

DMSO has been used in gel form instead of an aqueous solution for lateral epicondylitis (tennis elbow) and humero-scapular pain. In a placebo-controlled double-blind study, 157 patients received either 10% DMSO gel applied three times per day for 2 weeks or the inactive gel excipient.[30] Treatment was started within 72 h after the acute onset of symptoms and pain, and mobility of the joints was observed after 3, 7, and 14 days of treatment with DMSO as compared to the placebo gel recipients. Mild undesired events were seen in 8 of 77 DMSO patients and 3 of 80 patients treated with the placebo gel.[30]

One of the first studies to investigate the analgesic potential of DMSO in rheumatoid arthritis, osteoarthritis, tenosynovitis, and other articular disorders came from Japan in 1967.[31] These investigators reported using 50% aqueous solution of DMSO topically for 3 days on 318 patients and reported DMSO significantly relieved joint pain when assessed 14 days later, using a variety of tests involving articular symptoms and signs, such as spontaneous pain, tenderness on pressure, pain on motion, local heat, grip strength, circumference of joint, and range of joint motion.[31] They concluded that DMSO improved grip strength, relieved pain, and increased the range of joint motion but had no effect on joint swelling as compared to placebo. DMSO seemed more effective in cases of short duration. The only unfavorable side effect was temporary local irritation of the skin, which disappeared after the last DMSO application.[31]

As reports of the analgesic action of topical DMSO in arthritis and musculoskeletal disorders became known, other investigators looked into the possibility of using topical DMSO for surface tissue damage. One of these reports examined incisional pain after thoracotomy.[32] DMSO (60% concentrations) was applied directly to the closed wound incision extending 10 in. on either side. The skin was then permitted to dry for 10 min, and a sterile dressing was applied. Six hours later, the dressing was removed, and the same DMSO amount and concentration were reapplied. The applications were repeated every 6 h for 3 days. Thereafter, from 4 to 6 days, 70% DMSO was applied to the healing wound, and the treated patients as a group were able to move about more easily both in and out of bed, to resume early motion in the arm and shoulder in the operative side, and, in general, to enjoy a more rapid and less complicated postoperative course than patients who had not been treated with DMSO.[32] Also, it was observed that treated patients showed fewer gastrointestinal complications, such as nausea, vomiting, ileus, and constipation. The investigators, surgeons, and nurses providing the daily DMSO treatments were impressed with the apparent effectiveness of DMSO in relieving the usual postoperative thoracotomy pain.[32]

Patients with painful furuncular otitis were treated with 90% topical DMSO by applying DMSO into the auditory meatus and to the adjacent outer ear using a dripping-wet cotton-tipped applicator.[33] Patients were treated once daily or twice daily in serious cases. Excellent and dramatic results with complete remission of symptoms were achieved in 28 cases, and marked improvement was observed in 52 cases, while 22 cases remained unimproved. Many of the patients treated with DMSO had been refractory to cortisone and antibiotic creams. A distinctly longer-lasting effect could be achieved when antibiotics such as tetracycline and erythromycin dissolved in DMSO were applied.[33]

Not all studies using topical DMSO for rotator cuff and tendinitis pain showed positive results. In a year-long study involving 102 patients presenting with either medial or lateral epicondylitis (tennis elbow) or rotator cuff tendonitis were treated with topical applications of DMSO. Beneficial effects were assessed with respect to improvement in pain, tenderness or swelling, and increase in the range of motion. Forty patients treated with topical 70% DMSO aqueous solution did not report any more benefit from the drug than patients who received a 5% DMSO aqueous placebo solution.[34]

A review of DMSO in otological conditions by Freeman[35] in 1976 claimed that a thorough review by him of the literature revealed no evidence that topical application of the drug was useful and concluded that although DMSO was not ototoxic, double-blind studies on patients with otological infections demonstrated that 90% DMSO applied within the external auditory canal had no anti-inflammatory, antibacterial, or analgesic properties, and there was no evidence that it potentiated the action of other medicines when mixed preparations were used. To support this conclusion, Freeman[35] cited 11 references in his review, most of which in fact supported the opposite conclusion from that derived by Freeman,[35] who referred to "many double-blind studies" that demonstrated the lack of activity by DMSO but never once cited any of these reports or references.

Such shoddy and scientifically lazy conclusions failing to cite any report in support of its tenet would haunt DMSO for years as the shroud of its mystique and sometimes enthusiastic hyperbole by its supporters grew exponentially to the degree of putting into question the credibility of many evidence-based and well-performed experiments, both basic and clinical, that showed either positive or negative findings for the actions of this molecule on a variety of models and systems.

Complex regional pain syndrome (CRPS), formerly known as reflex sympathetic dystrophy and causalgia, is a chronic systemic disease characterized by intense neuropathic pain out of proportion to the severity of the injury.[36] CRPS can result from even a mild tissue trauma to an extremity, and when it does, it is typically characterized by swelling and severe pain of the extremity and by dramatic color and temperature skin changes.[37] There is no specific test that can diagnose CRPS, and diagnosis is often made by the exclusion of similar pain conditions.

CRPS is divided into two types: CRPS-1, which lacks demonstrable nerve lesions, and CRPS-2, formerly known as causalgia, where the pain is generally neuropathic and more severe than type 1 and where evidence of nerve damage is present.[38]

The specific cause of CRPS is not known for certain, but dysfunction or damage to the peripheral or central nervous system (CNS) may follow a nerve injury after trauma or surgery.[39]

Another theory is that CRPS is caused by an immune response that leads to the changes seen in the skin (redness, warmth, swelling, and pain) involving proinflammatory mediators such as TNF-alpha, interleukins 1β, 2, 6, 8, and C-reactive protein,[40–42] as well as increased systemic levels of proinflammatory neuropeptides, including substance P and bradykinin.[43] Other findings indicate that when there are no clinical signs of peripheral nerve damage, posttraumatic focal CRPS-1 is nonetheless associated with significant loss of C-fibers and A-δ fibers in the affected area.[44]

About 50,000 new cases of CRPS are estimated to occur annually in the United States alone (Bruehl), generally following surgery, simple fractures, crush injuries, and sprains.[38] When a limb is affected, loss of strength and decreased active range of motion can disable the individual, resulting in continuous intense pain and psychological stress. In fact, the pain can be so intense in CRPS that it has the unfortunate honor of being described as the most painful syndrome or disease, scoring highest on the McGill pain scale (42 out of a possible 50), and considerably above such events as amputation of a digit and childbirth (see Figure 3.1).

After many randomized-controlled trials published, there is no clinical consensus on how to treat or manage CRPS.[45]

Pathophysiologically, oxidative damage by free radicals have been implicated in the development of CRPS.[46] In addition, pain sensation that increases with time,

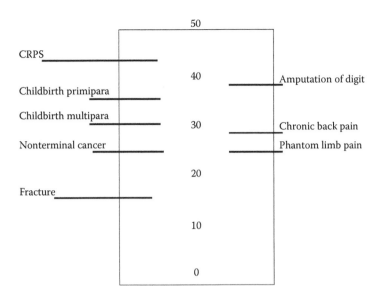

FIGURE 3.1 The McGill Pain Index is a subjective measure used clinically to compare pain perception from 0 to 50 in patients and provide a guideline for analgesic efficacy. CRPS is nearly at the 50-point mark, the highest pain sensation felt by humans. The analgesic least likely to do harm and provide good efficacy in the treatment of CRPS was reported to be DMSO. (From Quisel, A. et al., *J. Fam. Pract.*, 54(7), 599, 2005.)

also called *wind-up*, and CNS sensitization are important neurologic processes that have been reported to be involved in the unfolding of CRPS.[47] In this regard, it is interesting to note that elevated CNS glutamate levels promote wind-up and CNS sensitization.[47]

In addition, there is experimental evidence that demonstrates that N-methyl-D-aspartate (NMDA) receptors are localized in peripheral nerves and appear to be activated following CRPS.[48] Moreover, there is compelling evidence that the NMDA receptor has significant association in the CNS sensitization process involving CRPS.[49] This finding has led to the use of NMDA receptor blockers and free radical scavengers as a primary target for neuropathic pain[48] that can lead to central sensitization and *wind-up* phenomena such as that seen in CRPS.[50]

Good to excellent results in pain control of patients with CRPS type 1 have been shown after the topical application of 50% DMSO over the affected area.[51] The results indicated significantly favorable pain management, particularly when the CRPS was associated with inflammatory symptoms.[51] The mechanism by which DMSO is able to provide pain relief in one of the most difficult to treat pain conditions remains unclear.

A clue may lie in the fact that DMSO acts as a powerful free radical scavenger, and it is well known that free radicals following injury can be toxic to tissue and introduce acute or chronic pain sensation. DMSO is also an NMDA and glutamate receptor antagonist, and both substances can activate hyperalgesia and neuropathic pain.

One of the cardinal features of inflammatory states is that normally innocent stimuli can produce severe pain. This may be one of the mechanisms targeted by DMSO as an anti-inflammatory agent. Finally, DMSO has the ability to block C-fiber and small fiber pain fibers after its topical application. These actions may explain to some degree the number of studies that have tested DMSO in hundreds of patients where controlling pain and the consequences of CRPS were achieved.[23,26,51–53]

To compare the effects of two free radical scavengers, topical 50% DMSO and *N*-acetylcysteine (NAC) were used for the treatment of CRPS-1, in a randomized, double-dummy controlled, double-blind trial.[3] Treatments were conducted in 146 patients over a period of 24 months. Patients were randomized into two treatment groups: one was instructed to apply DMSO 50% five times daily to the affected extremity, and the second was treated with NAC 600 mg effervescent tablets three times daily, and both treatments were compared to each other and to a placebo. Significant differences were found for subscores of lower extremity function, in favor of DMSO treatment.[3] Subgroup analysis showed more favorable results for DMSO for warm CRPS-1 than for NAC.[3]

A study by Gaspar et al.[54] evaluated the topical treatment of 50% DMSO cream in type 1 CRPS over a period of 15 days, 1 month, 3 months, 6 months, and 1 year in 29 patients. The evaluation was performed using primarily a visual analogue scale (VAS) as the main efficacy variable that measures the patient's pain intensity subjectively. Secondary measures of efficacy by DMSO included the quality of life after treatment and the physical mobility of the affected limbs, involving range of motion, strength, and overall limb function. The results showed that topical application of 50% DMSO significantly diminished pain intensity approaching absence of pain, according to the VAS score at each period of observation. Secondary measures of efficacy were also recorded as higher scores on the quality of life questionnaire with a general improvement on limb rigidity, mobility, and function.[54] Thus, this study by Gaspar and his team[54] complies with the *QFW values* for the quality of easing pain (Q), restoring or significantly improving anatomic–physiological function (F), and general well-being (W) expressed by the patients following treatment.

One of the first studies to recognize the usefulness of DMSO in the treatment of CRPS was reported by Goris[55] in 1985. Goris[55] showed that DMSO as an anti-inflammatory and free radical scavenger agent could reduce the inflammation associated with CRPS and demonstrated, in an open study, a reduction of the CRPS symptoms using 50% DMSO cream. These results led to a prospective, randomized, and double-blind study on 32 patients following an accidental trauma and the development of acute CRPS.[56] Patients showed increased temperature, redness, and pain in the affected extremity, coupled to a limited range of motion of the limb. A VAS was used to measure the redness, swelling, and limited active range of motion of the affected extremity. DMSO 50% cream applied five times daily was compared to a placebo cream for 2 months. After 2 months, DMSO showed significant ($p < 0.01$) improvement of the VAS measures as compared to the placebo cream, and the authors recommended further use of DMSO for CRPS.[56] Side effects from the DMSO applications were mild scaling of the skin in the affected areas, which resolved in time.[56]

More recently, topical application of 50% DMSO three to five times daily was given as an anti-inflammatory therapy to 74 patients with CRPS and a mean disease duration of 17 weeks.[57] The DMSO treatment was supplemented with physical therapy and acetaminophen as an oral analgesic. Of the 74 patients with early CRPS, 55 (74%) patients significantly showed improvement in their pain score. Although no attempt was made to examine whether analgesia was obtained with DMSO or acetaminophen, the authors concluded that this package treatment should become standard therapy for patients presenting with early CRPS.[57]

A review of the most promising treatments for CRPS-1 at the present time concluded that early treatment of this syndrome is more likely to respond to treatment.[58] Of all the therapies for CRPS considered, including a wide range of analgesics, ganglionic blockers, steroids, adrenoceptor antagonists, acupuncture, calcitonin, and the glutamate blocker ketamine, the authors concluded that the therapy *least likely to do harm and supported by evidence of efficacy* was topical 50% DMSO cream[58] (see Figure 3.1).

In dealing with the subject of pain in humans, it must be remembered that pain is one of the most subjective of conditions and consequently one of the most difficult to test objectively by any means. Many extrinsic and intrinsic factors such as varying pain thresholds, gender, age, level of stress, culture, health status, mental conditioning, and other factors make pain a unique ailment in assessing and successfully treating it.

Pharmaceutical-grade DMSO has been used extensively as a penetration enhancer over the past 30 years demonstrating a favorable safety and efficacy profile. DMSO enhances the transdermal delivery of both hydrophilic and lipophilic medications that can provide localized drug delivery.

NSAIDs are widely used in the treatment of pain associated with a variety of indications, including arthritic conditions, but their usefulness is often limited by dose-dependent adverse events, such as gastrointestinal disturbances, cardiovascular events, and renal toxicity. The risk of such effects could be reduced by the use of topical formulations, which offer the potential to deliver analgesic concentrations locally, at the site of inflammation, while minimizing systemic concentrations.

In view that DMSO can potentiate the pharmacological activity of other agents and carry many of these compounds to a biological system, a number of DMSO mixtures have been described for a variety of medical conditions. In managing osteoarthritis, a condition with no cure characterized by pain, stiffness in at least one joint, and impairment or loss of function of the affected joint(s), diclofenac sodium topical solution 1.5% in 45% DMSO (TDiclo) was approved by the FDA in 2009 for the treatment of osteoarthritis. This NSAID–DMSO preparation now offers an alternative to single-use (NSAID) agents by reducing the NSAIDs risk of adverse events through DMSO's ability to enhance the penetration of diclofenac and thereby lower the dose of diclofenac.[59–61]

A recent multicenter, randomized, blinded, Phase III clinical study lasting 4–12 weeks of TDiclo for knee or hand osteoarthritis in 280 patients over the age of 75 reported that TDiclo is an appropriate treatment choice for osteoarthritis, with relatively few patients experiencing gastrointestinal or rarely cardiovascular and renal/urinary adverse events.[62] Results from this study suggest that TDiclo may offer an alternative to oral NSAID therapy for osteoarthritis of the knee or hand, particularly for patients at increased risk for serious systemic adverse events from oral NSAIDs.[62]

Prostate cancer is the most common cancer in men. Although it generally affects older men, it causes significant mortality when it metastasizes. The main therapy for metastatic prostate cancer involves androgen manipulation, chemotherapy, and radiotherapy.[63]

Prostate cancer pain is difficult to treat because the pathophysiology of pain in cancer patients is complex and remains poorly understood, making intractable pain from cancer a challenge for both patients and clinicians.[64]

Infusion of DMSO–sodium bicarbonate (DMSO–SB) was used to treat 18 patients with metastatic prostate cancer.[65] After a 90-day follow-up, patients treated with DSMO–SB showed significant improvement in clinical symptoms, including pain, blood and biochemistry tests, and quality of life.[65] There were no major side effects from the treatment. This study strongly suggested that therapy with DMSO–SB infusions could provide a rational alternative to conventional treatment for patients with metastatic prostate cancer.[65]

Another approach for the palliative treatment of advanced prostate cancer involved a subcutaneous implanted device that uses DMSO as a solubilizing excipient. The device was approved by the FDA in 2000. The implanted device contains a solution of 104 mg of DMSO and 72 mg of leuprolide acetate[66] (Viadur), a synthetic form gonadotropin-releasing hormone that can decrease blood levels of the male hormone testosterone. Androgen deprivation using leuprolide acetate and DMSO as an excipient has been shown to have a salutary effect in men with advanced prostate cancer.[67] In two open-label multicenter studies, long-term efficacy and safety using Viadur implanted device for 1 year was reported in a total of 122 patients.[68]

Among the many antiviral agents used against herpes simplex virus (HSV), idoxuridine and adenine arabinoside have shown minimal therapeutic benefit. However, when combined with DMSO, this treatment has been successful in acyclovir-resistant strains of HSV. In a double-blind, randomized, patient-initiated treatment study at five medical centers, 301 immunocompetent patients experiencing a recurrence of herpes labialis were treated with topical 15% idoxuridine (IDU) in 80% DMSO.[69] Using this drug combination, the mean duration of pain was significantly reduced by 1.3 days and the mean healing time to loss of crust by 1.7 days.[69] When only patients with classic herpes lesions (vesicle, ulcer, or crust formation) were considered, a greater drug effect using the idoxuridine–DMSO combination was seen for lowering the duration of pain by 2.6 days and increasing the mean healing time to normal skin by 2.3 days, a significant difference when compared to controls.[69]

Idoxuridine–DMSO (IDU) has also been used for herpes zoster and for the prevention of post herpetic neuralgia and compared to the antiviral agent acyclovir. After 4 days of topical treatment, IDU showed better control of vesicle formation, itching, and pain than oral acyclovir or placebo.[70]

These studies support the concept of DMSO as a penetrant and enhancer of other pharmaceuticals and suggest that additional combinations of DMSO and proven active medications for many illnesses should be further studied with the goal of lowering the dose of the active agent and consequently their systemic toxicity by localizing the treatment to a body part. This pharmacotherapeutic approach has not been exploited to a greater potential, especially with newer active agents for disease control that are replacing older, less efficacious, and more toxic medications. The effects of DMSO as a neuroprotective agent in the brain and spinal cord are discussed in more detail in Chapters 8 and 9.

INTERSTITIAL CYSTITIS

Interstitial cystitis (IC) is a painful chronic bladder condition that causes debilitating bladder pain that can be associated with urinary urgency, frequency, and nocturia. It is characterized by inflammation, erosion, and thinning of the bladder epithelial cells, and its hallmark symptom is pelvic and perineal pain. IC affects about three to eight million women and one to four million men in the United States, reducing the quality of life and social function of those affected. IC often overlaps overactive bladder (OAB), which can make diagnosis of either one or both difficult.[71] It is generally accepted that the hallmark for IC is pain and abdominal tenderness, whereas for OAB, it is urgency, frequency, and nocturia. However, both conditions can share these complaints.

The cause of IC is not clear, and it is still a disease diagnosed by exclusion since many other conditions resemble the symptoms of IC. There is no definitive test to identify IC, but other conditions mimicking IC can be ruled out by urinalysis, urine culture, cystoscopy, biopsy of the bladder wall, and, in men, laboratory examination of prostatic secretions.

IC is an understudied disorder lacking diagnostic and strategic treatment modalities, most likely from the limited research funding available to investigators.[72] Oral or intravesical therapies are the mainstay of treatment, while surgical procedures are reserved for refractory cases. This condition usually warrants a multidisciplinary approach for optimum outcome.[73]

In 1978, Shirley and his colleagues[74] introduced DMSO as a treatment for IC and reported dramatic results. They treated 213 patients with a variety of inflammatory conditions involving the lower genitourinary tract. These conditions included intractable IC, radiation cystitis, chronic prostatitis, and chronic female trigonitis.[74]

In this series by Shirley and his colleagues, the use of intravesical instillation of DMSO resulted in an excellent or good response in 54% of the female patients with no serious side effects noted. The treatments, however, are not curative and frequently require long-term maintenance with periodic installation of DMSO.

This initial report was followed by more studies from the same group and by others,[75,76] confirming the effectiveness of intravesical DMSO in controlling the symptoms of IC. Some reports showed 93% improvement in IC symptoms compared to 35% in the placebo group.[76] These early reports were followed by independent research studies from many countries, and by the end of 1978, DMSO was the first drug treatment approved by the FDA for the symptomatic relief of IC. Presently, DMSO remains the sole intravesical instillation agent for IC. The brand name for DMSO use in IC is RIMSO-50.

In clinical practice, the urinary bladder is filled with 50% DMSO for 15 min as a treatment for IC (Figure 3.2). This approach has the advantage that it reduces side effects from systemic application by achieving a high concentration of the drug

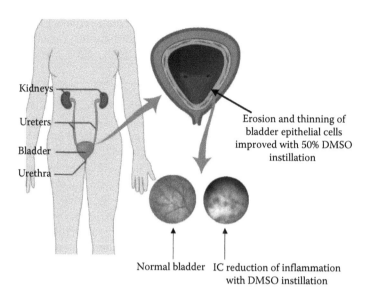

Kidneys

Ureters

Bladder

Urethra

Erosion and thinning of
bladder epithelial cells
improved with 50% DMSO
instillation

Normal bladder   IC reduction of inflammation
with DMSO instillation

FIGURE 3.2 (**See color insert.**) DMSO instillation for interstitial cystitis.

locally in the bladder. The initial disadvantage of bladder instillation was that it required a health-care provider to perform the procedure on patients during a clinic visit.

By 1987, patients did not need to go to a clinic to have DMSO instilled in their bladders, but they could be taught to self-administer DMSO at home using special kits.[77] No complications have been noted when the patient is taught how to self-administer DMSO.

The effectiveness of intravesical instillation of DMSO can be summarized as follows:

1. It reduces bladder inflammation and pain.[78,79]
2. It helps relax the bladder and pelvic detrusor muscles, thus reducing muscle spasm[80]; detrusor fibrosis is found in approximately 53% of IC patients and may be a major factor in treatment failure.[81]
3. It relieves pain by reducing small nerve fiber conduction of pain fibers, by diminishing inflammation, and by lowering substance P (a neuropeptide associated with pain) from bladder nerves.[23,82,83]
4. It helps reduce scar tissue by preventing collagen formation that promotes scarring.[84]
5. It reduces erosion and thinning of the bladder wall by limiting inflammation[78,79] (Figure 3.2).

The contractility of the bladder is attenuated after the intravesical treatment with DMSO, and it is reasonable to assume that the relaxant effect of DMSO is related to its therapeutic effect in relieving the inflammatory-related symptoms associated with IC.[85,86]

When IC is refractory to DMSO instillation, DMSO *bladder cocktails* have been tried with some success. These cocktails entail mixing DMSO with heparin, steroids, hyaluronic acid, analgesics, and other substances.[87,88]

GASTROINTESTINAL DISORDERS

Despite the presence of inflammatory changes in many gastrointestinal disorders, not much research has been reported using DMSO as an anti-inflammatory agent in these conditions. Most of the studies using DMSO in inflammatory bowel disease have been done outside the United States.

One use of DMSO in gastrointestinal disorders has aimed at Crohn's disease with secondary amyloidosis (AL), which develops in approximately 1% of patients with Crohn's disease. Oral administration of DMSO in 15 patients was found to be an effective treatment for amyloid A AL complicating gastrointestinal involvement and the early stage of renal dysfunction.[89] DMSO was given in three equal doses after meals at a dose of 3–20 g/day in a 33% solution with water or juice for 8 weeks. This treatment was observed to improve the renal function in 5 out of 10 renal AL patients when DMSO was initiated early in the disease process but had no effect on those patients with severe and/or advanced renal dysfunction.[89] In six patients, specific improvements were seen in gastrointestinal AL, gastrointestinal symptoms, including diarrhea, and protein-losing gastroenteropathy.[89]

Salim[90] used DMSO as a free radical scavenger on patients with recurrent ulcerative colitis and reported that 51% of patients receiving sulfasalazine and prednisolone alone showed good symptomatic control after 2 weeks of this regimen, and when DMSO was added to this treatment, 84% of patients showed similar symptomatic benefit.[90] This author concluded that oxygen-derived free radicals may be involved in the pathologic process of ulcerative colitis and that blocking these free radicals promoted the protection of the colon against recurrence of attacks.[90]

AUTOIMMUNE DISORDERS

In 1980, Pestronk and Drachman[91] reported that humoral immune mechanisms play an important part in the development of myasthenia gravis (MG). MG is a chronic autoimmune disorder in which circulating antibodies cause progressive weakness of the skeletal muscles by blocking acetylcholine receptors at the postsynaptic neuromuscular junction. The disease most commonly begins by weakening the muscles that control eye movements and eyelids, an effect that produces blurring, double vision, and eyelid drooping. In time, other muscle groups are affected, making chewing, swallowing, and breathing a life-threatening situation.

In the 1980 report, Pestronk and Drachman[91] noted that in the course of testing immunosuppressive drugs that could treat experimental autoimmune MG, they discovered that DMSO used as a vehicle to dissolve some of these agents itself produced a rapid and sustained fall of anti-AChR antibody titer. This observation led to their confirmation of this finding when DMSO was used on an experimental rat model of MG. It was seen that DMSO suppressed anti-AChR antibody levels by an average of 53%–76%, an effect that was similar whether DMSO was given by oral, rectal, or intraperitoneal routes.[92] DMSO treatment also suppressed the anti-AChR antibody response to a weak primary antigenic stimulus in the MG rat model.[92]

More recent studies on experimental autoimmune cystitis reveals that DMSO can impair effector T cells in a dose-dependent manner *in vitro*.[93] Effector T cells secrete inflammatory cytokines such as interleuli-6 in the bladder, which is considered to be responsible for the development and propagation of painful bladder symptoms in IC.[94]

There have been no human trials using DMSO for any autoimmune disorder, including MG.

RESPIRATION

DMSO has been shown to be an effective respiratory stimulant in experimental hypoxic states and following brain trauma in nonhuman primates[95,96] (see also Chapter 2). These findings led Klein and his colleagues[97] to use DMSO in patients with acute respiratory distress syndrome (ARDS). ARDS is a severe lung syndrome characterized by inflammation and edema of the lung parenchyma causing hypoxemia that can lead to multiple organ failure and death.[98,99] Most often, ARDS can result from sepsis or trauma, but shock and substances that will damage the alveolar epithelium can bring about a reaction.[98]

Klein et al.[97] reported that based on the properties that were known at the time for DMSO, such as a mucolytic respiratory stimulant, and its ability to lower tissue oxygen requirements, 10% DMSO was given intravenously to three near-death patients who had not responded to conventional treatments. All three patients responded to the DMSO administration within an hour, showing a significant rise in paO_2 from the 60s to high 80s and 90s and oxygen saturation increasing substantially to 95% within 8 h. The authors concluded that DMSO as a treatment for ARDS merited further investigation in view of the difficulty in treating this syndrome.

AMYLOIDOSIS AND SCLERODERMA

Primary systemic AL is characterized by the deposition of insoluble amyloid proteins in various organs contributing to the dysfunction of these organs. The median survival of patients with AL is 12 months, and if heart failure is present, mean survival is 6 months after the onset of symptoms.[99] Thus, the ability to dissolve the amyloid deposits or block their production would be expected to improve the affected organ function and extend survival.

The beneficial effects of DMSO on amyloid deposition were first reported by Isobe and Osserman[100] in 1976, and since then, a number of reports have confirmed the usefulness of DMSO therapy on organ damage caused by AL.[89,101] There has been some controversy that DMSO may inhibit amyloid production or act principally as an anti-inflammatory agent or free radical scavenger in organ damage associated with AL rather than a dissolver of amyloid fibrils.[102]

By contrast, single-case studies, though anecdotal, are occasionally of interest because they can provide clues as to the effectiveness of DMSO in disorders such as

AL, where a major clinical trial is not feasible. Such a case study was described of a patient with multiple myeloma who developed pulmonary infiltrates caused by AL and was given transdermal DMSO for 8 weeks, which resulted in a dramatic regression of the pulmonary infiltrates shown by x-ray analysis together with the stabilization of arterial blood gases.[103]

This report received confirmation in another case study of pulmonary AL treated with 10 mL DMSO per day for 2 years 104 (sutani). In a disorder that is known to progress rather aggressively within months of symptom onset, DMSO treatment resulted in significant reduction of pulmonary nodules and slowdown of disease progression.[104]

Similarly, it was reported that seven patients with secondary AL showed an unequivocal improvement of renal function following 3–6 months of 7–15 g/day of continuous oral DMSO treatment, which resulted in a 30%–100% rise of creatinine clearance and a decline in amyloid subunits in the urine that suggested breakdown or mobilization of the amyloid fibrils out of the tissue into urine.[105] This improvement in renal function was maintained as long as DMSO treatment was continued.[105]

Scleroderma is a connective tissue disease that involves changes in the skin, blood vessels, muscles, and internal organs, and it is thought to be caused by the accumulation of collagen.[106] It should be noted that DMSO application has been reported to reduce collagenous fibers in animals and in scleroderma patients.[107,108]

Scherbel[109] reported that cutaneous ulcers in scleroderma (systemic sclerosis) were seen to resolve after 2 weeks following topical applications of 35% DMSO applied three times daily. Patients receiving 35% DMSO had the topical dose increased to 46% after 2 days, and DMSO applications were reduced as ulcer healing appeared. As ulcers healed during a 6-month observation period, joint motion, grip strength, and skin softening were recorded to improve in 19 of 22 patients who participated in the study.[109]

However, a prospective, randomized, double-blind trial that compared topical therapy using 0.85% normal saline, and 70% DMSO for the treatment of digital ulcers in 84 patients with systemic sclerosis failed to show any difference between saline control and DMSO application with respect to total number of open ulcers, total surface area of open ulcers, average surface area per open ulcer, and the number of infected or inflamed ulcers.[110]

THROMBOEMBOLIC EVENTS

Myocardial infarction (MI) and ischemic stroke are leading causes of death and disability worldwide. Both conditions share a common pathologic process that results from unstable and vulnerable atherosclerotic plaques that form on the inside walls of coronary, peripheral, and cerebral arteries.[111] These plaques can rupture from high shear stress and trigger a thromboembolic event leading to an MI or ischemic stroke.[112]

Atherosclerotic plaques are composed mainly of extracellular lipid-rich material inside and outside of cells together with connective tissue components such as cholesterol and accumulated macrophages from the bloodstream. Coronary artery disease (CAD) generally precedes MI sometimes by decades, but the harm results when the coronary arteries become partially or totally blocked by atherosclerotic plaque formation. Rupture of an atherosclerotic plaque can activate platelets to form a clot that can dislodge and travel to a coronary or a cerebral artery where partial coronary occlusion can induce anginal pain, and a total occlusion can result in an MI or an ischemic stroke.

Thus, apart from their role in primary hemostasis, platelet activation is also critical in atherothrombosis where platelet adhesion and aggregation in the local microenvironment stimulate the release of their own inflammatory mediators to promote life-threatening cardiovascular and cerebrovascular thrombotic events.[113]

In addition to their role in acute thrombus formation, platelets are able to foster a complex inflammatory response that influences atherosclerotic lesion progression and induces plaque rupture.[114] Platelets involved in vascular inflammation appear to interact with endothelial cells, macrophages, and smooth muscle cells to create a pathologic fibroproliferative state.[115]

The route of entry of platelets into the atherosclerotic plaque and their exact location inside the plaque are, however, not completely understood. The key role played by platelets in atherothrombosis has led to the search of effective and safe antiplatelet agents that can block platelet aggregation.

Platelet-mediated thrombosis is dependent on three steps: (1) activation, (2) aggregation, and (3) adhesion, and each of these steps is a potential therapeutic target for the development of antiplatelet agents. Only platelet adhesion antagonists have not been approved for clinical use.

Antiplatelet treatment is a clinically crucial mainstay in acute and long-term secondary thromboembolic prevention. Antiplatelet drugs that work on different receptor sites of the platelet have been found to be extremely effective in preventing cardiovascular and cerebrovascular events and their recurrence, especially in high-risk patients and patients undergoing a percutaneous coronary intervention.[115,116]

These antiplatelet drugs include platelet-activating inhibitors such as aspirin that can irreversibly block cyclooxygenase-1 (COX-1) and the formation of prostaglandin PGH_2, the precursor of the powerful platelet aggregant and vasoconstrictor, thromboxane A_2.[117] It has been reported that PGH_2 and thromboxane A_2 receptors are found to be increased in patients with acute MI.[118]

The benefit of aspirin used alone or in combination with other antiplatelet drugs has been extensively studied as a primary and secondary prevention of thrombotic events in patients at high risk of cardiovascular and cerebrovascular embolism.[119,120]

The main problem with using aspirin to prevent the recurrence rate of ischemic events, aside from its bleeding potential, is that it targets only the COX-1 pathway and therefore cannot take aim at other pivotal signaling pathways, which can also lead to coronary and cerebrovascular thrombosis.

A second group of platelet activation inhibitors currently approved for clinical use are the thienopyridines such as ticlopidine, clopidogrel, and prasugrel, which target the blockade of the platelet ADP receptor $P2Y_{12}$.[121] It is well known that $P2Y_{12}$ receptor-mediated signaling processes play a central role in platelet activation and aggregation.[122] ADP is a potent platelet-aggregating agent that was originally shown to induce myocardial ischemia and arrhythmias when injected experimentally into the coronary arteries of pigs.[123]

Clopidogrel is currently the $PY2_{12}$ receptor antagonist of choice, although its broad interindividual variability in patients resistant to this antiplatelet drug and its delayed onset of action that can take 4–5 days before reaching its maximum antiplatelet effect has resulted in a considerable number of adverse cardiovascular and ischemic events.[124] These adverse events in patients treated with clopidogrel have reduced its clinical expectations as a safe antiplatelet agent.[124,125]

Despite this concern, the antagonism of the $P2Y_{12}$ receptor with clopidogrel has been shown to be clinically effective in the prevention of MI, ischemic stroke, and vascular death.[126] Nevertheless, there is hope that a more clinically effective and safer $P2Y_{12}$ antagonist will be found to prevent the incidence of ischemic and cardiovascular events.[126]

To increase its efficacy, clopidogrel has been combined with aspirin and found more effective than aspirin alone in acute coronary syndrome and stroke.[127–129] It is also reasonable to speculate that dual antiplatelet therapy such as the combination of clopidogrel–aspirin is more effective than either monotherapy because it targets *two* antiplatelet pathways, for example, ADP-induced platelet aggregation (clopidogrel) and thromboxane A_2 platelet aggregation (aspirin).

However, there is a rising concern that dual therapy using clopidogrel–aspirin increases the risk of bleeding due to the powerful anti thrombotic effect of using two anticoagulants than using aspirin monotherapy.[130] More importantly, the possible synergy of this dual therapy has been reported to be no better than aspirin alone with respect to lowering the rate of MI, stroke, or death from cardiovascular causes.[122,130]

Other antiplatelet agents used clinically include thrombin receptor antagonists such as the phosphodiesterase (PDE) inhibitors dipyridamole and cilostazol.[131]

Since thrombin is the most potent platelet activator known, there is hope that thrombin antagonists represent a promising novel strategy to reduce ischemic events without increasing the risk of bleeding or cardiovascular events, as other antiplatelet agents are prone to do.[132,133]

In this sense, a monotherapy that could antagonize platelet aggregation induced by more than one signaling pathway might offer improved efficacy and better safety than dual antiplatelet therapy because it would avoid the powerful anti thrombotic effect that is typically seen from two strong anti thrombotic agents that can lead to life-threatening bleeding complications as shown by clopidogrel–aspirin.[134]

DMSO has been shown to possess several unique pharmacological activities that could identify it as an antiplatelet monotherapeutic drug for use against CAD and stroke.

DMSO combines four crucial properties that have been consistently sought in a drug to help prevent primary and secondary thromboembolic events: (1) *platelet deaggregation*, (2) *tissue factor antagonism*, (3) *free radical scavenging*, and (4) *anti-inflammation*.

These properties resemble those observed with aspirin, which is also an effective anti-inflammatory and antiplatelet agent that blocks the production of prostaglandins by blocking the COX-1 pathway and consequently the synthesis of thromboxane A_2. As will be seen later, DMSO, unlike aspirin, also blocks tissue factor, a precursor of the powerful platelet aggregator thrombin, and additionally shows consistent activity as a free radical scavenger.

These activities by DMSO target two different signaling pathways known to induce platelet aggregation, namely, the prostaglandin cascade and thrombin activation. It remains to be seen in future clinical trials of DMSO whether targeting two different signaling pathways involved in platelet aggregation enhances its pharmacotherapeutic potential to prevent or reverse coronary thrombosis and stroke events.

This section will examine how these activities reported by DMSO could serve to provide patients at risk or at recurrence of CAD and stroke with significant short- or long-term prevention.

PLATELET DEAGGREGATION AND FREE RADICAL SCAVENGING

An early report by White[135] in 1971 reported the antiaggregatory property by DMSO for platelets incubated in platelet-rich plasma. This report was followed by an extended study from the same author and his colleagues,[136] that showed how DMSO might exert its antiaggregation activity on platelets. An antibiotic ionophore A23187 was used to produce rapid, irreversible concentration-dependent platelet aggregation *in vitro*.[136]

It was observed that the elevation of the cyclic 3′,5′-adenosine monophosphate (cAMP) level in platelets inhibited the response to all aggregating substances used in that study, including the A23187 ionophore. It was found that 10 mM DMSO added to the platelet-rich plasma was able to deaggregate these platelets ostensibly by increasing the level of cAMP in platelets following A23187-induced aggregation.[136]

Although it was not entirely clear in 1974 how DMSO was able to influence platelet cAMP *in vivo*, one suggestion was that DMSO altered platelet membrane fluidity to indirectly increase intracellular cAMP.[137]

Görög and Kovács[138] reported soon after the studies by White that topically applied DMSO was effective against platelet thrombi induced by ADP, a powerful platelet aggregator, when placed in the microcirculation of the hamster's cheek pouch.[138] These investigators also observed that 10% DMSO topically applied to a platelet-occluded venule preparation isolated from rat mesentery was able to inhibit thrombus formation in those venules when flow was experimentally restored.[138]

One of the more curious reports showing the antiaggregatory activity of DMSO was shown by Pace et al.[139] in 1983. This team, working for a major pharmaceutical, was investigating potential agents for the treatment of ischemic cardiac disease on a canine model when they made the mistake of using DMSO as a diluent for the experimental platelet deaggregators. Like good scientists, they observed the strange finding that most of the experimental agents tested had almost identical excellent antiaggregatory properties when dissolved in DMSO, at which point, when they tested DMSO by itself, they realized to their dismay that when 1 g/mL of DMSO was injected into a partially stenosed left circumflex branch of the left main coronary artery, the low coronary blood flow stemming from the stenosis was reversed.[139]

These findings were later confirmed by Levett and his team,[140] who infused 50% DMSO intravenously using a similar canine model of myocardial ischemia obtained by the ligation of the left descending coronary artery. They reported that cardiac output increased significantly, and both systemic and cerebrovascular resistance decreased in the DMSO-treated animals as compared to controls.[140]

On the question of how DMSO may deaggregate platelets, it has been shown that DMSO can increase cAMP accumulation in platelets by inhibiting its breakdown enzyme, the cyclic nucleotide PDE, a reaction that was observed in human peripheral blood lymphocytes *in vitro*.[137] This action on PDEs by DMSO could be similar to that of cilostazol, a promising therapeutic drug for patients at risk of ischemic stroke and major adverse cardiac events.[141] Cilostazol exerts its antiaggregation activity by selectively inhibiting PDE type 3.[142]

It should be noted that, at the present time, platelets are known to possess three PDEs, PDE2, PDE3, and PDE5, of the 60 different platelet isoforms described in mammalian species.[143,144] It is not yet known which of these PDEs DMSO targets for the inhibition of platelet aggregation.

PDEs are fundamental for platelet function, and their inhibition can dampen platelet aggregation by increasing cAMP and cyclic guanosine 3′–5′-monophosphate (cGMP), an action that limits the levels of intracellular nucleotides.[145]

Increasing cAMP by DMSO also implies that DMSO may serve an important function in manipulating signal transduction mediated by second messenger molecules such as cAMP. Inhibition of platelet signal transduction may suppress platelet activation regardless of the initial stimulus.[143] As mentioned before, DMSO could achieve this not only by directly inhibiting platelet-activating second messengers like cAMP and cGMP but also by amplifying the action of the endothelium-derived PGI$_2$, which can indirectly increase intracellular platelet cAMP.[143]

The ability to inhibit PDE enzymes has great clinical significance because it can prolong or enhance the effects of physiological processes mediated by cAMP, especially where cAMP plays a role in coronary heart disease. For example, it is generally recognized that prostacyclin (PGI$_2$) is the most potent inhibitor of platelet aggregation and that this action involves the production of cAMP.[146,147]

PGI$_2$-induced inhibition of platelet aggregation is mediated by the increase in cellular cAMP level through the activation of membrane adenylate cyclase, a process that starts as PGI$_2$ binds to its receptors on the platelet surface.[148]

It is known from the early studies of LaHann and Horita[149] that DMSO is able to stimulate the antiplatelet prostanoid PGE$_1$ and that at low doses in some *in vitro* systems, it also blocks prostaglandin biosynthesis and platelet aggregation induced by arachidonic acid.[22] These findings imply that low or high doses of DMSO can have a stimulating effect on specific prostaglandins like PGE$_1$ or totally block prostaglandin biosynthesis, including PGI$_2$ and thromboxane A$_2$.[22] Blocking thromboxane A$_2$ myocardial receptors could prevent MI and sudden cardiac death.[150]

By contrast, one *in vitro* study that examined the effect of DMSO on PGI$_2$ synthesis concluded that DMSO suppressed PGI$_2$ production in cultured aortic endothelial cells by interfering with the release of arachidonic acid in the prostaglandin cascade. This study did not investigate what action, if any, DMSO might have exerted on thromboxane A$_2$ production since a high dose of DMSO, like aspirin, would inhibit not only PGI$_2$ but also thromboxane synthesis by blocking the COX-1 arachidonic acid pathway.

In a recent study by Asmis et al.,[151] this is precisely what DMSO appears to do. The incubation of platelet-rich plasma with DMSO 0.5% (6.3 Mm) for 30 min significantly decreased total COX-1 activity by 36% compared to control. When DMSO was compared to the COX-1-specific inhibitor SC-560, a similar inhibition of 33% of total COX activity was observed with SC-560 as compared to control.[151] When DMSO and SC-560 were *combined*, no further reduction in COX-1 activity was seen, an indication that the residual COX-2 activity was not inhibited by either agent and that DMSO acted as a selective inhibitor of COX-1.[151] This is an important finding because the inhibition of COX-2 by some NSAIDs has been associated with an increased risk of renal failure, heart attack, thrombosis, and stroke through an increase in thromboxane A_2, which reduces the synthesis of PGI_2.[152,153]

In the study by Asmis et al.,[151] when the exogenous thromboxane A_2 analogue U46619 was combined with DMSO, the inhibitory effect of DMSO on platelet aggregation was prevented, implying that DMSO has the ability to block the prothrombotic release of thromboxane A_2.

The collective evidence from all these studies is that DMSO, like aspirin, is an antiplatelet agent with unique activities in blocking prostaglandin biosynthesis and the production of the powerful platelet aggregator thromboxane A_2, a cornerstone of acute coronary syndrome and stroke prevention. Like aspirin, whose optimal oral dose is recommended to be not greater than 75–81 mg/day, DMSO is in need of a clinical study that establishes its optimal dose and route of administration so it can best serve as a long-term antiplatelet agent in the control and prevention of acute coronary syndromes and stroke.

The data presented here on the effects of DMSO as a treatment for coronary stenosis and cardiac ischemia strongly suggest that this drug could be an effective anticlot-forming, cardioprotective, and stroke-preventing agent with similar actions as aspirin but void of the potential side effects such as hemorrhage, gastric pain, nausea, and skin rashes often seen with aspirin.

TISSUE FACTOR AND INFLAMMATION

The third and fourth properties expressed by DMSO as a potential antithrombotic agent are its inhibition of tissue factor expression and its anti-inflammatory activity.

Atherothrombotic vascular disorders often underlie CAD, stroke, and peripheral arterial disease and involve coagulation-induced thrombogenic molecules that can generate inflammatory reactions.

In the coagulation cascade, thrombin promotes platelet adhesion and aggregation by activating protease-activated receptors on the cell membrane of the platelet.[154]

Because thrombin plays a central role in arterial thrombogenesis, the goal of most treatment regimens is to block thrombin generation or inhibit its activity.

Direct thrombin inhibitors are a new class of anticoagulants that are used to prevent arterial or venous thrombosis. By reducing the bound thrombin activation of platelets, direct thrombin inhibitors can also be used as antiplatelet agents for such conditions as acute coronary syndromes and stroke.[155] They act by inactivating free thrombin or the thrombin that is bound to fibrin.[155]

Inactivation of fibrin-bound thrombin can be achieved by blocking upstream molecules in the coagulation cascade such as tissue factor. This approach is as effective as direct thrombin inhibition in preventing thrombin-induced platelet aggregation.[154]

Tissue factor is a cell surface glycoprotein that interacts with plasma factor VIIa to activate the coagulation cascade and generate thrombin in a stepwise activation of a series of proenzymes.[156]

The best known function of tissue factor is its role in blood coagulation where together with factor VIIa it forms a common pathway that leads to the formation of thrombin.[157]

Thrombin plays a dual action role where it normally contributes to signaling platelets for hemostasis and, abnormally, where it triggers platelet activation and arterial thrombosis that can result in heart attacks and stroke.[158] Thrombin has a large array of functions, but its primary role is the conversion of fibrinogen to fibrin, the primary building block of a hemostatic plug.[159,160] Although tissue factor is essential for hemostasis, its aberrant upregulation can also lead to atherothrombosis via the formation of thrombin.

The exact role of tissue factor in thrombosis remains unclear, but it has been found present in atherosclerotic plaques, and recent studies suggest that tissue factor–positive microparticles are the most abundant source in plaques.[161]

Rupture of high-risk vulnerable plaques with the subsequent tissue factor exposure is responsible for coronary thrombosis, acute MI, and sudden cardiac death.[162,163] Moreover, plaque thrombogenicity is directly correlated to the presence of elevated tissue factor content.[164] When a plaque ruptures, tissue factor on the plaque surface triggers thrombus formation, leading to downstream arterial occlusion.[165]

Thrombosis and inflammation are two major consequences of blood coagulation, and both can promote each other. Tissue factor can induce both thrombosis and inflammation through coagulant mediators factor VIIa, factor Xa, and fibrin production, all of which are proinflammatory.[166]

It would seem that if a drug could be found that suppressed tissue factor, vulnerable atherosclerotic plaque formation and platelet activation leading to arterial thrombosis could be dampened and possibly prevented.

Emerging evidence in a key research finding by Camici et al.[167] showed that DMSO was able to suppress tissue factor expression and activity as well as thrombus formation in a mouse carotid artery injury model. Tissue factor expression was seen to increase 19- and 24-fold in human aortic endothelial cells stimulated with tissue necrosis factor-a (TNF-α) or thrombin, respectively.[167] What was remarkable about this study was that DMSO was seen to inhibit the tissue factor induction of TNF-α and thrombin in a concentration-dependent manner. In the same study, in addition to blocking tissue factor mRNA expression, DMSO also inhibited the proliferation and migration of vascular smooth muscle cells, which occurs in vascular injury and atherogenesis after signal transduction pathways are activated.[168] No toxic effects by DMSO on the aortic endothelial cells were observed even at the highest doses used of 12.6 mmol/L,[157] a finding previously supported in human corneal endothelial cells.[169] The findings from this study prompted the suggestion that DMSO has the potential to be applied to drug-eluting stents in order to inhibit platelet aggregation, particularly since DMSO was also shown to suppress paclitaxel- and rapamycin-induced tissue factor expression, which are known to represent a cause for stent thrombosis in drug-eluting stents.[170,171]

DMSO was found to suppress in a concentration-dependent manner tissue factor expression and activity in response to TNF-α or thrombin exposure in human endothelial cells, monocytes, and vascular smooth muscle cells *in vitro*.[167] Moreover, it was additionally observed that DMSO prevented the proliferation and migration of vascular smooth muscle cells from human aorta, an outcome that could have important clinical application in treating coronary thrombosis and MI.[167]

In summary, DMSO combines antiplatelet and anti-inflammatory properties much like aspirin, a drug of choice for the prevention of cardiovascular disease and stroke,[172] but unlike aspirin, DMSO adds two more properties that can additionally block alternate signaling pathways that lead to coronary artery syndromes. These two additional properties, anti-free radical effect on platelet aggregation[173,174] and anti thrombin activity on thromboembolism,[151,167] place DMSO in a class by itself as a potential treatment for thrombogenesis of any etiology.

Clinical studies to investigate these actions by DMSO seem highly warranted.

CLINICAL TOXICOLOGY

The DMSO toxicology in animals is discussed in Chapter 2. A thorough human toxicology study was reported by Brobyn[5] in 1967 due to the wide and ever-increasing use of DMSO not only under clinical supervision but also as a home remedy where DMSO could be purchased over the counter by consumers.

A short-term study (14 days) and a long-term study (90 days) were undertaken by Brobyn,[5] who reported his findings on 213 healthy human volunteers. These volunteers were periodically examined by specialists after being administered 1 g/kg/day 80% DMSO daily. This dose was considered 3–30 times higher than the typical DMSO dose used or recommended. The specific clinical examinations included complete physical examinations with laboratory examinations of blood and urine, ophthalmologic, dermatologic, cardiologic, neurologic, pulmonary function, and bone marrow studies.[5]

The results of this study concluded that no significant adverse effects from either the short-term or long-term administration of DMSO were observed after a very extensive toxicology study was conducted.[5] Minor side effects of DMSO administered daily included skin irritation, nausea, sedation, and diarrhea, which resolved following discontinuation of DMSO.[5] DMSO at 3–30 times the usual treatment dose in humans appeared as a relatively very safe drug, and in particular, no lens changes that were seen to occur in some animal species were detected in any subject after extensive ophthalmological examinations after 3-month follow-up.[5] This absence of ocular changes after high and prolonged administration of DMSO has been fully confirmed in other studies.[175–177] There has not been another major toxicological study on topical DMSO following Brobyn's[5] report, but thousands of papers dealing with DMSO and its various routes administration, including intravenous, oral, dermal, and intravesical, have reported that adverse reactions are relatively mild and usually can occur in relation to its concentration and its mode of administration.[177–179]

Following intravenous administration, the best-documented side effect from DMSO is intravascular hemolysis after intravenous infusion of 40% solution or greater, which can result in urinary excretion of hemoglobin.[180] Dark urine is sometimes a marker of kidney damage, but despite a dose-dependent transient hemolysis with resultant hemoglobinuria, no injury or alteration of renal function after DMSO administration occurs.[181] The hemolysis appears related to osmotic changes exerted by DMSO on erythrocytes. In one case report, 50 mL intravesical instillation of 50% DMSO was used successfully to stop chronic, profuse hematuria due to eosinophilic cystitis, a condition characterized by dysuria, frequency, and hematuria.[182]

Intravascular hemolysis by DMSO after intravenous administration can be prevented by using less than a 30% DMSO solution.[178] However, in order to avoid fluid overload and hypernatremia, DMSO should not be given at highly diluted rates of infusion, that is, 10% or less.[183] Fluid overload can be avoided if the DMSO solution is infused between 25% and 35% concentration.[184,185]

An annoying side effect of DMSO administration by any route includes a garlic-like breath odor due to the pulmonary excretion of its breakdown product to dimethyl sulfide.[179] This odor produced from DMSO can potentially interfere with double-blinded clinical trials when intravenous or topical DMSO is used because the sulfide odor will easily identify the patient receiving this treatment. However, the odor problem caused by DMSO administration can be neutralized with scouting equipment placed in the room that absorbs the dimethyl sulfide and maintains the room relatively odor free.

DMSO is a relatively safe drug. Side effects to DMSO are common, but these are usually minor and are related to the concentration of DMSO used and the route of administration.[186–188]

REFERENCES

1. Jacob SW, Bischel M, Herschler RJ. Dimethyl sulfoxide (DMSO): A new concept in pharmacotherapy. *Curr Ther Res Clin Exp.* 1964;6:134–135.
2. Forouzanfar T, Kemler M, Kessels AG, Köke AJ, van Kleef M, Weber WE. Comparison of multiple against single pain intensity measurements in complex regional pain syndrome type I: Analysis of 54 patients. *Clin J Pain.* 2002;18(4):234–237.
3. Tran de QH, Duong S, Bertini P, Finlayson RJ. Treatment of complex regional pain syndrome: A review of the evidence. *Can J Anaesth.* 2010;57:149–166.
4. Perez RS, Zuurmond WW, Bezemer PD, Kuik DJ, van Loenen AC, de Lange JJ, Zuidhof AJ. The treatment of complex regional pain syndrome type I with free radical scavengers: A randomized controlled study. *Pain.* 2003;102(3):297–307.
5. Brobyn RD. The human toxicology of dimethyl sulfoxide. *Ann N Y Acad Sci.* 1975;243:497–506.
6. Rokyta R, Fricová J. Ontogeny of the pain. *Physiol Res.* 2012;61(Suppl 1):S109–S122.
7. Schaible HG. Peripheral and central mechanisms of pain generation. *Handb Exp Pharmacol.* 2007;177:3–28.
8. Cervero F. Sensory innervation of the viscera: Peripheral basis of visceral pain. *Physiol Rev.* 1994;74(1):95–138.
9. McMahon SB. Mechanisms of cutaneous, deep and visceral pain. In *Textbook of Pain*, 3rd edn, eds. P.D. Wall and R. Melzack, Churchill Livingstone, Edinburgh, U.K., 1994; pp. 128–153.
10. Burgess PR, Perl FR. Myelinated afferent fibres responding specifically to noxious stimulation of the skin. *J Physiol.* 1967;190:541–562.

11. Kerr FWL. Neuroanatomical substrates of nociception in the spinal cord. *Pain.* 1975;1:325–356.
12. Clark D, Hughes J, Gasser HS. Afferent function in the group of nerve fibers of slowest conduction velocity. *Am J Physiol.* 1935;114:69–76.
13. Collins WF, Nulsen FE, Randt CT. Relation of peripheral nerve fiber size and sensation in man. *Arch Neurol.* 1960;3:381–385.
14. Torebjork HE, Hallin RG. Perceptual changes accompanying controlled preferential blocking of A and C fibre responses in intact human skin nerves. *Exp Brain Res.* 1973;16:321–332.
15. Murinson BB, Griffin JW. C-fiber structure varies with location in peripheral nerve. *J Neuropathol Exp Neurol.* 2004;63(3):246–254.
16. Craig KD, Weiss SM. Vicarious influences on pain-threshold determinations. *J Pers Soc Psychol.* 1971;19:53–59.
17. Dworkin RH, O'Connor AB, Backonja M, Farrar JT, Finnerup NB, Jensen TS, Kalso EA et al. Recommendations for the pharmacological management of neuropathic pain: An overview and literature update. *Mayo Clin Proc.* 2010;85(3 Suppl):S3–S14.
18. Acquadro C, Berzon R, Dubois D, Leidy NK, Marquis P, Revicki D, Rothman M; PRO Harmonization Group. Incorporating the patient's perspective into drug development and communication: An ad hoc task force report of the Patient-Reported Outcomes (PRO) Harmonization Group meeting at the Food and Drug Administration, February 16, 2001. *Value Health.* 2003;6(5):522–531.
19. Forouzanfar T, Koke A, van Kleef M, Weber W. Treatment of complex regional pain syndrome type 1. *Eur J Pain.* 2002;6:105–122.
20. Beecher HK. The powerful placebo. *J Am Med Assoc.* 1955;159(17):1602–1606.
21. Colloca L, Miller FG. The nocebo effect and its relevance for clinical practice. *Psychosom Med.* 2011;73(7):598–603.
22. Panganamala RV, Sharma HM, Heikkila RE, Geer JC, Cornwell DG. Role of hydroxyl radical scavengers dimethyl sulfoxide, alcohols and methional in the inhibition of prostaglandin biosynthesis. *Prostaglandins.* 1976;11(4):599–607.
23. Evans MS, Reid KH, Sharp JB Jr. Dimethylsulfoxide (DMSO) blocks conduction in peripheral nerve C fibers: A possible mechanism of analgesia. *Neurosci Lett.* 1993;150(2):145–148.
24. Funk CD. Prostaglandins and leukotrienes: Advances in eicosanoid biology. *Science.* 2001;294:1871–1875.
25. Duggan JM. Aspirin ingestion and perforated peptic ulcer. *Gut.* 1972;13(8):631–633.
26. Becker DP, Young HF, Nulsen FE, Jane JA. Physiological effects of dimethyl sulfoxide on peripheral nerves: Possible role in pain relief. *Exp Neurol.* 1969;24(2):272–276.
27. Paul MM. Interval therapy with dimethyl sulfoxide. *Ann N Y Acad Sci.* 1967;141(1):586–598.
28. Lockie LM, Norcross BM. A clinical study on the effects of dimethyl sulfoxide in 103 patients with acute and chronic musculoskeletal injuries and inflammations. *Ann N Y Acad Sci.* 1967;141(1):599–602.
29. Blumenthal LS, Fuchs M. The clinical use of dimethyl sulfoxide on various headaches, musculoskeletal, and other general medical disorders. *Ann N Y Acad Sci.* 1967;141(1):572–585.
30. Kneer W, Kühnau S, Bias P, Haag RF. Dimethylsulfoxide (DMSO) gel in treatment of acute tendopathies. A multicenter, placebo-controlled, randomized study. *Fortschr Med.* 1994;112(10):142–146.
31. Matsumoto J. Clinical trials of dimethyl sulfoxide in rheumatoid arthritis patients in Japan. *Ann N Y Acad Sci.* 1967;141(1):560–568.
32. Penrod DS, Bacharach B, Templeton JY. Dimethyl sulfoxide for incisional pain after thoracotomy: Preliminary report. *Ann N Y Acad Sci.* 1967;141(1):493–495.

33. Asen H. Dimethyl sulfoxide: A new concept in the treatment of ENT diseases. *Ann NY Acad Sci.* 1967;141(1):451–457.
34. Percy EC, Carson JD. The use of DMSO in tennis elbow and rotator cuff tendonitis: A double-blind study. *Med Sci Sports Exerc.* 1981;13(4):215–219.
35. Freeman GR. DMSO in otology. *Laryngoscope.* 1976;86(7):921–929.
36. Chung OY, Bruehl S. Complex regional pain syndrome. *Curr Treat Options Neurol.* 2003;5:499–511.
37. Bruehl S. An update on the pathophysiology of complex regional pain syndrome. *Anesthesiology.* 2010;113(3):713–725.
38. Harden RN, Bruehl S, Galer BS, Saltz S, Bertram M, Backonja M, Gayles R, Rudin N, Bughra M, Stanton-Hicks M. Complex regional pain syndrome: Are the IASP diagnostic criteria valid and sufficiently comprehensive? *Pain.* 1999;83:211–219.
39. Parkitny L, McAuley JH, Di Pietro F, Stanton TR, O'Connell NE, Marinus J, van Hilten JJ, Moseley GL. Inflammation in complex regional pain syndrome: A systematic review and meta-analysis. *Neurology.* 2013;80(1):106–117.
40. Alexander GM, van Rijn MA, van Hilten JJ, Perreault MJ, Schwartzman RJ. Changes in cerebrospinal fluid levels of pro-inflammatory cytokines in CRPS. *Pain* 2005;116:213–219.
41. Maihofner C, Handwerker HO, Neundörfer B, Birklein F. Mechanical hyperalgesia in complex regional pain syndrome: A role for TNF-alpha? *Neurology* 2005;65:311–313.
42. Schinkel C, Gaertner A, Zaspel J, Zedler S, Faist E, Schuermann M. Inflammatory mediators are altered in the acute phase of posttraumatic complex regional pain syndrome. *Clin J Pain.* 2006;22(3):235–239.
43. Birklein F, Schmelz M, Schifter S, Weber M. The important role of neuropeptides in complex regional pain syndrome. *Neurology.* 2001;57:2179–2184.
44. Oaklander AL, Rissmiller JG, Gelman LB, Zheng L, Chang Y, Gott R. Evidence of focal small-fiber axonal degeneration in complex regional pain syndrome-I (reflex sympathetic dystrophy). *Pain.* 2006;120:235–243.
45. Cossins L, Okell RW, Cameron H, Simpson B, Poole HM, Goebel A. Treatment of complex regional pain syndrome in adults: A systematic review of randomized controlled trials published from June 2000 to February 2012. *Eur J Pain.* 2013;17(2):158–173.
46. Zollinger PE, Tuinebreijer WE, Breederveld RS, Kreis RW. Can vitamin C prevent complex regional pain syndrome in patients with wrist fractures? A randomized, controlled, multicenter dose-response study. *J Bone Joint Surg Am.* 2007;89(7):1424–1431.
47. Correll GE, Maleki J, Gracely EJ, Muir JJ, Harbut RE. Subanesthetic ketamine infusion therapy: A retrospective analysis of a novel therapeutic approach to complex regional pain syndrome. *Pain Med.* 2004;5(3):263–275.
48. Collins S, Sigtermans MJ, Dahan A, Zuurmond WW, Perez RS. NMDA receptor antagonists for the treatment of neuropathic pain. *Pain Med.* 2010;11(11):1726–1742.
49. Kiefer RT, Rohr P, Ploppa A et al. A pilot open-label study of the efficacy of subanesthetic isomeric *S*(+)-ketamine in refractory CRPS patients. *Pain Med.* 2008;9(1):44–54.
50. Sunder RA, Toshniwal G, Dureja GP. Ketamine as an adjuvant in sympathetic blocks for management of central sensitization following peripheral nerve injury. *J Brachial Plex Peripher Nerve Inj.* 2008;25(3):22.
51. van Dieten HE, Perez RS, van Tulder MW, de Lange JJ, Zuurmond WW, Ader HJ, Vondeling H, Boers M. Cost effectiveness and cost utility of acetylcysteine versus dimethyl sulfoxide for reflex sympathetic dystrophy. *Pharmacoeconomics.* 2003;21(2):139–148.
52. Repine JE, Pfenninger OW, Talmage DW, Berger EM, Pettijohn DE. Dimethyl sulfoxide prevents DNA nicking mediated by ionizing radiation or iron/hydrogen peroxide-generated hydroxyl radical. *Proc Natl Acad Sci USA.* 1981;78(2):1001–1003.
53. Rosenblum WI, El-Sabban F. Dimethyl sulfoxide (DMSO) and glycerol, hydroxyl radical scavengers, impair platelet aggregation within and eliminate the accompanying vasodilation of, injured mouse pial arterioles. *Stroke.* 1982;13(1):35–39.

54. Gaspar M, Bovaira M, Carrera-Hueso FJ, Querol M, Jiménez A, Moreno L. Efficacy of a topical treatment protocol with dimethyl sulfoxide 50% in type 1 complex regional pain syndrome. *Farm Hosp.* 2012;36(5):385–391.

55. Goris RJ. Treatment of reflex sympathetic dystrophy with hydroxyl radical scavengers. *Unfallchirurg.* 1985;88(7):330–332.

56. Zuurmond WW, Langendijk PN, Bezemer PD, Brink HE, de Lange JJ, van loenen AC. Treatment of acute reflex sympathetic dystrophy with DMSO 50% in a fatty cream. *Acta Anaesthesiol Scand.* 1996;40(3):364–367.

57. van Eijs F, Geurts JW, Van Zundert J, Faber CG, Kessels AG, Joosten EA, van Kleef M. Spinal cord stimulation in complex regional pain syndrome type I of less than 12-month duration. *Neuromodulation.* 2012;15(2):144–150.

58. Quisel A, Gill JM, Witherell P. Complex regional pain syndrome: Which treatments show promise? *J Fam Pract.* 2005;54(7):599–603.

59. Simon LS, Grierson LM, Naseer Z, Bookman AA, Zev Shainhouse J. Efficacy and safety of topical diclofenac containing dimethyl sulfoxide (DMSO) compared with those of topical placebo, DMSO vehicle and oral diclofenac for knee osteoarthritis. *Pain.* 2009;143(3):238–245.

60. Bookman AAM, Williams KSA, Shainhouse JZ. Effect of a topical diclofenac solution for relieving symptoms of primary osteoarthritis of the knee: A randomized controlled trial. *CMAJ.* 2004;171(4):333–338.

61. Barkin RL. Topical nonsteroidal anti-inflammatory drugs: The importance of drug, delivery, and therapeutic outcome. *Am J Ther.* February 22, 2012; Epub ahead of print.

62. Roth SH, Fuller P. Pooled safety analysis of diclofenac sodium topical solution 1.5% (w/w) in the treatment of osteoarthritis in patients aged 75 years or older. *Clin Interv Aging.* 2012;7:127–137.

63. Mostofi FK, Murphy GP, Mettlin C, Sesterhenn IA, Batsakis JG, Khaliq SU, Nadimpalli V, Tahan S, Siders DB, Kollin J. Pathology review in an early prostate cancer detection program: Results from the American Cancer Society-National Prostate Cancer Detection Project. *Prostate.* 1995;27(1):7–12.

64. Hoang BX, Levine SA, Shaw DG, Tran DM, Tran HQ, Nguyen PM, Tran HD, Hoang C, Pham PT. Dimethyl sulfoxide as an excitatory modulator and its possible role in cancer pain management. *Inflamm Allergy Drug Targets.* 2010;9(4):306–312.

65. Hoang BX, Le BT, Tran HD, Hoang C, Tran HQ, Tran DM, Pham CQ et al. Dimethyl sulfoxide-sodium bicarbonate infusion for palliative care and pain relief in patients with metastatic prostate cancer. *J Pain Palliat Care Pharmacother.* 2011;25(4):350–355.

66. Moul JW, Civitelli K. Managing advanced prostate cancer with Viadur (leuprolide acetate implant). *Urol Nurs.* 2001;21(6):385–388, 393–394.

67. Marks LS. Luteinizing hormone-releasing hormone agonists in the treatment of men with prostate cancer: Timing, alternatives, and the 1-year implant. *Urology.* 2003;62(6 Suppl 1):36–42.

68. Fowler JE Jr; Viadur Study Group. Patient-reported experience with the Viadur 12-month leuprolide implant for prostate cancer. *Urology.* 2001;58(3):430–434.

69. Spruance SL, Stewart JC, Freeman DJ, Brightman VJ, Cox JL, Wenerstrom G, McKeough MB, Rowe NH. Early application of topical 15% idoxuridine in dimethyl sulfoxide shortens the course of herpes simplex labialis: A multicenter placebo-controlled trial. *J Infect Dis.* 1990;161(2):191–197.

70. Aliaga A, Armijo M, Camacho F, Castro A, Cruces M, Díaz JL, Fernández JM, Iglesias L, Ledo A, Mascaró JM. A topical solution of 40% idoxuridine in dimethyl sulfoxide compared to oral acyclovir in the treatment of herpes zoster. A double-blind multicenter clinical trial. *Med Clin (Barc).* 1992;98(7):245–249.

71. Elliott CS, Payne CK. Interstitial cystitis and the overlap with overactive bladder. *Curr Urol Rep.* 2012;13(5):319–326.

72. You S, Yang W, Anger JT, Freeman MR, Kim J. 'Omics' approaches to understanding interstitial cystitis/painful bladder syndrome/bladder pain syndrome. *Int Neurourol J.* 2012;16(4):159–168.

73. Quillin RB, Erickson DR. Management of interstitial cystitis/bladder pain syndrome: A urology perspective. *Urol Clin North Am.* 2012;39(3):389–396.

74. Shirley SW, Stewart BH, Mirelman S. Dimethyl sulfoxide in treatment of inflammatory genitourinary disorders. *Urology.* 1978;11:215.

75. Ek A, Engberg A, Frödin L, Jönsson G. The use of dimethyl-sulfoxide (DMSO) in the treatment of interstitial cystitis. *Scand J Urol Nephrol.* 1978;12(2):129–131.

76. Perez-Marrero R, Emerson LE, Feltis JT. A controlled study of dimethyl sulfoxide in interstitial cystitis. *J Urol.* 1988;140(1):36–39.

77. Biggers RD. Self-administration of dimethyl sulfoxide (DMSO) for interstitial cystitis. *Urology.* 1986;28(1):10–11.

78. Parkin J, Shea C, Sant GR. Intravesical dimethyl sulfoxide (DMSO) for interstitial cystitis—A practical approach. *Urology.* 1997;49:105–107.

79. Stewart BH, Branson AC, Hewitt CB, Kiser WS, Straffon RA. The treatment of patients with interstitial cystitis, with special reference to intravesical DMSO. *J Urol.* 1972;107(3):377–380.

80. Shiga KI, Hirano K, Nishimura J, Niiro N, Naito S, Kanaide H. Dimethyl sulphoxide relaxes rabbit detrusor muscle by decreasing the Ca^{2+} sensitivity of the contractile apparatus. *Br J Pharmacol.* 2007;151(7):1014–1024.

81. Richter B, Hesse U, Hansen AB, Horn T, Mortensen SO, Nordling J. Bladder pain syndrome/interstitial cystitis in a Danish population: A study using the 2008 criteria of the European Society for the Study of Interstitial Cystitis. *BJU Int.* 2010;105(5):660–667.

82. Kushner L, Chiu PY, Brettschneider N, Lipstein A, Eisenberg E, Rofeim O, Moldwin R. Urinary substance P concentration correlates with urinary frequency and urgency in interstitial cystitis patients treated with intravesical dimethyl sulfoxide and not intravesical anesthetic cocktail. *Urology.* 2001;57(6 Suppl 1):129.

83. Rossberger J, Fall M, Peeker R. Critical appraisal of dimethyl sulfoxide treatment for interstitial cystitis: Discomfort, side-effects and treatment outcome. *Scand J Urol Nephrol.* 2005;39:73–77.

84. Sant GR. Intravesical 50% dimethyl sulfoxide (Rimso-50) in treatment of interstitial cystitis. *Urology.* 1987;29(4 Suppl):17–21.

85. Stewart BH, Shirley SW. Further experience with intravesical dimethyl sulfoxide in the treatment of interstitial cystitis. *J Urol.* 1976;116(1):36–38.

86. Stewart BH, Persky L, Kiser WS. The use of dihyl sulfoxide (DMSO) in the treatment of interstitial cystitis. *J Urol.* 1967;98(6):671–672.

87. Ghoniem GM, McBride D, Sood OP, Lewis V. Clinical experience with multiagent intravesical therapy in interstitial cystitis patients unresponsive to single-agent therapy. *World J Urol.* 1993;11(3):178–182.

88. Hung MJ, Chen YT, Shen PS, Hsu ST, Chen GD, Ho ES. Risk factors that affect the treatment of interstitial cystitis using intravesical therapy with a dimethyl sulfoxide cocktail. *Int Urogynecol J.* 2012;23(11):1533–1539.

89. Amemori S, Iwakiri R, Endo H, Ootani A, Ogata S, Noda T, Tsunada S et al. Oral dimethyl sulfoxide for systemic amyloid A amyloidosis complication in chronic inflammatory disease: A retrospective patient chart review. *J Gastroenterol.* 2006;41(5):444–449.

90. Salim AS. Role of oxygen-derived free radical scavengers in the management of recurrent attacks of ulcerative colitis: A new approach. *J Lab Clin Med.* 1992;119(6):710–717.

91. Pestronk A, Drachman DB. Dimethyl sulphoxide reduces anti-receptor antibody titres in experimental myasthenia gravis. *Nature.* 1980;288(5792):733–734.

92. Pestronk A, Teoh R, Sims C, Drachman DB. Effects of dimethyl sulfoxide on humoral immune responses to acetylcholine receptors in the rat. *Clin Immunol Immunopathol.* 1985;37(2):172–178.

93. Kim R, Liu W, Chen X, Kreder KJ, Luo Y. Intravesical dimethyl sulfoxide inhibits acute and chronic bladder inflammation in transgenic experimental autoimmune cystitis models. *J Biomed Biotechnol.* 2011;2011:937061.

94. Lotz M, Villiger P, Hugli T, Koziol J, Zuraw BL. Interleukin-6 and interstitial cystitis. *J Urol.* 1994;152(3):869–873.

95. de la Torre JC, Rowed DW. DMSO: A new respiratory stimulant? *J Clin Pharmacol.* 1974;14(7):345–353.

96. de la Torre JC, Rowed DW, Kawanaga HM, Mullan S. Dimethyl sulfoxide in the treatment of experimental brain compression. *J Neurosurg.* 1973;38(3):345–354.

97. Klein HA, Samant S, Herz B, Pearlman H. Dimethyl sulfoxide in adult respiratory distress syndrome: Abstract of preliminary report. *Ann N Y Acad Sci.* 1983;411:389–390.

98. Mann A, Early GL. Acute respiratory distress syndrome. *Missouri Med.* 2012;109(5):371–375.

99. Kyle RA, Greipp PR. Amyloidosis (AL). Clinical and laboratory features in 229 cases. *Mayo Clin Proc.* 1983;58(10):665–683.

100. Isobe T, Osserman EF. Effects of DMSO on Bence-Jones proteins, amyloid fibrils and casein-induced amyloidosis. In *Amyloidosis*, eds. O. Wagelius, A. Pasternak. Academic Press, London, U.K., 1976, p. 247.

101. Malek RS, Wahner-Roedler DL, Gertz MA, Kyle RA. Primary localized amyloidosis of the bladder: Experience with dimethyl sulfoxide therapy. *J Urol.* 2002;168(3):1018–1020.

102. van Rijswijk MH, Ruinen L, Donker AJ, de Blécourt JJ, Mandema E. Dimethyl sulfoxide in the treatment of AA amyloidosis. *Ann N Y Acad Sci.* 1983;411:67–83.

103. Iwasaki T, Hamano T, Aizawa K, Kobayashi K, Kakishita E. A case of pulmonary amyloidosis associated with multiple myeloma successfully treated with dimethyl sulfoxide. *Acta Haematol.* 1994;91(2):91–94.

104. Sutani A, Tabe K, Nagata M, Kuramitsu K, Shimizu Y, Sakamoto Y. Treatment of pulmonary amyloidosis with dimethyl sulfoxide—A case report. *Nihon Kokyuki Gakkai Zasshi.* 2004;42(9):825–830.

105. Ravid M, Shapira J, Lang R, Kedar I. Prolonged dimethylsulphoxide treatment in 13 patients with systemic amyloidosis. *Ann Rheum Dis.* 1982;41(6):587–592.

106. Hinchcliff M, Varga J. Systemic sclerosis/scleroderma: A treatable multisystem disease. *Am Fam Physician.* 2008;78(8):961–968.

107. Gries G, Bublitz G, Lindner J. The effect of dimethyl sulfoxide on the components of connective tissue. (Clinical and experimental investigations). *Ann N Y Acad Sci.* 1967;141(1):630–637.

108. Frommhold W, Bublitz G, Gries G. The use of DMSO for the treatment of postirradiation subcutaneous plaques. *Ann N Y Acad Sci.* 1967;141(1):603–612.

109. Scherbel AL. The effect of percutaneous dimethyl sulfoxide on cutaneous manifestations of systemic sclerosis. *Ann N Y Acad Sci.* 1983;411:120–130.

110. Williams HJ, Furst DE, Dahl SL, Steen VD, Marks C, Alpert EJ, Henderson AM, Samuelson CO Jr, Dreyfus JN, Weinstein A. Double-blind, multicenter controlled trial comparing topical dimethyl sulfoxide and normal saline for treatment of hand ulcers in patients with systemic sclerosis. *Arthritis Rheum.* 1985;28(3):308–314.

111. Corti R., Farkouh ME, Badimon, JJ. The vulnerable plaque and acute coronary syndromes. *Am J Med.* 2002;113:668–680.

112. Gijsen F, van der Giessen A, van der Steen A, Wentzel J. Shear stress and advanced atherosclerosis in human coronary arteries. *J Biomech.* 2013;46(2):240–247.

113. Davi G, Pathono G. Platelet activation and atherothrombosis. *N Engl J Med.* 2007;357:2482–2494.

114. Picker SM. Platelet function in ischemic heart disease. *J Cardiovasc Pharmacol.* 2013;61(2):166–174.

115. Picker SM. Antiplatelet therapy in the prevention of coronary syndromes: Mode of action, benefits, drawbacks. *J Cardiovasc Pharmacol.* 2012;61(2):166–174.

116. Sibbing D, Orban M, Massberg S. Potent P2Y12 receptor inhibitors in patients with acute coronary syndrome. Agents, indications, issues to consider in clinical practice. *Hamostaseologie.* 2013;33(1):9–15.

117. Ruggeri ZM. Platelets in atherothrombosis. *Nat Med.* 2002;8:1227–1234.

118. Yuhki K, Kashiwagi H, Kojima F, Kawabe J, Ushikubi F. Roles of prostanoids in the pathogenesis of cardiovascular diseases. *Int Angiol.* 2010;29(2 Suppl):19–27.

119. Antithrombotic Trialists' Collaboration. Collaborative meta-analysis of randomised trials of antiplatelet therapy for prevention of death, myocardial infarction, and stroke in highrisk patients. *BMJ.* 2002;324:71–86.

120. Patrono C, Garc Rodr uez LA, Landolfi R, Baigent C. Low-dose aspirin for the prevention of atherothrombosis. *N Engl J Med.* 2005;353:2373–2383.

121. Savi P, Herbert JM. Semin Thromb Hemost. Clopidogrel and ticlopidine: P2Y12 adenosine diphosphate-receptor antagonists for the prevention of atherothrombosis. *Semin Thromb Hemost.* 2005;31(2):174–183.

122. Thomson RM, Anderson DC. Aspirin and clopidogrel for prevention of ischemic stroke. *Curr Neurol Neurosci Rep.* 2013;13(2):327.

123. Jørgensen L. ADP-induced platelet aggregation in the microcirculation of pig myocardium and rabbit kidneys. *J Thromb Haemost.* 2005;3(6):1119–1124.

124. Ferreiro JL, Angiolillo DJ. Clopidogrel response variability: Current status and future directions. *Thromb Haemost.* 2009;102(1):7–14.

125. Trenk D, Kristensen SD, Hochholzer W, Neumann FJ. High on-treatment platelet reactivity and P2Y12 antagonists in clinical trials. *Thromb Haemost.* 2013;109(5):834–845.

126. Bates ER, Lau WC, Angiolillo DJ. Clopidogrel-drug interactions. *Am Coll Cardiol.* 2011;57(11):1251–1263.

127. The Clopidogrel in Unstable Angina to Prevent Recurrent Events Trial Investigators. Effects of clopidogrel in addition to aspirin in patients with acute coronary syndromes without ST-segment elevation. *N Engl J Med.* 2001;345:494–502.

128. Markus HS, Droste DW, Kaps M, Larrue V, Lees KR, Siebler M, Ringelstein EB. Dual antiplatelet therapy with clopidogrel and aspirin in symptomatic carotid stenosis evaluated using Doppler embolic signal detection: The Clopidogrel and Aspirin for Reduction of Emboli in Symptomatic Carotid Stenosis (CARESS) trial. *Circulation.* 2005;111(17):2233–2240.

129. Kreutz RP, Nystrom P, Kreutz Y, Miao J, Kovacs R et al. Inhibition of platelet aggregation by prostaglandin E1 (PGE1) in diabetic patients during therapy with clopidogrel and aspirin. *Platelets.* 2013;24(2):145–150.

130. Bhatt DL, Fox KA, Hacke W, Berger PB, Black HR, Boden WE, Cacoub P et al. Clopidogrel and aspirin versus aspirin alone for the prevention of atherothrombotic events. *N Eng J Med.* 2006;354(16):1706–1717.

131. Jeong YH, Park Y, Kim IS, Yun SE, Kang MK, Hwang SJ, Kwak CH, Hwang JY. Enhanced platelet inhibition by adjunctive cilostazol to dual antiplatelet therapy after drug-eluting stent implantation for complex lesions. *Thromb Haemost.* 2010;104(6):1286–1289.

132. Ueno M, Ferreiro JL, Angiolillo DJ. Mechanism of action and clinical development of platelet thrombin receptor antagonists. *Expert Rev Cardiovasc Ther.* 2010;8(8):1191–200.

133. Yang J, Xu K, Seiffert D. Challenges and promises of developing thrombin receptor antagonists. *Recent Pat Cardiovasc Drug Discov.* 2010;5(3):162–170.

134. Diener HC, Bogousslavsky J, Brass LM, Cimminiello C, Csiba L, Kaste M, Leys D, Matias-Guiu J, Rupprecht HJ; MATCH Investigators. Aspirin and clopidogrel compared with clopidogrel alone after recent ischaemic stroke or transient ischaemic attack in high-risk patients (MATCH): Randomised, double-blind, placebo-controlled trial. *Lancet*. 2004;364(9431):331–337.

135. White JG. Platelet microtubules and microfilaments: Effects of cytochalasin B on structure and function. In *Platelet Aggregation*, ed. J. Caen. Masson, Paris, France, 1971, p. 30.

136. White JG, Rao GH, Gerrard JM. Effects of the ionophore A23187 on blood platelets I. Influence on aggregation and secretion. *Am J Pathol*. 1974;77(2):135–149.

137. Hynie S, Lanefelt F, Fredholm BB. Effects of ethanol on human lymphocyte levels of cyclic AMP. In vitro: Potentiation of the response to isoproterenol, prostaglandin E2 or adenosine stimulation. *Acta Pharmacol Toxicol (Copenh)*. 1980;47(1):58–65.

138. Görög P, Kovács IB. Antiarthritic and antithrombotic effects of topically applied dimethyl sulfoxide. *Ann N Y Acad Sci*. 1975;243:91–97.

139. Pace DG, Kovacs JL, Klevans LR. Dimethyl sulfoxide inhibits platelet aggregation in partially obstructed canine coronary vessels. *Ann N Y Acad Sci*. 1983;411:352–356.

140. Levett JM, Johns LM, Grina NM, Mullan BF, Kramer JF, Mullan JF. Effects of dimethyl sulfoxide on systemic and cerebral hemodynamic variables in the ischemic canine myocardium. *Crit Care Med*. 1987;15(7):656–660.

141. Rogers KC, Faircloth JM, Finks SW. Use of cilostazol in percutaneous coronary interventions. *Ann Pharmacother*. 2012;46(6):839–850.

142. Yang Y, Luo J, Kazumura K, Takeuchi K, Inui N, Hayashi H, Ohashi K, Watanabe H. Cilostazol suppresses adhesion of human neutrophils to HUVECs stimulated by FMLP and its mechanisms. *Life Sci*. 2006;79(7):629–636.

143. Gresele P, Momi S, Falcinelli E. Anti-platelet therapy: Phosphodiesterase inhibitors. *Br J Clin Pharmacol*. 2011;72(4):634–646.

144. Hidaka H, Asano T. Human blood platelet 3′:5′-cyclic nucleotide phosphodiesterase. Isolation of low-Km and high-Km phosphodiesterase. *Biochim Biophys Acta*. 1976;429:485–497.

145. Rondina MT, Weyrich AS. Targeting phosphodiesterases in anti-platelet therapy. *Handb Exp Pharmacol*. 2012;210:225–238.

146. Marcus AJ, Safier LB. Thromboregulation: Multicellular modulation of platelet reactivity in hemostasis and thrombosis. *FASEB J*. 1993;7(6):516–522.

147. Vane JR, Anggård EE, Botting RM. Regulatory functions of the vascular endothelium. *N Engl J Med*. 19905;323(1):27–36.

148. Moncada S, Palmer RM, Higgs EA. Nitric oxide: Physiology, pathophysiology, and pharmacology. *Pharmacol Rev*. 1991;43(2):109–142.

149. LaHann TR, Horita A. Effects of dimethyl sulfoxide (DMSO) on prostaglandin synthetase. *Proc West Pharmacol Soc*. 1975;18:81–82.

150. Dorn GW 2nd, Liel N, Trask JL, Mais DE, Assey ME, Halushka PV. Increased platelet thromboxane A2/prostaglandin H2 receptors in patients with acute myocardial infarction. *Circulation*. 1990;81(1):212–218.

151. Asmis L, Tanner FC, Sudano I, Luscher TF, Camici GG. DMSO inhibits human platelet activation through cyclooxygenase-1 inhibition. A novel agent for drug eluting stents? *Biochem Biophys Res Commun*. 2010;391:1629–1633.

152. Trelle S, Reichenbach S, Wandel S, Hildebrand P, Tschannen B, Villiger PM, Egger M, Jüni P. Cardiovascular safety of non-steroidal anti-inflammatory drugs: Network meta-analysis. *BMJ*. 2011;342:c7086.

153. Kearney PM, Baigent C, Godwin J, Halls H, Emberson JR, Patrono C. Do selective cyclo-oxygenase-2 inhibitors and traditional non-steroidal anti-inflammatory drugs increase the risk of atherothrombosis? Meta-analysis of randomised trials. *BMJ*. 2006;332(7553):1302–1308.

154. Martorell L, Martínez-González J, Rodríguez C, Gentile M, Calvayrac O, Badimon L. Thrombin and protease-activated receptors (PARs) in atherothrombosis. *Thromb Haemost*. 2008;99(2):305–315.

155. Xiao Z, Theroux P. Platelet activation with unfractionated heparin at therapeutic concentrations and comparisons with a low-molecular-weight heparin and with a direct thrombin inhibitor. *Circulation*. 1998;97:251–256.

156. Winstanley L, Chen R. New thrombin and factor Xa inhibitors for primary and secondary prevention of ischaemic stroke. *CNS Neurol Disord Drug Targets*. 2013;12(2):242–251.

157. Damman P, Woudstra P, Kuijt WJ, de Winter RJ, James SK. P2Y12 platelet inhibition in clinical practice. *J Thromb Thrombolysis*. 2012;33(2):143–153.

158. Sambrano GR, Weiss EJ, Zheng YW, Huang W, Coughlin SR. Role of thrombin signalling in platelets in haemostasis and thrombosis. *Nature*. 2001;413(6851):74–78.

159. Toschi V, Gallo R, Lettino M, Fallon JT, Gertz SD, Fernandez OA, Chesebro JH, Badimon L, Nemerson Y, Fuster V. Tissue factor modulates the thrombogenicity of human atherosclerotic plaques. *Circulation*. 1997;95:594–599.

160. Moons AH, Levi M, Peters RJ. Tissue factor and coronary artery disease. *Cardiovasc Res*. 2002;53:313–325.

161. Owens AP 3rd, Mackman N. Role of tissue factor in atherothrombosis. *Curr Atheroscler Rep*. 2012;14(5):394–401.

162. Mackman N. Role of tissue factor in hemostasis, thrombosis, and vascular development. *Arterioscler Thromb Vasc Biol*. 2004;24(6):1015–1022.

163. Mackman N. Role of tissue factor in hemostasis and thrombosis. *Blood Cells Mol Dis*. 2006;36(2):104–107.

164. Wolberg AS, Mast AE. Tissue factor and factor VIIa—Hemostasis and beyond. *Thromb Res*. 2012;129(Suppl 2):S1–S4.

165. Jude B, Zawadzki C, Susen S, Corseaux D. Relevance of tissue factor in cardiovascular disease. *Arch Mal Coeur Vaiss*. 2005;98(6):667–671.

166. Libby P, Simon DI. Inflammation and thrombosis: The clot thickens. *Circulation*. 2001;103(13):1718–1720.

167. Camici GG, Steffel J, Akhmedov A, Schafer N, Baldinger J, Schulz U, Shojaati K et al. Dimethyl sulfoxide inhibits tissue factor expression, thrombus formation, and vascular smooth muscle cell activation: A potential treatment strategy for drug-eluting stents. *Circulation*. 2006;114(14):1512–1521.

168. Gerthoffer WT. Mechanisms of vascular smooth muscle cell migration. *Circ Res*. 2007;100(5):607–621.

169. Bourne WM, Shearer DR, Nelson LR. Human corneal endothelial tolerance to glycerol, dimethylsulfoxide, 1,2-propanediol, and 2,3- butanediol. *Cryobiology*. 1994;31:1–9.

170. Steffel J, Latini RA, Akhmedov A, Zimmermann D, Zimmerling P, Luscher TF, Tanner FC. Rapamycin, but not FK-506, increases endothelial tissue factor expression: Implications for drug-eluting stent design. *Circulation*. 2005;112:2002–2011.

171. Stahli BE, Camici GG, Steffel J, Akhmedov A, Shoojati K, Graber M, Luscher TF, Tanner FC. Paclitaxel enhances thrombin-induced endothelial tissue factor expression via c-Jun terminal NH2 kinase activation. *Circ Res*. 2006;99:149–155.

172. Campbell CL, Smyth S, Montalescot G, Steinhubl SR. Aspirin dose for the prevention of cardiovascular disease: A systematic review. *JAMA*. 2007;297(18):2018–2024.

173. Violi F, Ghiselli A, Iuliano L, Alessandri C, Cordova C, Balsano F. Influence of hydroxyl radical scavengers on platelet function. *Haemostasis*. 1988;18(2):91–98.

174. Nishimura H, Rosenblum WI, Nelson GH, Boynton S. Agents that modify EDRF formation alter antiplatelet properties of brain arteriolar endothelium in vivo. *Am J Physiol*. 1991;261(1 Pt 2):H15–H21.

175. Shirley HH, Lundergan MK, Williams HJ, Spruance SL. Lack of ocular changes with dimethyl sulfoxide therapy of scleroderma. *Pharmacotherapy*. 1989;9(3):165–168.

176. Garcia CA. Ocular toxicology of dimethyl sulfoxide and effects on retinitis pigmentosa. *Ann N Y Acad Sci.* 1983;411:48–51.
177. Rubin LF. Toxicologic update of dimethyl sulfoxide. *Ann N Y Acad Sci.* 1983;411:6–10.
178. Jacob SW, de la Torre JC. Pharmacology of dimethyl sulfoxide in cardiac and CNS damage. *Pharmacol Rep.* 2009;61(2):225–235.
179. Santos NC, Figueira-Coelho J, Martins-Silva J, Saldanha C. Multidisciplinary utilization of dimethyl sulfoxide: Pharmacological, cellular, and molecular aspects. *Biochem Pharmacol.* 2003;65(7):1035–1041.
180. Waller FT, Tanabe CT, Paxton HD. Treatment of elevated intracranial pressure with dimethyl sulfoxide. *Ann N Y Acad Sci.* 1983;411:286–292.
181. Muther RS, Bennett WM. Effects of dimethyl sulfoxide on renal function in man. *JAMA.* 1980;244(18):2081–2083.
182. Sibert L, Khalaf A, Bugel H, Sfaxi M, Grise P. Intravesical dimethyl sulfoxide instillations can be useful in the symptomatic treatment of profuse hematuria due to eosinophilic cystitis. *J Urol.* 2000;164(2):446.
183. Marshall LF, Camp PE, Bowers S. Dimethyl sulfoxide for the treatment of intracranial hypertension: A preliminary trial. *Neurosurgery.* 1984;4(6):659–663.
184. Karaca M, Bilgin U, Akar M, de la Torre JC. Dimethyl sulfoxide lowers ICP after closed head trauma. *Eur J Clin Pharmacol.* 1991;40:113–114.
185. Kulah A, Akar M, Baykut L. Dimethyl sulfoxide in the management of patients with brain swelling and increased intracranial pressure after severe closed head injury. *Neurochirurgia.* 1990;33:177–180.
186. Jacob SW, Herschler R. Pharmacology of DMSO. *Cryobiology.* 1986;23:14–27.
187. Smith ER, Hadidian Z, Mason MM. The single—and repeated—dose toxicity of dimethyl sulfoxide. *Ann N Y Acad Sci.* 1967;141(1):96–109.
188. Willhite CC, Katz PI. Toxicology updates. Dimethyl sulfoxide. *J Appl Toxicol.* 1984;4(3):155–160.

4 DMSO in Genetics

PROTECTION FROM IONIZING RADIATION

One of the most important biological properties of dimethyl sulfoxide (DMSO) is its ability to protect cells from radiation damage.[1] DMSO has been used in a number of studies to repair radiation damage induced in a variety of mammalian models and *in vitro* preparations.[2] Such radiation damage to mammalian cells generally involves injury to deoxyribonucleic acid (DNA). DNA molecules are double-stranded helices running in opposite directions to each other. In mammals, most of the DNA is stored inside the cell nucleus in a dense string-like fiber called chromatin.

DNA is composed of building blocks called nucleotides that consist of a deoxyribose sugar bound on one side and a phosphate group bound to the other side of four nitrogenous bases. The bases are composed of adenine (A), thymine (T), guanine (G), and cytosine (C) (Figure 4.1). Phosphates and sugars that are adjacent to these nucleotides link to form a long polymer and the double helix, a process called transcription.

The outside strands of DNA are made up of the sugar and phosphate portions of the nucleotides, while the middle parts are made of the nitrogenous bases. The nitrogenous bases on the two strands of DNA pair up, purine with pyrimidine (A always with T, G always with C), and are held together by weak hydrogen bonds. As the keeper of genetic information in each living cell, the integrity and stability of DNA are essential to life.

In eukaryotic cells, when DNA is active, its code is copied to messenger RNA, which then carries the DNA information to the ribosomal factories within the cell cytoplasm in order to manufacture proteins. The proteins thus formed in the ribosome are made up of amino acids that are linked together to extend the growing protein chain, a process called translation.

In humans, DNA strand breaks are a common daily occurrence arising at a frequency of tens of thousands per cell per day from direct attack by intracellular metabolites and from spontaneous DNA decay.[3] Despite this daily assault on DNA, strand breaks generally can be easily fixed by a collection of repair processes available to the cell. These lesions to the DNA structure are detected by damage sensors that initiate various response reactions.[4]

These repair processes are vital to the integrity of its genome and to the survival of the organism. Strand breaks can also result from mechanical stress, ionizing radiation, and chemical damage, for example, from reactive oxygen species (ROS).

Key experiments in the 1960s showed that the ratios of nitrogen bases A–T and G–C are constant in all living things.

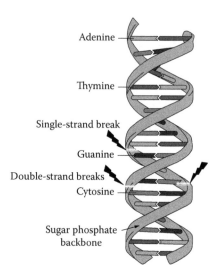

Adenine

Thymine

Single-strand break

Guanine

Double-strand breaks

Cytosine

Sugar phosphate
backbone

FIGURE 4.1 **(See color insert.)** Ionizing radiation can create two types of breaks: (1) in the sugar phosphate backbone or (2) in the base pairs of the DNA. These breaks can cause either single (bolt) or double DNA strand breaks (bolts). Direct DNA damage occurs when radiation particles physically break one or two strands of DNA. Indirect damage occurs when radiation creates harmful free radicals to damage DNA, a target of repair for DMSO. Single-strand DNA breaks are more prone to damage than double-strand DNA breaks because more toxic agents target single-strand DNA regions than double strands. Ionizing radiation of single-strand breaks are more easily repaired by the cell than cell repair of similar ionizing radiation of double-strand breaks, which can be lethal to the cell. Double-strand DNA breaks are the cause of many diseases, including various types of cancer.

Damage to DNA can create a loss of genetic instructions for producing new cells and impair the well-functioning and development of every living organism. However, DNA is not inert, and like any other biological molecule, it is continually the target of damage from the environment, which, if not repaired, can result in metabolic dysfunction, mutation, and potentially cell death. For this reason, all cells contain a range of repair pathways that have evolved to optimize their survival following damage to their DNA.

One of these repair pathways that follow genomic DNA damage is the DNA damage response, which coordinates DNA repair, cell cycle arrest, and ultimately cell death.[5]

This process is a complex signaling network involving cell cycle checkpoints and DNA repair/damage control pathways that work to minimize changes in normal metabolic function.

The DNA damage response is affected by an assortment of cellular components and processes, which include chromatin structure, DNA replication, cell growth, and cell cycle status. When the DNA damage response fails, genomic instability and several human syndromes, particularly cancer, can appear. Paradoxically, DNA-damaging agents are sometimes used in cancer therapy, and drugs targeting DNA damage response proteins, cell cycle, and cancer differentiation therapy are presently an exciting field being explored as potential new anticancer agents.

As reviewed later, DMSO has been shown to protect cells from genotoxic stress, including free radical damage from excessive ROS and hydroxyl radical formation. DMSO can additionally accelerate DNA strand break repair or prevent damage to DNA from ionizing radiation. However, it remains a challenge to point to one property under the DMSO umbrella of activities that might broadly explain the extraordinary biological actions of this simple polar molecule in providing protection to cells exposed to harmful stimuli.

DMSO PROTECTION OF SINGLE- AND DOUBLE-STRANDED DNA BREAKS

There is growing evidence that deficiencies in DNA damage signaling and repair pathways are fundamental to most, if not all, developing cancers in humans. The cell response to DNA damage is by activating the DNA damage response pathway, which involves cell cycle arrest and the transcriptional and posttranscriptional activation of a subset of genes associated with DNA repair. Transcriptional and posttranscriptional activations involve the increased expression of a gene at the level of DNA–RNA (transcription) and at the level of RNA between transcription and translation (posttranscriptional). Cell cycle arrest occurs when DNA damage is inflicted and continues until the damage is repaired. If the damage cannot be repaired, the cell commits suicide by ordering apoptosis or programmed cell death. In a way, apoptosis is often a defense reaction that can keep other cells from dying.

If the affected cells are unable to respond or repair DNA damage, genetic instability can result, which will increase the risk of disease, especially cancer development. The most dangerous type of DNA damage that can develop in mammalian cells occurs when DNA double-strand breaks are induced. DNA double-strand breaks are potentially lethal lesions that can be created by ionizing radiation from x-rays and gamma rays[6] (Figure 4.1). Besides ionizing radiation, DNA double-strand breaks can result from the excessive formation within the cell of free radicals, such as ROS. ROS can target mitochondria by creating intracellular oxidative stress damage.[6]

Although cells can adapt to mild forms of radiation damage, especially if only one strand of DNA is damaged, heavy doses of radiation can overwhelm the defensive process that cells possess to fix DNA from ionizing damage. However, cells have trouble fixing the DNA double-strand breaks due to the two DNA ends generated by the breaks becoming physically dissociated from one another, making the repair difficult to perform. Cells that cannot fix the DNA double-strand breaks by themselves and do not undergo apoptosis can become cancerous.

Mammalian cells have two methods to repair DNA double-strand breaks, by homologous recombination or by nonhomologous end joining (NHEJ). The first method uses a duplicate set of DNA as a template to repair the damaged DNA, and the second method (NHEJ) is able to repair the break directly without outside influence.[7] The first method of DNA repair, homologous recombination, is more accurate than NHEJ because it uses the undamaged DNA strand from its template to make sure the repair is done correctly and no nucleotides are lost in the transaction as it often happens with NHEJ. How the cell determines whether homologous recombination or NHEJ will be used to repair a break is not entirely clear.

It is estimated that one DNA double-strand break in a key gene is enough to kill the cell outright or cause the cell to trigger the more complex process of apoptosis.[8]

The double-stranded DNA breaks can be repaired by the mammalian cell using powerful methods to repair this damage as soon as it happens.[9] To understand how DMSO may help repair double-stranded DNA breaks, a review of the mammalian cell's response to this type of genotoxic damage is in order. In the first place, the pathologic forms of double-stranded DNA breaks are repaired by NHEJ because they often occur when a nearby homology donor is not available.[10] Common causes of double-stranded DNA breaks are ionizing radiation (Figure 4.1) and ROS from an overactive oxidative metabolism.[11]

Ionizing radiation from x-rays and gamma rays passes through human bodies every day at a rate of 300 million ionizing particles/h, and as they do, they create free radicals from water along their path. Radiation damage can strike DNA inside the living cell in two ways: first, by ionizing the DNA atoms and changing them to ions, and second, by ionizing water molecules to produce free radicals that can react with and damage DNA molecules. Free radicals formed in this manner can damage DNA by inducing single- and double-strand breaks. These breaks are easily fixed by the cells using NHEJ, which requires specific enzymes such as a nuclease to resect the damaged DNA, a polymerase to fill in new DNA, and a ligase to restore integrity to the DNA strands.[10] When these repair enzymes are unavailable, repair agents such as DMSO can be used with relative success, as discussed later.

A second major cause of double-stranded DNA breaks occurs when ROS are formed. ROS can form from environmental pollutants such as polychlorinated biphenyls (PCBs), which are industrial chemicals used as dielectrics in capacitors, as lubricants, and as flame retardants in cooling fluids.[12]

Despite the fact that PCB production in the United States was halted in the late 1970s, it is estimated that more than a million tons of PCB particles have escaped into the environment worldwide where they persist and accumulate through the food chain in animal tissues, including humans.[13] Studies have reported a causal link between exposure to PCBs and non-Hodgkin lymphoma, a frequently fatal form of cancer[14] as well as other cancers, including liver and biliary cancer,[15] breast cancer,[16] and skin cancer.[17]

Double-stranded DNA breaks have also been linked to malignant brain tumors such as gliomas, meningiomas, and medulloblastomas.[18]

These findings are of intense interest with respect to the properties DMSO is observed to exert as an agent that can help repair DNA double-strand breaks and its potential as a therapeutic and chemosuppressor of various forms of cancer arising primarily from DNA damage.

The radioprotective properties of DMSO have been known for a long time and were first reported by Ashwood-Smith[19] in 1961 following the work on the ability of some sulfoxides to suppress the effect of irradiation. A series of other reports soon followed showing that the application of DMSO to newborn rat skin protected them from damage from x-ray exposure.[20]

A study on the fruit fly (*Drosophila melanogaster*) given DMSO via the abdomen showed a significant reduction of x-ray-induced mortalities and sex-linked recessive lethal mutations of male sperm.[21] That report concluded that DMSO provided radio-protection to the fruit fly at both the whole-body and genetic levels.[21] However, that conclusion was at odds with another experiment that showed DMSO did not protect mouse testes subjected to x-irradiation.[22]

A common malady among elderly people exposed to solar radiation was shown to have been completely prevented in mice after x-ray radiation (1000 radians) when DMSO was prophylactically applied topically to the eyes.[23]

These early studies were rarely followed up by pharmaceutical firms or academic institutions with human trials even when the therapeutic potential to neutralize or lessen ionizing radiation damage from solar, nuclear reactions, radioactive materials, x-rays, and other external sources of beta and gamma rays to human and animal body parts seemed not only obvious but also medically warranted. This was particularly baffling especially because the sources for radiation damage were strongly linked to many types of cancer even in the early 1950s following the nuclear bombings of Hiroshima and Nagasaki in 1945 and a study of the populations affected by this radiation exposure.

To our knowledge and perplexity, the military forces in the United States, who would have an extra interest in protecting its more vulnerable personnel from radiation exposure and damage from any source, never showed an interest in either funding extracurricular investigations with DMSO or employing its own research facilities to study radiation prevention and treatment.

The exact explanation for the radioprotective effects of DMSO on tissue exposed to ionizing radiation remains unclear, but the effects are generally attributed to the suppression of DNA damage caused by the indirect action of radiation and the formation of hydroxyl radicals, which DMSO is able to scavenge and neutralize.[1,2,20]

However, this conclusion was questioned by more recent experiments on golden hamster embryo cells treated with DMSO after exposure to gamma rays. Although gamma ray damage to DNA single-strand breaks yielded hydroxyl radicals, the principal protective effects of DMSO were observed to be not due to the scavenging of hydroxyl radicals, but rather due to the scavenging of H^+ atoms or other ions, which created protective DMSO radicals.[24,25]

The finding that DMSO may act to protect golden hamster embryonic cells from radiation damage by scavenging H^+ ions does not negate that in other models, using other conditions to produce DNA strand breaks, DMSO can also be effective in its protection of DNA strand breaks through the inhibition of free radicals, including those cytotoxic molecules known to damage DNA including ROS and hydroxyl radicals.

Chinese hamster ovary (CHO) cells are a cell line derived from the ovary of the Chinese hamster used routinely in biological and medical research in genetic studies and gene expression.

CHO are the preferred host cell lines in biological research and mutagenesis studies because of their rapid growth, ease of manipulation, and high production of recombinant therapeutic proteins such as monoclonal antibodies, growth factors, hormones, blood factors, interferons, and enzymes.[26]

Ionizing radiation-sensitive mutants of CHO cell lines that are deficient in DNA repair of single- and double-strand breaks are a preferred model to unravel the complexities of DNA repair and to screen potential therapies that can aid in the DNA repair process.

DMSO was tested for its protective effects against the induction and rejoining of DNA double-strand breaks and inactivation using CHO cells exposed to both high- and low-linear energy transfer (LET) radiations.[27] The cells were exposed under aerobic conditions as monolayers to either low-LET photons or high-LEt alpha particle radiation. Gamma irradiation in the presence of 0.5 mol/dm³ DMSO, 90%, was observed to be rejoined within a 3 h incubation. It was inferred that this action by DMSO was due to the scavenging of free radicals near the double-strand breaks after gamma radiation, but this was not the case after alpha particle radiation.[27] It seemed likely that the severity of damage reduced by DMSO was due to minimizing the formation of hydroxyl-induced sugar/base assault forming near the vicinity of the double-strand breaks.[27]

The frequency of human chromosomal aberrations induced by irradiation was examined in the presence of 0.5 M DMSO.[28] The number of aberrations of human chromosomes 1–4, 7, 11, and 12 in the presence of DMSO was found significantly reduced following irradiation as calculated by DNA content.[28]

These studies suggest that DMSO may exert two viable protective mechanisms in repairing single and double DNA strand breaks by its ability to capture both the damaging H^+ ions and hydroxyl radicals formed after ionizing radiation.

In a series of studies from Kyoto University in Japan, investigators tested different concentrations of DMSO on CHO cells irradiated with x-rays. At 1.28 mM DMSO, cells were treated for 2 h before and during irradiation.[29] These studies showed that DMSO was able to suppress some of the lethal consequences of irradiation such as chromosomal aberrations, but the DMSO dose was found to be toxic.[24,25] Further studies by this group found that the treatment of CHO cells with 64 mM DMSO for 1 h was nontoxic and suppressed bystander signal between irradiated and nonirradiated cells.

Bystander signals refer to the phenomenon in which unirradiated cells exhibit irradiated effects as a result of signals received from nearby irradiated cells. This blocking action on bystander signals by DMSO therefore suggested that DMSO suppressed signal transduction in the irradiated cells since they did not appear to emit radiation to unirradiated neighboring cells in the vicinity.[30,31]

DMSO has been used as a synchronizing agent in CHO cells, since there are few methods available that can arrest the cell cycle efficiently and reversibly in the CHO cell line.[32] The reason is that these cells show poor growth inhibition exerted by confluence. It has been shown that incubation with DMSO induces a G_1 cell cycle arrest in adherent Chinese hamster CHO K-1 cells[33] (Figure 4.2). These results indicate that DMSO not only is able to arrest CHO cells in G_1 cell cycle but also stimulates cell differentiation, inducing CHO cells to acquire some features of differentiated skin cells.[32] DMSO exerts its induced growth arrest primarily by stimulating contact inhibition-dependent growth arrest, which is reported to be mediated by high levels of the cyclin-dependent kinase inhibitor p27.[34] Transitions

Theoretical treatment of cancer stem cells
by DMSO differentiation therapy

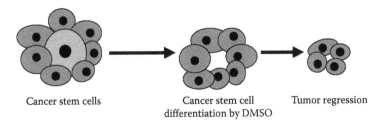

Cancer stem cells   Cancer stem cell   Tumor regression
                           differentiation by DMSO

FIGURE 4.2 Cytodifferentiation of cancer stem cells converted to noncancerous growth by DMSO's ability to arrest cell cycle at G_1 phase. (From Fiore, M. et al., *Mutagenesis*, 17(5), 419, September 2002; Chang, C.K. et al., *J. Surg. Res.*, 95, 181, 2001; Wang, C.C. et al., *PLoS ONE*, 7(4), e33772, 2012.)

between cell cycles are catalyzed by a family of cyclic-dependent kinases and inhibited by a host of proteins purified from arrested cells.[34]

A more recent study by Kashino and his Kyoto University colleagues revealed an elegant explanation of how DMSO may exert its radioprotective effects on irradiated CHO cells.[35] Their results put to question the general assumption that the radioprotective effects of DMSO is due to the suppression of DNA damage induced by the indirect action of radiation. Indirect action of radiation can impair or damage cells by creating free radical molecules that are highly reactive due to the presence of unpaired electrons on the molecule. Free radicals that are thus formed are capable of creating chemical reactions within the cell that can induce damage to the genetic code. DNA strand breaks can then encode a mutated protein with oncogenic (cancer) potential. The mutation may be passed on during cell division, perhaps leading to a malignant cell or some other mutation. In some cases, a mutation may remain dormant for years and even forever.[36]

The recent findings by Kashino et al.[35] concluded that DMSO appears to repair double-strand DNA breaks, first, by blocking hydroxyl radicals created by the radiation from inhibiting NHEJ. Second, DMSO may protect from DNA double-strand breaks by affecting the conformational change of chromatin where DNA is stored, an action that would allow NHEJ to be activated for DNA repair.

Activation of NHEJ is known to be the major repair pathway of DNA double-strand breaks caused by irradiation, and any element that prevents the repair process will inhibit DNA repair. Although other mechanisms exist for DNA single-strand breaks, only homologous recombination and NHEJ are able to repair DNA double-strand breaks. This is critical because if genotoxic products generated by radiation can inhibit NHEJ, for example, unrepaired or misrepaired lesions lead to chromosomal aberrations and possible carcinogenesis.

If DMSO can prevent the inhibition of NHEJ by genotoxic molecules such as hydroxyl radicals or by affecting chromatin conformational changes from blocking NHEJ, as the Kyoto studies suggest, the targeted cells stand a better chance of surviving from radiation damage and from possible mutagenesis leading to malignancies.

A deeper understanding of how DMSO aids DNA repair mechanisms could clarify its potential value as a therapeutic agent in oncology where tumors with DNA repair defects are the rule.

DMSO IN CELLULAR DIFFERENTIATION

Differentiated cells are distinguished from other cells in the same organism by the transition of a cell from one cell type to another, and in that sense, cells of one type never make the products characteristic of another type. Cancer cells, for example, generally synthesize proteins that are not characteristic of the tissue type they come from, and consequently when a tumor makes proteins, those proteins are not normally made by the tissue that the tumor has invaded. That means that cancer cells can give rise to disturbed patterns of protein synthesis, and some of these proteins can target DNA during replication and create mutagens, such as chemical carcinogens.

In cytopathology, the level of cellular differentiation is used as a measure of cancer progression. *Grade* is a marker of how differentiated a cell in a tumor has become.

To understand how differentiated cancer cells differ from their normal counterpart, four cellular dysfunctions need to be considered. First, whereas normal cell proliferation is strictly regulated, cancer cells are not. Second, the differentiation process is disturbed in cancer cells. Third, chromosomes are destabilized so that cancer cell growth is rapid and frequent. Fourth, apoptosis or regulated cell death does not work properly in cancer cells.

A useful anticancer agent may target one or several of these cellular dysfunctions or the mechanisms that promote their disorganized behavior.

There is now considerable evidence that the production of cancer cells is generally linked to lesions that generate abnormalities in protein synthesis or in nucleic acids that direct the course of protein synthesis and regulate all cell activities. Any genetic disease implies a defect in nucleic acids, and although some genetic diseases are classified as protein abnormalities, there is always an inherent defect in the nucleotides that compose DNA and RNA.[37]

It is at the treatment level of chemical carcinogens and mutagens that DMSO as shown preclinical usefulness as a potential anticancer agent. Many carcinogens including nitrosamines are known to react strongly with DNA to produce strand break lesions in the base pairs or sugar phosphate backbone[38] (Figure 4.1).

In 1956, British scientists Peter Magee and John Barnes reported that dimethylnitrosamine produced liver tumors in rats.[39] Nitrosamines are considered to be mutagens, and of the more than 300 nitrosamines known, 90% of them have been found to be carcinogenic in a wide variety of experimental animals, a feature that suggests that they are also carcinogenic in humans.[40] Many nitrosamines are organ specific. For example, dimethylnitrosamine causes liver cancer in experimental animals as shown by Magee and Barnes,[40] whereas some of the tobacco-specific nitrosamines cause lung cancer.[41]

In the late 1970s, extensive attention was focused on the use of sodium nitrite as a preservative for cured meats due to their production of nitrosamines and the potential for inducing cancer. It was found that antioxidants such as ascorbic acid (vitamin C) could inhibit nitrosamine formation, and a law was passed that required cured meat to be treated with ascorbic acid, or its isomer, erythorbic acid, a cheaper version of vitamin C. Other antioxidants, like vitamin E, were also found to inhibit nitrosamine formation in cured meats. It is not known whether DMSO, a powerful antioxidant, was considered as an additive to cured meats since its cost and safety would have been a keen attraction to meat manufacturers. The effect of DMSO in inhibiting nitrosamines is reviewed later.

Animals exposed to cancer-producing nitrosamines and treated with several polyene antimycotics combined with DMSO showed marked antineoplastic activity in animals treated with the carcinogenic agent diethyl nitrosoamine. The 5-month survival of the animals was 76% as compared to 35% with the untreated control group.[42]

DMSO has been used to induce differentiation in a variety of human mesenchymal stem cells.[43,44] In one recent study, DMSO was shown to differentiate cultured P19 embryonic carcinoma cells into beating cardiomyocyte-like cells. The differentiation by DMSO 1% was accelerated by a differentiation-inducing factor-1 (Br-DIF-1), which turned the cancer cells into spontaneously beating cardiomyocyte-like cells, partly by enhancing the expression of the T-type Ca^{2+} channel gene.[42] These results suggest that accelerating DMSO-induced differentiation of embryonic cancer cells by using Br-DIF-1 as a supplement is able to spontaneously change cancer cells into beating cardiomyocyte-like cells, while lowering the concentration of DMSO to safer levels.[45]

A similar P19 embryonic carcinoma cell line derived from stem cells confirmed the ability of DMSO to differentiate these cells into cardiomyocytes. Experiments such as these suggest the possibility that differentiated cancer cells derived from stem cells are a promising intervention for cardiac repair and that DMSO may be a key agent in this transformation.[46]

Just as important is the exciting possibility of induction by DMSO of cardiomyocytes from embryonic stem cells that could be used for transplantation to repair the heart after a heart attack or in conditions where left ventricular failure impairs the heart's pumping capacity. Although this clinical approach to a damaged heart is still in its research infancy, it raises optimism that one day it could become the greatest revolution in cardiovascular medicine in a century. Work has already begun in attempting to make this heart repair a reality, and preliminary clinical findings have indicated the feasibility of this approach.

The versatility of DMSO can be seen when other cancer cell differentiators are compared to DMSO. When oxytocin was used as a P19 cancer cell differentiator, aggregated P19 cells could be differentiated into cardiomyocytes, but monolayer cells did not differentiate and express specific cardiac muscle marker genes. However, when DMSO was used instead of oxytocin, both aggregates and monolayer cells could be differentiated into cardiomyocytes.[47]

A study explored the preventive effects of DMSO on experimental hepatic fibrosis induced by dimethylnitrosamine in rats.[48] DMSO 2 mL/kg daily for 4 weeks was given orally in the drinking water, and at the end of the observation period, DMSO was seen to prevent the dimethylnitrosamine-induced body and liver weight loss with no major side effects. Significantly, DMSO also suppressed the induction of hepatic fibrosis and the expression of mRNA for type-1 collagen in the liver as determined by histological evaluation and reduced hepatic hydroxyproline.[48] This study also determined that DMSO inhibited lipopolysaccharide-induced tumor necrosis factor-alpha (TNF-alpha) and nitric oxide production as reflected by reduced TNF-alpha mRNA levels.[48]

TNF-alpha is a pro-inflammatory cytokine that can be produced by a wide variety of cells and is involved in an assortment of pathological processes and a wide spectrum of human diseases. TNF-alpha is associated with chronic inflammation and autoimmune disorders as well as malignant disease and tumor development.[49]

TNF-alpha is often used experimentally to produce cell damage. This was the case in a study where TNF-alpha was added to human keratinocyte cultures to induce toxic production of metalloproteinase-9 and metalloproteinase-9 mRNA expression.[50] As metalloproteinases formed in the keratinocyte culture, an inflammatory environment was created that simulated impaired chronic wound healing. In situations where inflammation is allowed to continue, TNF-alpha and its product metalloproteinase-9 can persist in wounds and impede normal healing. Following DMSO 1% v/v administration to the inflammatory keratinocytic preparation, metalloproteinase-9 was inhibited, and metalloproteinase-9 mRNA expression levels were significantly lowered.[50] The results imply that DMSO blocked TNF-alpha induction of inflammatory metalloproteinase molecules possibly at the transcription level.[50]

The pathological activities of TNF-alpha are not restricted to inflammation. Its association with many disorders is reflected by its correlation with an increased risk of morbidity and mortality.[51] However, much research has focused on inhibiting the effects of TNF-alpha in such diseases as rheumatoid arthritis and Crohn's disease.

For this reason, a number of anti-TNF-alpha inhibitors such as infliximab and adalimumab have been approved by the FDA for prescription use in the United States for the treatment of rheumatoid arthritis and Crohn's disease, which constitute a billion-dollar industry despite their reported serious adverse effects that include non-Hodgkin's lymphoma, heart failure, demyelinating disease, and systemic infections, including tuberculosis.[52,53]

Curiously, despite its relative cost-effectiveness, safety, and efficacy as a TNF-alpha blocker and powerful anti-inflammatory agent, DMSO has not yet been tested in clinical trials against these debilitating disorders.

DMSO was studied as an anticancer agent by one of the great cell biologists of the twentieth century, Charlotte Friend. Among the many contributions to the field of cancer research made by Charlotte Friend, while working at Mt. Sinai Hospital in New York, was to isolate the first virus shown to cause leukemia in adult Swiss mice. Her original observations had an enormous impact on the growing field of leukemia and sarcoma tumor viruses.[54] The last 20 years of Friend's career was spent studying cell differentiation induced by Friend leukemia virus made possible with the use of DMSO.

The Friend leukemia virus was originally perceived as a disorder resulting from a defect in cell maturation. It was discovered that these leukemic cells were not subject to control by normal regulators such as erythropoietin or hemoglobin. Investigators using the Friend leukemia virus model reported that DMSO increased the infectivity of some viruses, but when Friend tried to increase the virus infectivity with DMSO, she discovered instead that this solvent could induce differentiation of Friend erythroleukemia cells with a loss of their malignant potential.[55] This monumental discovery has led to the hope that less toxic forms of cancer therapy using DMSO or similar agents may be achieved based on the induction of cytodifferentiation. Friend and her coworkers published a series of papers showing induced cytodifferentiation of Friend leukemia cells to normal erythrocyte maturation, thus confirming the role of DMSO in the differentiation process.[56,57] The experiments by Friend et al. were reproduced by others who reported erythroid cell–transforming spleen focus-forming virus exposed to DMSO-induced erythroid differentiation of these cells.[58,59]

Treating malignant tumors through the induction of cell differentiation as shown in a number of experiments using DMSO thus became an attractive concept in oncological research that still continues to this day.

Friend speculated that the loss of malignancy potential for the leukemia cells might reflect a toxic action by DMSO, which could injure these cells so they were not able to replicate and grow malignant after a prolonged (96 h) exposure to 280 mM DMSO.[55] However, this assumption was never proven. Despite these encouraging *in vitro* results, *in vivo* experiments to test the chemotherapeutic potential of DMSO in infected Swiss mice using a 7% DMSO solution in the drinking water neither prolonged the life of the infected mice nor improved their disease course when compared to nontreated controls.[55] The 1975 study by Friend and Scher[55] was a kiss of death for DMSO investigations as an anticancer agent for nearly a decade because after Friend's negative DMSO findings, scientists decided that this research avenue was not worth pursuing. On the other hand, the negative rodent findings by Friend and Scher[55] have not, to the best of our knowledge, been replicated. If another rat experiment using DMSO in the drinking water is attempted, a minimum 28% or higher solution of DMSO in the drinking water for several weeks may be a more realistic dose than a 7% solution as used by Friend in her study.[55] The reason is that DMSO is quickly excreted from the body after oral ingestion, and a cumulative therapeutic serum concentration is unlikely to result from a dose of 10% or less. The trick is to get rats to drink the same amount of water daily at a comparable rate to untreated controls using the bitter-tasting DMSO solution mixed into the drinking water. This was achieved using 28% DMSO in the drinking water to which several drops of an artificial sweetener had been added. Rats were observed to drink their daily 100 mL water ration for several months consistently in both DMSO and untreated controls.[60]

This trick of neutralizing the bitter taste of DMSO in water may have been used in another experiment. For example, a study was done to determine whether DMSO is able to potentiate anticancer activity when given in conjunction with other antitumor drugs against a rodent tumor.[61] DMSO was given in the drinking water at nontoxic doses ranging from 25% to 32% mixed with other anticancer agents. It was found that continuous DMSO ingestion from daily water drinking potentiated the antitumor agents' ability to significantly reduce tumor growth and prolong the survival time of tumor-bearing rats.[61]

Despite the initial setback in 1975, interest in DMSO as an anticancer agent resumed less than a decade later when it was reported that DMSO was seen to exert antiviral activity in a number of studies where it appeared to block viral replication[62] (see also Chapter 5).

These tissue culture findings, whether positive or negative, point out the difficulty of bridging the gap between relatively simple *in vitro* culture preparations and complex *in vivo* models that display a myriad of uncontrollable variables and permutated responses to chronic drug administration.

Viewed from another angle, the basis for cytodifferentiation and growth arrest of erythroleukemia cells by DMSO developed by Friend has been used as a roadway map to develop other polar molecules similar in structure to DMSO for anticancer activity of solid tumors and blood cancers.[63]

The use of DMSO as a differentiation-inducing agent in conjunction with conventional chemotherapeutic agents or radiation therapy for the treatment of solid tumors needs to be clinically tested because such an approach could become a second- or third-line therapy in patients with advanced cancer.

Another possible way DMSO could be used as a cell differentiation–inducing agent is by turning nonviral cancer tumors into nonaggressive growths. This concept has begun to receive attention for its simplicity and variety of cell differentiation–inducing agents presently available.

CANCER STEM CELLS AND DIFFERENTIATION

Cancer stem cells are cancer cells found within tumors or blood cancers that possess characteristics associated with normal stem cells, specifically, the ability to generate all cell types found in a particular cancer sample. Cancer stem cells have been implicated in the formation of many cancers since they can self-propagate with more cancer cells and they can also differentiate into a variety of cell types. Studies have shown that populations of cancer stem cells have been isolated from solid tumors from such organs as brain, prostate, colorectum, pancreas, and lung.[64–68]

Differentiation therapy is an attractive alternative to chemotherapy and radiation, which can indiscriminately kill any cell whether cancerous or not. It is well known that chemotherapy often comes with side effects such as fatigue, hair loss, and extreme nausea. Currently, a new more serious side effect of long-term chemotherapy has been found that can alter brain function for years after posttreatment. Chemotherapy-induced cognitive impairment, or *chemobrain*, has been well established in the literature.[69] Chemobrain is a common term used by cancer survivors to describe thinking and memory problems that can occur after cancer treatment. Chemobrain has also been called chemo fog or chemotherapy-related cognitive impairment.

In any case, cancer stem cells are resistant to chemotherapy and radiation therapy.[70] The way differentiation therapy works is not by killing cells but by turning cancer stem cells into some other cell types that are not cancerous.[71]

Exposure of DMSO to cancer stem cells could disrupt the cycle of self-renewal in cancer stem cells and promote these cells to become a new noncancerous cell type, such as skin cells (Figure 4.2). The reason why this therapeutic approach is feasible has to do with some key actions by DMSO on cell cycle activity. Previous studies have demonstrated that DMSO can modulate the activator protein-1 (AP-1) activity and lead to cell cycle arrest at the G1 phase.[72] AP-1 is a nuclear transcription complex that regulates genetic expression from a variety of stimuli, including cytokines, growth factors, and stress.[73] AP-1 is composed of Jun–Jun and Jun–Fos protein dimers that are involved in various biological mechanisms, including normal cell growth, differentiation, and tumor development.[74]

Moreover, an important cell culture finding was recently reported showing DMSO as the key stimulator of the tumor suppressor protein called human liver DnaJ-like protein (HLJ1).[75] Tumor suppression was observed when DMSO was able to activate AP-1 through JunB and JunD proteins in highly invasive human lung adenocarcinoma cell line.[75] HLJ1 is a novel tumor suppressor protein recently discovered that inhibits cancer cell cycle progression, proliferation, invasion, and tumorigenesis and is significantly correlated with relapse and survival and with prognosis in non-small-cell lung carcinoma patients.[76]

As a response to DMSO's tumor-suppressing action, mRNA and protein expressions of HLJ1 were observed to increase in the adenocarcinoma cell lines. Targeted induction of HLJ1 represents a promising approach for cancer therapy. These collective findings imply that DMSO could be a major primary treatment or ligand for the development of anticancer drugs that induce HLJ1 expression where activation of cell migration and cell proliferation pathways can be arrested in cancer cells.[75,77]

This is an idea that has not yet been experimentally tried and obviously requires more basic and clinical exploration. But the reward of finding a nontoxic, active, and cost-effective anticancer agent such as DMSO or its use as an adjuvant to chemotherapy or radiation therapy may be an attractive lead to identify a novel therapeutic approach in the search for more effective treatments of invasive cancers.

The importance of cancer research and search for effective treatments cannot be overstated. Although many new treatments discovered in the last few decades have lowered the incidence of cancer deaths, cancer is still the second most common cause of death in the United States, killing about 600,000 people/year, exceeded only by heart disease. Cancer is estimated by the World Health Organization (WHO) to account for 7.6 million deaths worldwide.

Since the collective evidence presented in this chapter indicates that DMSO can cause cancer cells to become benign or to arrest their growth, it is important that DMSO research into this important field be stimulated with government or private funding.

There are still a number of challenges that need to be addressed in order to translate the research and understanding of cancer stem cells to clinical applications using differentiation-induced therapy as reported for DMSO. These challenges should focus on signaling pathways (which are also shared by normal stem cells), regulation of differentiation, genomics, and proteomics.[75,77]

REFERENCES

1. Ashwood-Smith MJ. Radioprotective and cryoprotective properties of DMSO. In *Dimethyl Sulfoxide*. Jacob SW, Rosenblum E, Wood DC (eds.). Marcel Dekker, Inc., New York, 1971; pp. 147–187.
2. Raaphorst GP, Azzam EI. Fixation of radiation-induced potentially lethal damage by anisotonic treatment and its modification by DMSO or BrdUrd in V79 cells. *Radiat Environ Biophys*. 1985;24(3):175–184.
3. Caldecott KW. Single-strand break repair and genetic disease. *Nat Rev Genet*. August 2008;9(8):619–631.

4. Lindsey-Boltz LA, Sancar A. RNA polymerase: The most specific damage recognition protein in cellular responses to DNA damage? *Proc Natl Acad Sci USA.* 2007;104:13213–13214.

5. Fortini P, Ferretti C, Dogliotti E. The response to DNA damage during differentiation: Pathways and consequences. *Mutat Res.* 2013;Mar–Apr:743–744.

6. Blackwood JK, Rzechorzek NJ, Bray SM, Maman JD, Pellegrini L, Robinson NP. End-resection at DNA double-strand breaks in the three domains of life. *Biochem Soc Trans.* February 1, 2013;41(1):314–320.

7. Kinashi Y, Takahashi S, Kashino G, Okayasu R, Masunaga S, Suzuki M, Ono K. DNA double-strand break induction in Ku80-deficient CHO cells following boron neutron capture reaction. *Radiat Oncol.* September 5, 2011;6:106.

8. Rich T, Allen RL, Wyllie AH. Defying death after DNA damage. *Nature.* October 12, 2000;407(6805):777–783.

9. Goodsell DS. The molecular perspective: Double-stranded DNA breaks. *Oncologist.* May 2005;10(5):361–362.

10. Lieber MR. The mechanism of double-strand DNA break repair by the nonhomologous DNA end-joining pathway. *Annu Rev Biochem.* 2010;79:181–211.

11. Friedberg EC, Walker GC, Siede W, Wood RD, Schultz RA, Ellenberger T. *DNA Repair and Mutagenesis.* ASM Press, Washington, DC, 2006; p. 1118.

12. Tanabe S. PCB problems in the future: Foresight from current knowledge. *Environ Pollut.* 1988;50:5–28.

13. National Toxicology Program (NTP). *Report on Carcinogens, Substance Profiles,* 11th edn. Polychlorinated Biphenyls (PCBs), U.S. Department of Human Health and Human Services, 2004 (CAS no. 1336-36-3).

14. Freeman MD, Kohles SS. Plasma levels of polychlorinated biphenyls, non-Hodgkin lymphoma, and causation. *J Environ Public Health.* 2012;2012:258981.

15. Brown DP. Mortality of workers exposed to polychlorinated biphenyls—An update. *Arch Environ Health.* 1987;42(6):333–339.

16. Recio-Vega R, Velazco-Rodriguez V, Ocampo-Gómez G, Hernandez-Gonzalez S, Ruiz-Flores P, Lopez-Marquez F. Serum levels of polychlorinated biphenyls in Mexican women and breast cancer risk. *J Appl Toxicol.* 2011;31(3):270–278.

17. Gallagher RP, Macarthur AC, Lee TK et al. Plasma levels of polychlorinated biphenyls and risk of cutaneous malignant melanoma: A preliminary study. *Int J Canc.* 2011;128(8):1872–1880.

18. Johnson NM, Lemmens BB, Tijsterman M. A role for the malignant brain tumour (MBT) domain protein LIN-61 in DNA double-strand break repair by homologous recombination. *PLoS Genet.* March 2013;9(3):e1003339.

19. Ashwood-Smith MJ. The radioprotective action of dimethyl sulphoxide and various other sulphoxides. *Int J Radiat Biol.* 1961;3:41–48.

20. Dod JL, Shewell L. Local protection against x-irradiation by dimethylsulphoxide. *Brit J Radiol.* 1968;41:950.

21. Barnett BM. Radioprotective effects of dimethyl sulfoxide in *Drosophila melanogaster. Radiat Res.* 1972;51:134.

22. Ashwood-Smith MJ. Inability of dimethyl sulphoxide to protect mouse testes against the effect of x-irradiation. *Int J Radiat Biol.* 1961;3:101–105.

23. Hagemann RF, Evans TC, Riley EF. Modification of radiation effect on the eye by topical application of dimethyl sulfoxide. *Radiat Res.* 1970;44:368–372.

24. Watanabe M, Suzuki M, Suzuki K, Hayakawa Y, Miyazaki T. Radioprotective effects of dimethyl sulfoxide in golden hamster embryo cells exposed to gamma rays at 77 K. II. Protection from lethal, chromosomal, and DNA damage. *Radiat Res.* October 1990;124(1):73–78.

25. Miyazaki T, Hayakawa Y, Suzuki K, Suzuki M, Watanabe M. Radioprotective effects of dimethyl sulfoxide in golden hamster embryo cells exposed to gamma rays at 77 K. I. Radical formation as studied by electron spin resonance. *Radiat Res.* October 1990;124(1):66–72.

26. Walsh G. Biopharmaceutical benchmarks 2010. *Nat Biotechnol.* 2010;28:917–924.

27. deLara CM, Jenner TJ, Townsend KM, Marsden SJ, O'Neill P. The effect of dimethyl sulfoxide on the induction of DNA double-strand breaks in V79-4 mammalian cells by alpha particles. *Radiat Res.* October 1995;144(1):43–49.

28. Cigarrán S, Barrios L, Caballín MR, Barquinero JF. Effect of DMSO on radiation-induced chromosome aberrations analysed by FISH. *Cytogenet Genome Res.* 2004;104(1–4):168–172.

29. Koyama S, Kodama S, Suzuki K et al. Radiation-induced long-lived radicals which cause mutation and transformation. *Mutat Res.* 1998;421:45–54.

30. Kashino G, Prise KM, Suzuki K et al. Effective suppression of bystander effects by DMSO treatment of irradiated CHO cells. *J Radiat Res.* 2007;48:327–333.

31. Harada T, Kashino G, Suzuki K et al. Different involvement of radical species in irradiated and bystander cells. *Int J Radiat Biol.* 2008;84:809–814.

32. Fiore M, Zanier R, Degrassi F. Reversible G(1) arrest by dimethyl sulfoxide as a new method to synchronize Chinese hamster cells. *Mutagenesis.* September 2002;17(5):419–424.

33. Fiore M, Degrassi F. Dimethyl sulfoxide restores contact inhibition-induced growth arrest and inhibits cell density-dependent apoptosis in hamster cells. *Exp Cell Res.* August 25, 1999;251(1):102–110.

34. Polyak K, Kato J-Y, Solomon MJ, Sherr CJ, Massague J, Roberts JM, Koff A. P27[Kip1], a cyclin inhibitor, links transforming growth factor-β and contact inhibition to cell cycle arrest. *Genes Dev.* 1994;8:9–22.

35. Kashino G, Liu Y, Suzuki M, Masunaga S, Kinashi Y, Ono K, Tano K, Watanabe M. An alternative mechanism for radioprotection by dimethyl sulfoxide; possible facilitation of DNA double-strand break repair. *J Radiat Res.* 2010;51(6):733–740.

36. Khanna K, Jackson SP. DNA double-strand breaks: Signalling, repair and the cancer connection. *Nat Genet.* 2001;27:247–254.

37. Budden T, Bowden NA. The role of altered nucleotide excision repair and UVB-induced DNA damage in melanomagenesis. *Int J Mol Sci.* January 9, 2013;14(1):1132–1151.

38. Magee PN, Barnes JM. Carcinogenic nitroso compounds. *Adv Canc Res.* 1967;10:163–246.

39. Magee PH, Barnes JM. The production of malignant primary hepatic tumors in the rat by feeding dimethylnitrosamine. *Br J Canc.* 1956;10:114–122.

40. Bartsch H, Montesano R. Relevance of nitrosamines to human cancer. *Carcinogenesis.* November 1984;5(11):1381–1393.

41. Russo P, Cardinale A, Margaritora S, Cesario A. Nicotinic receptor and tobacco-related cancer. *Life Sci.* November 27, 2012;91(21–22):1087–1092.

42. Ibragimova VKh, Aliev DI. [Radioprotective and antineoplastic activity of polyene antibiotics combined with dimethyl sulfoxide]. *Antibiot Khimioter.* 2002;47(9):3–8.

43. Okura H, Matsuyama A, Lee CM, Saga A, Kakuta-Yamamoto A, Nagao A. Cardiomyoblast-like cells differentiated from human adipose tissue-derived mesenchymal stem cells improve left ventricular dysfunction and survival in a rat myocardial infarction model. *Tissue Eng Part C Methods.* 2011;17:145–154.

44. Hegner B, Weber M, Dragun D, Schulze-Lohoff E. Differential regulation of smooth muscle markers in human bone marrow-derived mesenchymal stem cells. *J Hypertens.* 2005;23:1191–1202.

45. Seya K, Kanemaru K, Matsuki M et al. Br-DIF-1 accelerates dimethyl sulphoxide-induced differentiation of P19CL6 embryonic carcinoma cells into cardiomyocytes. *Br J Pharmacol.* February 2012;165(4):870–879.

46. Li L, Gong H, Yu H, Liu X, Liu Q, Yan G, Zhang Y, Lu H, Zou Y, Yang P. Knockdown of nucleosome assembly protein 1-like 1 promotes dimethyl sulfoxide-induced differentiation of P19CL6 cells into cardiomyocytes. *J Cell Biochem.* December 2012;113(12):3788–3796.

47. Fathi F, Murasawa S, Hasegawa S, Asahara T, Kermani AJ, Mowla SJ. Cardiac differentiation of P19CL6 cells by oxytocin. *Int J Cardiol.* May 1, 2009;134(1):75–81.

48. Nakamuta M, Ohta S, Tada S, Tsuruta S, Sugimoto R, Kotoh K, Kato M, Nakashima Y, Enjoji M, Nawata H. Dimethyl sulfoxide inhibits dimethylnitrosamine-induced hepatic fibrosis in rats. *Int J Mol Med.* November 2001;8(5):553–560.

49. Zidi I, Mestiri S, Bartegi A, Amor NB. TNF-alpha and its inhibitors in cancer. *Med Oncol.* June 2010;27(2):185–198.

50. Majtan J, Majtan V. Dimethyl sulfoxide attenuates TNF-α-induced production of MMP-9 in human keratinocytes. *J Toxicol Environ Health A.* 2011;74(20):1319–1322.

51. Strieter R, Kunkel S, Bone R. Role of tumor necrosis factor-alpha in disease states and inflammation. *Crit Care Med.* 1993;21(10 Suppl):S447–S463.

52. Hochberg MC, Lebwohl MG, Plevy SE et al. The benefit/risk profile of TNF-blocking agents: Findings of a consensus panel. *Semin Arthritis Rheum.* 2005;34:819–836.

53. Ljung T, Karlén P, Schmidt D et al. Infliximab in inflammatory bowel disease: Clinical outcome in a population based cohort from Stockholm County. *Gut.* 2004;53:849–853.

54. Friend C. Leukemia of adult mice caused by transmissible agent. *Ann NY Acad Sci.* 1957;68:522–532.

55. Friend C, Scher W. Stimulation by dimethyl sulfoxide of erythroid differentiation and hemoglobin synthesis in murine virus-induced leukemic cells. *Ann NY Acad Sci.* January 27, 1975;243:155–163.

56. Scher W, Tsuei D, Friend C. The structural basis for steroid modulation of DMSO-stimulated erythrodifferentiation. *Leuk Res.* 1980;4(2):217–229.

57. Tsuei D, Haubenstock H, Revoltella R, Friend C. Virus production and hemoglobin synthesis in variant lines of dimethyl sulfoxide-treated Friend erythroleukemia cells. *J Virol.* July 1979;31(1):178–183.

58. Dube SK, Pragnell IB, Kluge N, Gaedicke G, Steinheider G, Ostertag W. Induction of endogenous and of spleen focus-forming viruses during dimethylsulfoxide-induced differentiation of mouse erythroleukemia cells transformed by spleen focus-forming virus. *Proc Natl Acad Sci USA.* May 1975;72(5):1863–1867.

59. Gaedicke G, Abedin Z, Dube SK, Kluge N, Neth R, Steinheider G, Weimann BJ, Ostertag W. Control of globin synthesis during DMSO-induced differentiation of mouse erythroleukemic cells in culture. *Hamatol Bluttransfus.* 1974;14:278–287.

60. de la Torre JC. Unpublished observations.

61. Thuning CA, Fanshaw MS, Warren J. Mechanisms of the synergistic effect of oral dimethyl sulfoxide on antineoplastic therapy. *Ann NY Acad Sci.* 1983;411:150–160.

62. Wagner EK, Aguilar JS, Roy D et al. Dimethyl sulfoxide blocks herpes simplex virus-1 productive infection in vitro acting at different stages with positive cooperatively. *BMC Infect Dis.* 2002;2(9):1–10.

63. Marks PA, Breslow R. Dimethyl sulfoxide to vorinostat: Development of this histone deacetylase inhibitor as an anticancer drug. *Nat Biotechnol.* January 2007;25(1):84–90.

64. Singh SK, Clarke ID, Terasaki M et al. Identification of a cancer stem cell in human brain tumors. *Canc Res.* 2003;63:5821–5828.

65. Collins AT, Berry PA, Hyde C, Stower MJ, Maitland NJ. Prospective identification of tumorigenic prostate cancer stem cells. *Canc Res.* 2005;65:10946–10951.

66. Dalerba P, Dylla SJ, Park IK et al. Phenotypic characterization of human colorectal cancer stem cells. *Proc Natl Acad Sci USA.* 2007;104:10158–10163.

67. Li C, Heidt DG, Dalerba P et al. Identification of pancreatic cancer stem cells. *Canc Res.* 2007;67:1030–1037.

68. Ho MM, Ng AV, Lam S, Hung JY. Side population in human lung cancer cell lines and tumors is enriched with stem-like cancer cells. *Canc Res.* 2007;67:4827–4833.
69. Scherling CS, Smith A. Opening up the window into "chemobrain": A neuroimaging review. *Sensors (Basel).* March 6, 2013;13(3):3169–3203.
70. Alamgeer M, Peacock CD, Matsui W, Ganju V, Watkins DN. Cancer stem cells in lung cancer: Evidence and controversies. *Respirology.* April 16, 2013;18:757–764.
71. Garvalov BK, Acker T. Cancer stem cells: A new framework for the design of tumor therapies. *J Mol Med (Berl).* February 2011;89(2):95–107.
72. Chang CK, Albarillo MV, Schumer W. Therapeutic effect of dimethyl sulfoxide on ICAM-1 gene expression and activation of NF-κB and AP-1 in septic rats. *J Surg Res.* 2001;95:181–187.
73. Hess J, Angel P, Schorpp-Kistner M. AP-1 subunits: Quarrel and harmony among siblings. *J Cell Sci.* December 1, 2004;117(Pt 25):5965–5973.
74. Reddy SP, Mossman BT. Role and regulation of activator protein-1 in toxicant-induced responses of the lung. *Am J Physiol Lung Cell Mol Physiol.* 2002;283:L1161–L1178.
75. Wang CC, Lin S, Lai Y, Liu Y, Hsu Y, Chen JJ. Dimethyl sulfoxide promotes the multiple functions of the tumor suppressor HLJ1 through activator protein-1 activation in NSCLC cells. *PLoS ONE.* 2012;7(4):e33772.
76. Lin SY, Hsueh CM, Yu SL, Su CC, Shum WY, Yeh KC, Chang GC, Chen JJ. HLJ1 is a novel caspase-3 substrate and its expression enhances UV-induced apoptosis in non-small cell lung carcinoma. *Nucleic Acids Res.* October 2010;38(18):6148–6158.
77. Wang CC, Tsai MF, Hong TM et al. The transcriptional factor YY1 upregulates the novel invasion suppressor HLJ1 expression and inhibits cancer cell invasion. *Oncogene.* 2005;24:4081–4093.

5 DMSO in Basic Microbiology

DMSO IN BACTERIAL INFECTIONS

ERA OF ANTIMICROBIALS

Microbiology is the study of bacteria, fungi, parasites, and viruses; these are microorganisms that are too small to be seen with the naked eye. Microorganisms that are harmful to human health are referred to as pathogens, that is, organisms that are capable of causing disease. Most bacteria, however, are not pathogens, and it has been estimated that nonpathogenic or harmless bacteria make up over 70% of all bacteria. It is also estimated that about 100 trillion mostly benign bacteria—called normal flora—are found symbiotically in the human body, primarily in the oral mucosa, intestines, and surface of the skin. Bacterial cells consequently outnumber human cells by a factor of 10 to 1. Normal flora in the body has created the recently discovered human microbiome, an ecosystem that has been linked not only to the digestion of food but also to obesity, depression, asthma, Crohn's disease, and even cancer.[1]

Pathogenic bacteria are the cause of minor and deadly illnesses. Pathogens can be broken down into three groups consisting of cocci, bacilli, and spirilla. Cocci are subdivided into three groups, streptococci, diplococci, and staphylococci and are responsible for producing pus and causing boils and pustules. Bacilli notably produce diseases such as influenza and tuberculosis. Spirillum can be the cause of serious disease, including syphilis. Pathogens can spread to infect humans and animals through a range of vectors including air, water, soil, animals, and insects and by physical contact between the infected host and the noninfected victim. These organisms have wreaked havoc throughout the centuries by decimating a disproportionately large number of people from such epidemics and pandemics as the plague of Athens in 430 BC, the bubonic (black death) plague in 1338, the great Plague of London in 1665, the cholera pandemic of 1816, and the flu pandemic of 1918, to name but a few.

In the history of medicine, probably the greatest breakthrough of all time is the start of the antibiotic era and the control of infectious diseases that have been the leading cause of human morbidity and mortality for most of human existence. Prior to the antibiotic era, it was common to believe and easy to record that "the treatment is worse than the disease." Names like Koch, Pasteur, Domagk, Lister, and Ehrlich were instrumental in changing this cynical cliché by laying the groundwork for less toxic and more clinically effective antimicrobial therapy. Their work largely paved the way for the chance discovery of penicillin by Alexander Fleming in 1929.

Fleming made a keen observation noting that the *Penicillium notatum* mold accidentally cultured in a petri dish was able to kill several colonies of staphylococci bacteria that had also invaded the petri dish near the mold. Despite this monumental observation, penicillin languished on a laboratory shelf for a decade largely in part because Fleming did not seem to grasp the potential of his great discovery or the conviction to pursue it further. In support of this conclusion, it is well known that Fleming failed to undertake essential experiments where penicillin would be administered to animals infected with pathogens to see whether the animals would be protected and survive the systemic infection.[2] This was and still is a simple rule of thumb in experimental pharmacotherapeutics in 1929.

Credit for these animal experiments and penicillin's clinical effectiveness was finally realized in 1939 by a team at Oxford led by Ernst Chain, Howard Florey, who was the head of the laboratory where the work was done, and the chemist Norman Heatley.[3]

In 3 years, this team not only showed the bactericidal properties of penicillin in animals and humans but also purified, mass-produced, and made penicillin clinically available as an antibacterial agent so that, by 1944, hundreds of thousands of soldiers and civilians were able to survive wound infections sustained in the Second World War that would otherwise may have been fatal. The number of lives saved by penicillin continues to this day. Fleming, Florey, and Chain shared the Nobel Prize in Physiology or Medicine in 1945.

The brilliant Spanish physician and scientist Gregorio Marañon referred to penicillin as a miracle cure that opened a new era, full of hope for all humanity. He also prophetically warned that even a miracle cure had its limits. Presently, there are more than a thousand different preparations containing penicillin and over 150 antibacterial agents in the pharmacopoeia to control bacteria, viruses, molds, protozoa, and parasites.

At the end of 1945, no one could have predicted that the miracle cure that was penicillin would lose some or all of its effectiveness against evolving germs it had easily defeated in the past. The new post-antibiotic era of drug resistance where organisms are able to resist the effects of antibiotics to which they were previously sensitive was now in progress, not only for penicillin but also for many other newer antimicrobials developed after penicillin.

Antibiotic-resistant bacteria reveal the amount of sophistication and high adaptability these germs have been able to develop to changing environments. This reaction by disease-producing pathogens poses one of the most serious medical concerns of the twenty-first century because when it is encountered clinically, first-line antibiotic treatments lose their efficacy and second-line antibiotics that may have less efficacy, greater cost, and more safety concerns become the only line of defense. Bacteria that can resist many antibiotics are called multidrug resistant and are commonly referred to as superbugs. These superbugs are especially dangerous in prolonged hospital stays and in intensive care units where an increasing number of invasive measures and interventions and inadequate maintenance of hygiene standards pose a high risk for infection.[4]

Any adjuvant therapy or new approach that potentiates the activity of first-line antibiotics to prevent antibiotic resistance would help postpone the day when antibiotics are no longer generally efficacious.[5] This topic and the role played by dimethyl sulfoxide (DMSO) as an adjuvant in drug-resistant microbes is discussed in the section "How Bacteria Achieve Antibiotic Resistance" in Chapter 6.

DMSO AND PATHOGENS

When treating an infection, physicians may face a choice between using a bactericidal drug, a bacteriostatic drug, and a combination of the two. A bactericidal drug means that it kills bacteria, while a bacteriostatic drug prevents the growth of bacteria. The use of one or the other or both is dependent principally on the type of infection being treated and the achievement of a good clinical outcome with the least toxicity.[6]

The first observations concerning the bacteriostatic properties of DMSO were made by Jacob et al.[7] They reported that a 20% solution of DMSO showed bacteriostatic properties against *Escherichia coli*, *Staphylococcus aureus*, and *Pseudomonas bacilli* grown in cultures. Additionally, Jacob's group also observed that 1% DMSO solution altered the sensitivity of resistant tubercle bacilli (the cause of tuberculosis) to antibiotics. As will be seen in Chapter 6, hard-to-cure drug-resistant tuberculosis has become a serious and growing problem worldwide. This occurs when bacteria develop the ability to withstand multiple first-line antibiotic attack and relay that resistant ability to subsequent generations of bacterial offspring. Since that entire strain of resistant bacteria inherits this capacity to fend off the effects of various antibiotics, resistant bacteria can spread from one person to another. Even treatable forms of tuberculosis are particularly tricky to cure because drug-sensitive strains must be treated with a 6-month course of antibiotics. Tougher cases require costly long-term hospitalization and a regimen of multiple drugs that can last years. Patients die when this treatment is ineffective.

Perhaps one of the unusual properties of DMSO is its ability to penetrate biological membranes or act as a penetrant carrier for other even larger molecules. This property implies that enhancing the penetration of an antibiotic into a bacterial cell, for example, may reduce the concentration needed for the antibiotic to be effective and also reduce its potential toxicity.

As a bactericidal agent, DMSO's minimal inhibitory concentration is reported to be 50% for *S. aureus* and from 20% to 40% for beta-hemolytic streptococcus, *Corynebacterium* species, *E. coli*, and *Proteus* species.[8]

Pottz et al.[9] confirmed Jacob's team findings and further tested DMSO for bactericidal and bacteriostatic properties against a variety of microorganisms where they observed growth inhibition at concentrations of 40%–50% (Table 5.1).

Pottz et al.[9] found that the bactericidal concentration of DMSO required to inhibit the growth of *E. coli*, *A. cloacae*, *P. vulgaris*, *beta-hemolytic streptococci group A*, *Salmonella paratyphi B*, and *C. albicans* was 30%, and for *P. aeruginosa*, it was 10%. *D. pneumoniae* was most sensitive to DMSO and required only 5% concentration (Table 5.1). *S. aureus* and *Streptococcus faecalis* required a 40% and a 50% concentration respectively.

These microorganisms are responsible for many serious infections, which if left untreated can have organ damage consequences or be fatal to the host. For example,

TABLE 5.1
DMSO Bactericidal Activity

Test Microorganism	DMSO Minimal Inhibitory Concentration				
	0	10	20	40	50
E. coli	++++	++++	+	−	−
Aerobacter cloacae	++++	++++	+	−	−
Pseudomonas aeruginosa	++++	++++	++	−	−
Proteus vulgaris	++++	++++	++++	−	−
S. aureus**	++++	++++	++	−	−
Streptococcus pyogenes*	++++	++++	+	−	−
Salmonella schottmulleri	++++	++	++	−	−
Diplococcus pneumoniae	++++	−	−	−	−
Candida albicans	++++	+	+	−	−

Notes: *, beta-hemolytic streptococcus group A; **, coagulation positive; +, very mild growth; ++, mild growth; ++++, strong growth in blood agar or broth cultures; −, negative growth.[9]

certain strains of *E. coli*, which includes serotypes O157:H7, O104:H4 in contaminated food or water, can secrete Shiga toxigenic toxins that can result in bloody diarrhea and cause severe damage to the lining of the intestines and kidneys.[10]

A. cloacae is a rod-shaped bacterium generally found in normal gut flora that sometimes causes respiratory and urinary tract infections. Strep throat from beta-hemolytic streptococci group A is a common throat infection in school-aged children that can lead to a spectrum of more serious diseases including glomerulonephritis and rheumatic fever.[11]

S. paratyphi B and *S. faecalis* are contaminants in food and water that often infect the intestines and urinary tract. *P. aeruginosa* and *Candida* species are opportunistic pathogens commonly found in hospital intensive care units where they can cause stubborn infections in patients after surgery.[12] In this list of microbes, the most sensitive microorganism to DMSO appeared to be *D. pneumoniae* (now known as *Streptococcus pneumoniae*),[9] which is a major cause of pneumonia and a host of other infectious diseases such as meningitis, endocarditis, and brain abscesses.[13]

Shiga toxins from *Shigella dysenteriae* or *E. coli* inhibit cell-free protein synthesis resulting in multiple cytotoxic effects including neurotoxicity and enterotoxicity. Shiga toxin secreted by Shigella also binds to specific receptors on the cells' surface in order to attach and enter the cell where it can inactivate the 60S ribosomal subunit to inhibit protein synthesis and cause the destructive lesions seen in shigellosis. DMSO is reported to reduce the endocytosis that occurs after Shiga toxin release and prevent both the cytotoxicity and binding of the Shiga toxin, possibly by its ability to prevent membrane permeability or by inhibiting translocation of the toxin from the cell surface.[14]

It was assumed from Pottz et al.'s[9] original study that DMSO's antibacterial action may have been through the neutralization of penicillinase, an enzyme produced by some bacteria that provide resistance to beta-lactam antibiotics like penicillin. Further experimentation with DMSO to test the possibility that it neutralizes penicillinase's protective action on bacterial growth exposed to penicillin proved negative. Pottz et al.[9] concluded from their extensive findings that the killing action of DMSO on bacteria was not through neutralization of penicillinase. Pottz's study speculated that, instead, DMSO's antibacterial action may have been due to dissolution of susceptible microorganisms by some unknown mechanism exerted by DMSO.[9] This assumption was supported by microscopic analysis of *D. pneumoniae* smears, which showed that the majority of these microbes had dissolved into a sediment after contact with DMSO.[9]

It is known that the gram-positive microbe *S. aureus* can be cleared from murine lungs by the bactericidal action exerted by phagocytosis from resident alveolar macrophages without significant contribution from polymorphonuclear leukocytes (PMNs).[15] PMNs' bactericidal intrapulmonary killing activity is more prominent against gram-negative bacteria such as *P. aeruginosa*. Recent evidence indicates that actual phagocytosis of bacteria may not be necessary for intrapulmonary kill of gram-positive microbes and that the majority of inhaled staphylococci in mice are killed without being ingested by resident alveolar macrophages.[16,17]

One theory is that reactive oxygen metabolites, including the oxidant superoxide, and the radical hydroxyl anion are secreted by phagocytes to create the alveolar bactericidal activity. With this in mind, DMSO was used to test its ability as an antioxidant and hydroxyl radical scavenger to see whether the rate of pulmonary bactericidal activity against *P. aeruginosa* and *S. aureus* was depressed in infected mice. Intratracheal and aerosol delivery of these bacteria were given to mice.[18]

DMSO at doses of 50–165 mg/mouse was administered intraperitoneally 30 min before challenge with *S. aureus* and pseudomonas delivered by either intratracheal or aerosol route.[18] DMSO was observed to have no activity in inhibiting macrophage killing of *S. aureus* but did cause a dose-dependent inhibition of pseudomonas clearance from the lungs of treated mice,[18] supporting the assumption that alveolar bactericidal action against pseudomonas by PMNs is done through the production of reactive hydroxyl anions that are inhibited by the action of DMSO as a hydroxyl radical scavenger. Conversely, DMSO's lack of effect in allowing the clearance of *S. aureus* from infected murine lung suggests that phagocytic alveolar macrophages fail to produce reactive oxygen for bactericidal activity,[19] but this assumption has been challenged by others.[20]

The question arose whether using DMSO clinically for a variety of ailments might predispose the user to bacterial infections by inhibiting phagocyte bactericidal activity. This concern grew from an *in vitro* experiment where DMSO treatment of human and mouse neutrophils and mononuclear phagocytes caused a dose-dependent inhibition of the killing of *E. coli* and *Listeria monocytogenes*. However, when the experiment was repeated *in vivo using mice*, results showed that administration of DMSO prior to subjecting mice to *E. coli* and *L. monocytogenes* infection caused only a slight and nonsignificant decrease in the subsequent *in vitro* bactericidal activity of neutrophils and macrophages from those mice.[21] Repeated injections of mice with physiologically relevant doses of DMSO did not increase the lethality of either *E. coli* or *L. monocytogenes*, nor did it inhibit the clearance of a sublethal *Listeria* challenge from the spleen and liver of these mice. The findings indicated that clinical usage of DMSO does not predispose human subjects to bacterial infection.[21]

Bacterial endotoxin from gram-negative pathogens can directly injure lung endo-thelial cells in culture. This injury is unaffected by antioxidants such as superoxide dismutase or catalase because they are unable to enter cells. Bacterial endotoxins are prevented by DMSO, which penetrates cells to exert antioxidant activity.[22] Since allo-purinol can inhibit direct lung cell injury from endotoxin, it is probable that xanthine oxidase, which allopurinol is known to inhibit, may be a source of oxidant generation in lung endothelial cells. Current data suggest a two-stage oxidant process of lung cell injury where there is direct injury of the cell both by intracellular generation of toxic oxidants and triggering of an inflammatory response.[22] Inflammatory cells that have been activated to adhere to lung cells are able to enhance the injury by generat-ing and releasing extracellular oxidants, but this process can also be neutralized by DMSO's anti-inflammatory action.[22]

The inability of many antibiotics to enter the blood–brain barrier in mammals has been a limiting factor in the treatment of bacterial brain infections such as life-threatening meningitis and brain abcess.[23] Although pathogens can invade the brain in a variety of ways, the main pathway is through the bloodstream, where they can enter the subarachnoid space in places where the blood–brain barrier is vulnerable, for example, the choroid plexus. There is some controversy regarding the ability of DMSO to potentiate the penetration of molecules, including antibiotics, across the blood–brain barrier. Previous studies have shown that DMSO enhances the penetra-tion into the brain of the protein tracer horseradish peroxidase and of ketoconazole, an antifungal agent.[24,25] The ability of DMSO to carry molecules across the blood–brain barrier may depend on the size of the molecule to be delivered, its polarity, and molecular weight; however, the exact mechanism for achieving such penetration is unknown.[26] It has been suggested that molecules that exceed 70,000 Da (a unit of atomic mass) cannot be delivered by DMSO through the blood–brain barrier,[25] suggesting that DMSO may shrink the endothelial cells' tight junctions that make up the blood–brain barrier to allow small molecules to pass through. This idea was tested on seven healthy adult mares using trimethoprim (TMP)–sulfamethoxazole (SMZ), an effective antimicrobial combination against gram-negative organisms.[27] A solution of 40% DMSO was injected intravenously at 1 g/kg for 30 min followed immediately after by TMP–SMZ at a combined dosage of 44 mg/kg. The mares were also given the same treatment a week later without DMSO. The combination TMP–SMZ was 543 Da combined, well under the 70,000 Da theoretical limit of DMSO to enhance molecule penetration into the brain. It was found that DMSO was not able to enhance the penetration of TMP–SMZ into the cerebrospinal fluid but did delay the clearance of TMP from serum and increased TMP serum concentration, a finding that had no immediate explanation.[27]

These findings add to the unsettled controversy of how and what molecules DMSO is able to help penetrate the blood–brain barrier. However, it is clear that penetration of DMSO itself through the intact blood–brain barrier occurs and has been shown to use magnetic resonance spectroscopy.[28]

DMSO has been shown to enhance the killing activity of certain bactericidal such as peroxynitrite, whereas other hydroxyl radical scavengers cannot. Macrophages and other cells able to produce nitric oxide can generate the formation of peroxyni-trite, a strong and versatile oxidant and the reaction product of nitric oxide and superoxide, which are not strong oxidants themselves.[29] Peroxynitrite is highly bactericidal, killing *E. coli* in direct proportion to its concentration.[30] Three hydroxyl radical scavengers mannitol, ethanol, and benzoate did not significantly affect the killing potential of peroxynitrite on *E. coli* cultures, but 100 mM DMSO did enhance peroxynitrite-mediated killing. DMSO is a more efficient hydroxyl radical scavenger than the other three scavengers and increases the formation of nitrogen dioxide to enhance the killing power of peroxynitrite.[30]

Helicobacter pylori is a spiral, intragastric bacterium that can be found in at least a third to half of people worldwide, most of whom have no symptoms. The frequency of *H. pylori* infection varies from country to country with 70%–90% of adults harboring the microbe in underdeveloped countries to about 50% in the United States found carrying the bacterium in 60-year-olds.[31]

H. pylori can help the human body by helping to regulate the level of stomach acid secreted, and as it does, it creates a beneficial environment both for itself and its host. When this regulation of acid secretion is disturbed, a series of molecular changes in the stomach can occur where *H. pylori* strains become harmful to its host.

When that happens, the dark side of *H. pylori* is revealed as the causative organism in peptic ulcer disease and chronic gastritis. The *H. pylori* pathogenic strain is also linked to the development of duodenal ulcers and distal stomach adenocarcinoma. The discovery of *H. pylori* as the cause of peptic ulcer disease came about after Freedberg and Baron[32] found in 1940 *spirochetes* in about 40% of gastric biopsies. Even thought there were sporadic reports occasionally appearing in the medical literature about the association of *H. pylori* and peptic ulcers, the great majority of gastroenterologists believed that these ulcers arose principally from lifestyle factors or from physical and emotional stress due to acid secretion. The fact that *H. pylori* was capable of living in the acidic environment of the stomach was a major reason for most gastroenterologists to dismiss its potential pathogenicity because no one could envision how a bacterium could survive for long in such an environment.

Until 1984, anyone training in medicine was taught the axiomatic golden rule, "no acid, no ulcer," the intention being that peptic ulcers can occur only when the stomach secretes acid.[33] This long-standing and erroneous principle supported by no solid evidence of any kind became an unchallenged medical dictum in the teaching and management of gastric ulcers. The "no acid, no ulcer" rule also led to a financial boom for many pharmaceuticals that manufactured a hodgepodge of antacids. However, in 1983, a year before the continued pursuit of antacids for the treatment of ulcer disease, two Australian physicians, Robin Warren and Barry Marshall,[34] boldly and correctly identified and later successfully treated peptic and duodenal ulcers with antibiotics despite violent opposition by the medical establishment at the time. Their work not only cured a serious gastroduodenal disorder that afflicted millions of patients worldwide but also prevented a highly lethal cancer of the stomach. Marshall and Warren shared the Nobel Prize in 2005. The Warren–Marshall discovery has generated over 50,000 scientific publications to this date and has resulted in a significant decline of gastroduodenal ulcers in most Western countries.[35]

H. pylori is presently treated with a triple therapy consisting of two antibiotics, clarithromycin and amoxicillin, and a proton-pump inhibitor omeprazole.[36] This regimen has considerable side effects. Moreover, the increase in antibiotic resistance by this bacterium has led to the search for new therapeutic strategies, including anti-inflammatory agents and hydroxyl radical scavengers.

Searching for other options to treat *H. pylori*, an *in vitro* study removed *H. pylori* strains isolated from antral biopsy specimens and cultured these pathogens on sheep blood agar.[37] An agar diffusion test was used to screen the antibacterial effect of 10% DMSO and 1% allopurinol. It was found that therapeutically administered concentrations of DMSO and allopurinol had no growth-inhibiting effect on *H. pylori in vitro*.[37] No zones of inhibition were produced by either drug.[37] This study concluded that neither DMSO nor allopurinol at the doses used is suitable for *H. pylori* eradication treatment. A cautionary note was added to these negative findings by pointing out the limits of *in vitro* testing since it could not be ruled out that DMSO may work to interfere with adherence factors of *H. pylori* and, in this way, block this microbe from recolonization of the gastroduodenal epithelium and prevent duodenal ulcer relapse.[37]

The role of oxyradicals in the process of acute and chronic duodenal ulceration in the rat was examined by administering allopurinol and DMSO by gavage. To produce chronic duodenal ulceration, all rats received intramuscular reserpine injections (0.1 mg/kg) every day for 6 weeks. Pretreatment was given with dose-dependent DMSO and allopurinol. At least 70% of rats were protected against ulceration with 5% or 10% solutions of DMSO, and similar concentrations of allopurinol protected all animals. Neither agent was seen to protect rats from influencing the hyperchlorhydria.[38] These findings appeared to indicate that oxygen-derived free radicals played an important role in the formation and development of gastric ulcers and led to translationally examine DMSO and allopurinol in a clinical setting in patients diagnosed with *H. pylori*.[38]

In a randomized, controlled clinical study on 146 patients, cimetidine, placebo, and the free radical scavengers allopurinol and DMSO were tested to see their effects on *H. pylori* duodenal ulcer relapse, which is believed to be mediated by oxygen-derived free radicals.[38] DMSO was given orally at doses of 500 mg four times daily, allopurinol at 50 mg four times a day, and cimetidine 400 mg at bedtime to patients who had shown endoscopic healing from gastric mucosal infection with *H. pylori*.[38] Treatment of these agents continued for 1 year. During the follow-up endoscopy at 6 and 12 months, the cumulative relapse at 1 year was placebo 47%, cimetidine 24%, allopurinol 6%, and DMSO 6%. In the patients who relapsed, ulcer recurrence tended to occur early in those patients taking a placebo or cimetidine, while ulcer relapse tended to be evenly distributed over the year for the patients treated with DMSO or allopurinol. The conclusions from this study suggest that oxygen-derived free radicals are involved in the relapse of duodenal ulceration in patients infected with *H. pylori*.[38]

P. aeruginosa is a common opportunistic pathogen of humans, animals, and plants. This bacterium can colonize the respiratory tract in hospital-acquired infections particularly in immunocompromised or mechanically ventilated people.[39]

Because it thrives on most surfaces, this *P. aeruginosa* can colonize medical equipment, including in-dwelling catheters, causing cross-infections in hospitals and clinics. It has been shown that several *P. aeruginosa* virulence factors, both cell-associated and extracellular products, are potent inducers of proinflammatory mediators, such as the interleukins, a large group of cytokines mainly produced by PMNs and leukocytes during inflammation. In particular, interleukin-8 (IL-8) overexpression has been shown to induce ulcer formation in the cornea through neutrophil recruitment.[40] Interleukin-8 is often associated with inflammation. IL-8, also known as *neutrophil chemotactic factor*, has two primary functions. It induces chemotaxis in target cells, primarily neutrophils but also other granulocytes, causing them to migrate toward the site of infection, and also induces phagocytosis once they have arrived.[41] Chemotaxis is a process whereby bacteria and other cells move toward an object according to certain chemicals in their environment. This is fundamentally important for bacteria to move forward to find food or move away from a noxious source.

A relationship has been found between massive accumulation of neutrophils, mainly due to enhanced IL-8 levels, to contribute to lung infections from *P. aeruginosa*.[42]

The role of DMSO as an anti-inflammatory and free radical scavenger[43] was investigated in the possible inhibition of IL-8-mediated neutrophil recruitment induced by *P. aeruginosa* bacterial supernatant.[44] The test involved the question of IL-8 inhibition *in vitro* induced by *P. aeruginosa* in human bronchial epithelial cells. These cells produce IL-8 induced by *P. aeruginosa*, a process that was blocked by 1% exposure to DMSO as measured by the concentrations of RNA and protein present. These findings were extended to examine whether DMSO prevented the recruitment of neutrophils and IL-8 production induced by *P. aeruginosa* in dog trachea. It was seen that DMSO was able to prevent IL-8 production in the dog airway epithelial cells induced by *P. aeruginosa*. The mechanism for this inhibitory action by DMSO is not known but may be due to its anti-inflammatory effects and hydroxyl radical scavenger activity of recruited neutrophils since it has been shown that airway epithelial cells can form oxygen radicals and hydroxyl radicals after exposure to inflammatory mediators such as IL-8.[44]

Aside from their effects on airway epithelial cells, production of reactive oxygen species from bacterial toxins also contributes to the pathophysiology of intestinal inflammation. This is the case of the enterotoxic effects of toxin A generated by *Clostridium difficile*.

C. difficile is a slow-growing, spore-forming anaerobe that is found in the normal gut flora in 80% of healthy neonates and in the majority of people.[45] *C. difficile* can remain in check in the gastrointestinal tract normal flora until its destructive power is unleashed following antibiotic treatment that can eradicate other competing bacteria from the intestine, disrupt the normal microbiome, and result in colitis and diarrhea.[46]

Following antibiotic treatment and proliferation by *C. difficile*, two potent enterotoxins designated A and B are released by *C. difficile*, causing an acute inflammatory reaction and damage to the mucosal cells of the lamina propria in the large bowel.[46] These toxins are responsible for the enterocolitis and diarrhea that results from *C. difficile* infection.[46] After their release, toxin A or B or both can enter the cytosol of target cells where pseudomembranous colitis develops, characterized by loss of appetite, fatigue, offensive-smelling diarrhea, fever, and abdominal pain. In severe cases, life-threatening complications can develop, such as toxic megacolon.

C. difficile infection of the bowel in adults can be treated with oral vancomycin or preferably, with oral metronidazole, 500 mg three times/day for 2 weeks.[47] However, both medications have side effects, and in particular, metronidazole can cause additional pseudomembranous colitis, hives, and swelling of the face, lips, and throat.[48] Pediatric *C. difficile* infection appears to be on the rise, and some cases can be life-threatening.[49] The search for more effective and safer treatments for *C. difficile* has therefore become a medical priority.[50]

Infection with *C. difficile* also poses a socioeconomic problem for health-care costs. The estimated number of cases of *C. difficile*–associated disease exceeds 250,000 per year in the United States alone with total additional health-care costs approaching 1 billion dollars annually.[51]

The search for alternative therapy to *C. difficile* led Qiu et al.[52] to test the action of DMSO on reactive oxygen metabolites, which have been implicated in the pathogenesis of experimental colitis in both animal models and idiopathic inflammatory bowel disease of humans.[53] Reactive oxygen metabolites are mainly free radicals that appear to mediate damage to proteins and lipids in tissues targeted by *C. difficile*.[52] These free radicals include superoxide, hydroxyl ion, and hypochlorite, which may be involved in toxin A production generated by *C. difficle*.[52] Rats were subjected to a laparotomy, and two closed ileal loops were formed in each anima. Five microliters of toxin A was administered by intraluminal injection to each animal. Seven days prior to toxin A administration, rats were treated with 0.1%–5% DMSO or control buffer. Results of DMSO treatment are summarized in Figure 5.1.

Results of DMSO treatment and partial inhibition of toxin A support the assumption that hydroxyl and superoxide radicals contribute to the effects of toxin A–induced acute enteritis inflammation and microvascular injury to the rat ileum.[52] The finding that only partial inhibition of toxin A by DMSO was seen (Figure 5.1) suggests that other mediators such as prostaglandins and leukotrienes may also play a role in toxin A damage.[54]

These findings could be useful in the treatment of ulcerative colitis and Crohn's disease in humans since they are also characterized by reduced levels of antioxidant molecules that can neutralize inflammation generated by activated neutrophils and monocytes. Data to support this assumption have been reported showing oxidant damage in the inflamed mucosa of Crohn's patients but not in patients with prior Crohn's disease who had healing of the colon mucosa.[53]

DMSO IN VIRAL AND FUNGAL PATHOLOGY

One of the first antiviral uses of DMSO was a formulation for the treatment of oral herpes simplex in humans. Herpes simplex is an infection caused by a herpes simplex virus (HSV), which enters the body and forms blisters, itch, and pain and then heals and disappears. There are two types of herpes viruses: HSV-1 and HSV-2. Both types can cause lesions in the mouth and genitals, but HSV-1 is more predominant in cold sores of the mouth. Genital herpes affects the genitals, buttocks, or anal area.[55] Genital herpes is a sexually transmitted disease. Other herpes infections can affect the eyes, skin, or other parts of the body. Once an infection occurs, the virus

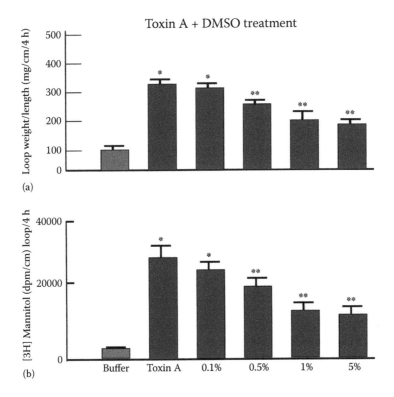

FIGURE 5.1 Inhibition of toxin A by dose-dependent DMSO treatment given orally 7 days before toxin exposure in rat-induced colitis. DMSO effective response is seen in concentrations between 0.1% and 5%. *Key*: Results are expressed as means ± SE of 8 loops/group. (a) Fluid-induced secretion of toxin A. (b) Mucosal permeability to mannitol. *$p \leq 0.05$ DMSO versus buffer control; **$p \leq 0.01$ DMSO versus toxin A alone. (Adapted from Qiu, B. et al., *Am. J. Physiol.*, 276(2 Pt 1), G485, 1999.)

spreads to nerve cells and stays in the body for the rest of a person's life. It may come back from time to time and cause symptoms or flares.

Many antiviral preparations have been introduced to alleviate the pain and healing of oral herpes, some better than others, including DMSO.[56,57]

One of the better antiviral preparations for herpes is a formulation consisting of a 5% solution of idoxuridine in various concentrations combined with DMSO. A small amount of the idoxuridine–DMSO solution was painted on the herpetic lesions of 16 patients, and details of the progress of the lesions were recorded daily by an observer in a double-blind manner.[58]

The duration of the attack until the arrest of lesion was 1.2 days in the group receiving idoxuridine–DMSO and 2.45 in those receiving only DMSO. Thus, there seemed to be a synergistic action when using idoxuridine–DMSO solution as opposed to DMSO alone. Both drugs were more effective than no treatment.[58] This study has been replicated many times following results from a study by Spruance et al.,[59] which showed significant acceleration of healing in herpes labialis lesions when

301 immunocompetent patients were treated with a solution of 15% idoxuridine–80% DMSO solution. Idoxuridine–DMSO preparation has been approved for herpes zoster in Canada and Europe.[60]

Another antiviral agent tested for the treatment of herpes simplex infection is topically administered acyclovir in polyethylene glycol (Acy-PEG). A comparison of this Acy-PEG ointment preparation was made versus acyclovir–DMSO ointment in excised human and guinea pig skin.[61] Results from this study established that Acy-PEG formulation penetrates human skin very slowly.[61] This result was attributed to PEG inefficacy since the preparation using acyclovir–DMSO ointment was significantly more effective. A more consistent interpretation of this study is that DMSO appears to exert antiviral activity by itself since it was found that acyclovir ointment alone had only limited therapeutic benefit.[61] There is, in fact, corroboration for this latter conclusion from another independent study done in Spain.

That study reports that although DMSO has been regarded as a useful excipient that can carry active antiviral agents across membranes to increase penetration into tissues of these agents, findings show that DMSO may express antiviral activity by itself.[62] Rabbit skin fibroblasts were used to grow the recombinant HSV-1 and Vero cells to analyze the action of DMSO.[62] DMSO was able to block productive infection in HSV-1 in Vero cells when administered after virus adsorption. DMSO appears to cause this block in HSV-1 production by acting at different sites with positive cooperativity, as suggested by the Hill numbers of the inhibition curves.[62] The results indicated that DMSO inhibits HSV-1 viral DNA synthesis *in vitro*, blocked the stability and maturation of the free virion, and also reduced the levels of a number of viral transcripts.[62] Thus, it was concluded from this study that DMSO may possess antiviral activity by previously undisclosed inhibitory mechanisms.

This important finding should be considered in relation to how herpes infections can progress to serious disorders. Although most of herpes infections are mild, they can lead to lethal encephalitis, and this virus accounts for 20% of all reported encephalitis cases.[63]

To our knowledge, there are no studies that have used DMSO either as an adjuvant or as a primary therapy in human viral encephalitis. Paradoxically, DMSO with acyclovir has been used for the prophylaxis and therapy of equine herpes virus type 1 (EHV-1) infection, which can cause myeloencephalopathy and death in horses.[64,65]

Since DMSO has been used as a solubilizing and penetrant carrier for other antiviral agents for such conditions as rabies and viral-induced myocarditis,[66,67] it is logical and justified to reappraise the potential antiviral activity by DMSO in the interest of drug safety since many of the antiviral agents on the market have adverse effects that could be avoided by considerably lowering their dose when combined with DMSO.

DMSO is a frequently used solvent and penetrant carrier through the skin for antifungal drugs. DMSO is recommended to dissolve antifungals with poor aqueous solubility, but recently, DMSO has gained attention for drugs that are currently prepared in water.

Using DMSO to prepare all approved antifungals would increase the consistency of the methods for drug preparation listed in the current Clinical and Laboratory Standards Institute guidelines and could help avoid precipitation of the antifungal agents, which would impair their efficacy.[68]

DMSO by itself has shown effectiveness against the strains of many fungi that cause skin, hair, and nail infections, including trichophyton, epidermophyton, and microsporum at dose-dependent concentrations of 1%–10%.[69] Trichophyton is a skin infection that causes athlete's foot, ringworm, jock itch, and similar dermatophyte infections of the nail, beard, skin, and scalp, while epidermophyton and microsporum affect hair, skin, and nails.[70]

DMSO was evaluated *in vitro* at a 2% concentration against an assortment of Candida species and was observed to slow the growth of all Candida isolates tested.[71]

Antifungal activity by DMSO was also observed recently at concentrations between 0.5% and 1% against six yeast species, including *Candida* species.[72]

The antifungal effects of DMSO are especially important in the geriatric population due to the higher prevalence of seborrheic dermatitis, mucosal and cutaneous candidiasis, tinea pedis, and onychomycosis found in this group of patients.[73] Treatment of these elderly people with dermatophytoses and yeast infections should depend on the patient's medication list to avoid potential drug interactions and whether topical or systemic therapy is indicated.[73]

REFERENCES

1. Turnbaugh PJ, Ley RE, Hamady M, Fraser-Liggett CM, Knight R, Gordon JI. The human microbiome project. *Nature.* 2007;449(7164):804–810.
2. Macfarlane G. *Alexander Fleming: The Man and the Myth.* Chatto and Windus, London, U.K., 1984.
3. Chain E, Florey HW, Adelaide MB, Gardner AD, Heatley NG, Jennings MA, Orr-Ewing J, Sanders AG. Penicillin as a chemotherapeutic agent. 1940. *Clin Orthop Relat Res.* 1993;295:3–7.
4. Kerwat K, Kerwat M, Graf J, Wulf H. Resistance to antibiotics and multiresistant pathogens. *Anasthesiol Intensivmed Notfallmed Schmerzther.* 2010;45(4):242–243.
5. Kalan L, Wright GD. Antibiotic adjuvants: Multicomponent anti-infective strategies. *Exp Rev Mol Med.* 2011;13:e5.
6. Pankey GA, Sabath LD. Clinical relevance of bacteriostatic versus bactericidal mechanisms of action in the treatment of Gram-positive bacterial infections. *Clin Infect Dis.* 2004;38(6):864–870.
7. Jacob SW, Bischel M, Herschler RJ. Dimethyl sulfoxide (DMSO): A new concept in pharmacotherapy. *Curr Ther Res Clin Exp.* 1964;6:134–135.
8. Kligman AM. Dimethyl sulfoxide, 2. *JAMA.* 1965;193:923–928.
9. Pottz GE, Rampey JH, Benjamin F. The effect of dimethyl sulfoxide (DMSO) on antibiotic sensitivity of a group of medically important microorganisms: Preliminary report. *Ann NY Acad Sci.* 1967;141(1):261–272.
10. Trotz-Williams LA, Mercer NJ, Walters JM, Maki AM, Johnson RP. Pork implicated in a Shiga toxin-producing *Escherichia coli* O157:H7 outbreak in Ontario, Canada. *Can J Public Health.* 2012;103(5):e322–e326.
11. Dale JB, Fischetti VA, Carapetis JR et al. Group A streptococcal vaccines: Paving a path for accelerated development. *Vaccine.* 2013;31(Suppl 2):B216–B222.
12. Jarvis WR, Martone W. Predominant pathogens in hospital infections. *J Antimicrob Chem.* 1992;29:19–24.
13. Siemieniuk R, Gregson D, Gill J. The persisting burden of invasive pneumococcal disease in HIV patients: An observational cohort study. *BMC Infect Dis.* 2011;11:314.

14. Jacewicz M, Keusch GT. Pathogenesis of Shigella diarrhea. VIII. Evidence for a trans-location step in the cytotoxic action of Shiga toxin. *J Infect Dis.* 1983;148(5):844–854.

15. Green GM, Kass EH. The role of the alveolar macrophage in the clearance of bacteria from the lung. *J Exp Med.* 1964;119:167–175.

16. Nathan CF, Murray HW, Cohn ZA. The macrophage as an effector cell. *N Engl J Med.* 1980;303:622–626.

17. Coonrod JD, Yoneda K. Detection and partial characterization of antibacterial factor(s) in alveolar lining material of rats. *J Clin Invest.* 1983;71:129–141.

18. Pesanti EL, Nugent KM. Modulation of pulmonary clearance of bacteria by antioxidants. *Infect Immun.* 1985;48(1):57–61.

19. De Chatelet LR, Mullikin D, McCall CE. The generation of superoxide anion by various types of phagocytes. *J Infect Dis.* 975;131:443–446.

20. Hoidal JR, Beall GD, Repine JE. Production of hydroxyl radical by human alveolar macrophages. *Infect Immun.* 1979;26:1088–1092.

21. Czuprynski CJ, Henson PM, Campbell PA. Effect of dimethyl sulfoxide on the in vitro and in vivo bactericidal activity of human and mouse neutrophils and mononuclear phagocytes. *Inflammation.* 1984;8(2):181–191.

22. Brigham KL. Oxidant stress and adult respiratory distress syndrome. *Eur Respir J Suppl.* 1990;11:482s–484s.

23. Barling RWA, Selkon JB. The penetration of antibiotics into cerebrospinal fluid and brain tissue. *J Antimicrob Chemother.* 1978;4:203–227.

24. Iwen PC, Miller NG. Enhancement of ketoconazole penetration across the blood-brain barrier of mice by dimethyl sulfoxide. *Antimicrob Agents Chemother.* 1986;30:617–618.

25. Broadwell RD, Salcman M, Kaplan RS. Morphologic effect of dimethyl sulfoxide on the blood brain barrier. *Science.* 1982;217:164–166.

26. Brayton CF. Dimethyl sulfoxide (DMSO): A review. *Cornell Vet.* 1986;76:61–90.

27. Green SL, Mayhew IG, Brown MP, Gronwall RR, Montieth G. Concentrations of tri-methoprim and sulfamethoxazole in cerebrospinal fluid and serum in mares with and without a dimethyl sulfoxide pretreatment. *Can J Vet Res.* 1990;54(2):215–222.

28. Delgado-Goñi T, Martín-Sitjar J, Simões RV, Acosta M, Lope-Piedrafita S, Arús C. Dimethyl sulfoxide (DMSO) as a potential contrast agent for brain tumors. *NMR Biomed.* 2013;26(2):173–184.

29. Pryor WA, Squadrito GL. The chemistry of peroxynitrite: A product from the reaction of nitric oxide with superoxide. *Am J Physiol.* 1995;268(5 Pt 1):L699–L722.

30. Zhu L, Gunn C, Beckman JS. Bactericidal activity of peroxynitrite. *Arch Biochem Biophys.* 1992;298(2):452–457.

31. Sachs G, Scott DR. *Helicobacter pylori*: Eradication or preservation. *F1000 Med Rep.* 2012;4:7–10.

32. Freedberg AS, Baron LE. The presence of spirochetes in human gastric mucosa. *Am J Dig Dis.* 1940;7:443–445.

33. Berkow R (ed.). *The Merck Manual of Diagnosis and Therapeutics, 16th Edition.* Merck & Co. Inc., Rahway, NJ, 1992.

34. Warren JR, Marshall B. Unidentified curved bacilli on gastric epithelium in active chronic gastritis. *Lancet.* 1983;i:1273–1275.

35. Ahmed M. 23 years of the discovery of *Helicobacter pylori*: Is the debate over? *Ann Clin Microbiol Antimicrob.* 2005;4:17–19.

36. Nakajima S, Graham DY, Hattori T, Bamba T. Strategy for treatment of *Helicobacter pylori* infection in adults. I. Updated indications for test and eradication therapy suggested in 2000. *Curr Pharm Des.* October 2000;6(15):1503–1514.

37. Ansorg R, von Recklinghausen G, Heintschel von Heinegg E. Susceptibility of *Helicobacter pylori* to simethicone and other non-antibiotic drugs. *J Antimicrob Chemother.* 1996;37(1):45–52.

38. Salim AS. The relationship between *Helicobacter pylori* and oxygen-derived free radicals in the mechanism of duodenal ulceration. *Intern Med.* 1993;32(5):359–364.

39. Venza I, Cucinotta M, Visalli M, De Grazia G, Oliva S, Teti D. *Pseudomonas aeruginosa* induces interleukin-8 (IL-8) gene expression in human conjunctiva through the recruitment of both RelA and CCAAT/enhancer-binding protein beta to the IL-8 promoter. *J Biol Chem.* 2009;284(7):4191–4199.

40. Xue LM, Willcox MD, Lloyd A, Wakefield D, Thakur A. Regulatory role of IL-1β in the expression of IL-6 and IL-8 in human corneal epithelial cells during *Pseudomonas aeruginosa* colonization. *Clin Exp Ophthalmol.* 2001;29:171–174.

41. Baggiolini M, Walz A, Kunkel SL. Neutrophil-activating peptide-1/interleukin-8, a novel cytokine that activates neutrophils. *J Clin Invest.* 1989;84:1045–1049.

42. Konig B, Vasil ML, Konig W. Role of haemolytic and non-haemolytic phospholipase C from *Pseudomonas aeruginosa* in interleukin-8 release from human monocytes. *J Med Microbiol.* 1997;46:471–478.

43. Jacob SW, de la Torre JC. Pharmacology of dimethyl sulfoxide in cardiac and CNS damage. *Pharmacol Rep.* 2009;61(2):225–235.

44. Massion PP, Lindén A, Inoue H, Mathy M, Grattan KM, Nadel JA. Dimethyl sulfoxide decreases interleukin-8-mediated neutrophil recruitment in the airways. *Am J Physiol.* November 1996;271(5 Pt 1):L838–L843.

45. Kyne L, Farrel RJ, Kelly CP. *Clostridium difficile. Gastroenterol Clin North Am.* 2001;30(3):753.

46. Lamont JT. How bacterial enterotoxins work: Insights from in vivo studies. *Trans Am Clin Climatol Assoc.* 2002;113(42):167.

47. Dineen SP, Bailey SH, Pham TH, Huerta S. *Clostridium difficile* enteritis: A report of two cases and systematic literature review. *World J Gastrointest Surg.* 2013;5(3):37–42.

48. Follmar KE, Condron SA, Turner II, Nathan JD, Ludwig KA. Treatment of metronidazole-refractory *Clostridium difficile* enteritis with vancomycin. *Surg Infect (Larchmt).* 2008;9:195–200.

49. Pant C, Deshpande A, Altaf MA, Minocha A, Sferra TJ. *Clostridium difficile* infection in children: A comprehensive review. *Curr Med Res Opin.* 2013;29(8):967–984.

50. Wilkins TD, Lyerly DM. *Clostridium difficile* testing: After 20 years, still challenging. *J Clin Microbiol.* 2003;41:531–534.

51. Kyne L, Hamel M, Polavaram R, Kelly C. Health care costs and mortality associated with nosocomial diarrhea due to *Clostridium difficile. Clin Infect Dis.* 2002;34:346–353.

52. Qiu B, Pothoulakis C, Castagliuolo I, Nikulasson S, LaMont JT. Participation of reactive oxygen metabolites in *Clostridium difficile* toxin A-induced enteritis in rats. *Am J Physiol.* 1999;276(2 Pt 1):G485–G490.

53. McKenzie SJ, Baker M, Buffinton G, Doe W. Evidence of oxidant-induced injury to epithelial cells during inflammatory bowel disease. *J Clin Invest.* 1996;98:136–141.

54. Triadafilopoulos G, Pothoulakis C, Weiss R, Giampaolo C, LaMont J. Comparative study of *Clostridium difficile* toxin A and cholera toxin in rabbit ileum. Role of prostaglandins and leukotrienes. *Gastroenterology.* 1987;92:1174–1180.

55. Studahl M, Lindquist L, Eriksson BM, Günther G, Bengner M, Franzen-Röhl E, Fohlman J, Bergström T, Aurelius E. Acute viral infections of the central nervous system in immunocompetent adults: Diagnosis and management. *Drugs.* 2013;73(2):131–158.

56. Field HJ, Vere Hodge RA. Recent developments in anti-herpesvirus drugs. *Brit Med Bull.* 2013;106:213–249.

57. Sehtman L. Dimethyl sulfoxide therapy in various dermatological disorders. *Ann NY Acad Sci.* January 27, 1975;243:395–402.

58. MacCallum FO, Juel-Jensen BE. Herpes simplex virus skin infection in man treated with idoxuridine in dimethyl sulphoxide. Results of double-blind controlled trial. *Brit Med J*. 1966;2(5517):805–807.

59. Spruance SL, Stewart JC, Freeman DJ, Brightman VJ, Cox JL, Wenerstrom G, McKeough MB, Rowe NH. Early application of topical 15% idoxuridine in dimethyl sulfoxide shortens the course of herpes simplex labialis: A multicenter placebo-controlled trial. *J Infect Dis*. 1990;161(2):191–197.

60. Aliaga A, Armijo M, Camacho F, Castro A, Cruces M, Díaz JL, Fernández JM, Iglesias L, Ledo A, Mascaró JM. A topical solution of 40% idoxuridine in dimethyl sulfoxide compared to oral acyclovir in the treatment of herpes zoster. A double-blind multicenter clinical trial. *Med Clin (Barc)*. 1992;98(7):245–249.

61. Freeman DJ, Sheth NV, Spruance SL. Failure of topical acyclovir in ointment to penetrate human skin. *Antimicrob Agents Chemother*. 1986;29(5):730–732.

62. Aguilar JS, Roy D, Ghazal P, Wagner EK. Dimethyl sulfoxide blocks herpes simplex virus-1 productive infection in vitro acting at different stages with positive cooperativity. Application of micro-array analysis. *BMC Infect Dis*. 24, 2002;2:9.

63. Whitley RJ, Soong SJ, Dolin R, Galasso GJ, Chien LT, Alford CA. Adenine arabinoside therapy of biopsy-proved herpes simplex encephalitis. National Institute of Allergy and Infectious Diseases collaborative antiviral study. *N Engl J Med*. 1977;297:289–294.

64. Henninger RW, Reed SM, Saville WJ, Allen GP, Hass GF, Kohn CW, Sofaly C. Outbreak of neurologic disease caused by equine herpesvirus-1 at a university equestrian center. *J Vet Intern Med*. 2007;21(1):157–165.

65. Wong D, Scarratt WK. Equine herpes myeloencephalopathy in a 12-year-old American quarter horse. *Vet Clin North Am Equine Pract*. 2006;22(1):177–191.

66. Ramya R, Mohana Subramanian B, Sivakumar V, Senthilkumar RL, Sambasiva Rao KR, Srinivasan VA. Expression and solubilization of insect cell-based rabies virus glycoprotein and assessment of its immunogenicity and protective efficacy in mice. *Clin Vaccine Immunol*. October 2011;18(10):1673–1679.

67. Yun SH, Lee WG, Kim YC, Ju ES, Lim BK, Choi JO, Kim DK, Jeon ES. Antiviral activity of coxsackievirus B3 3C protease inhibitor in experimental murine myocarditis. *J Infect Dis*. February 1, 2012;205(3):491–497.

68. Fothergill AW, Sanders C, Wiederhold NP. Comparison of MICs of fluconazole and flucytosine when dissolved in dimethyl sulfoxide or water. *J Clin Microbiol*. 2013;51(6):1955–1957.

69. Randhawa MA, Randhawa MA. The effect of dimethyl sulfoxide (DMSO) on the growth of dermatophytes. *Nihon Ishinkin Gakkai Zasshi*. 2006;47(4):313–318.

70. Fernández-Torres B, Carrillo AJ, Martín E, Del Palacio A, Moore MK, Valverde A, Serrano M, Guarro J. In vitro activities of 10 antifungal drugs against 508 dermatophyte strains. *Antimicrob Agents Chemother*. 2001;45:2524–2528.

71. Rodríguez-Tudela JL, Cuenca-Estrella M, Díaz-Guerra TM, Mellado E. Standardization of antifungal susceptibility variables for a semiautomated methodology. *J Clin Microbiol*. 2001;39(7):2513–2517.

72. Hazen KC. Influence of DMSO on antifungal activity during susceptibility testing in vitro. *Diagn Microbiol Infect Dis*. 2013;75(1):60–63.

73. Loo DS. Cutaneous fungal infections in the elderly. *Dermatol Clin*. 2004;22(1):33–50.

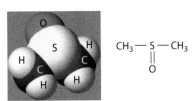

FIGURE 1.1 Structural 3D and 2D formulas of the DMSO molecule. Sulfur (S), oxygen (O), carbon (C), and hydrogen (H).

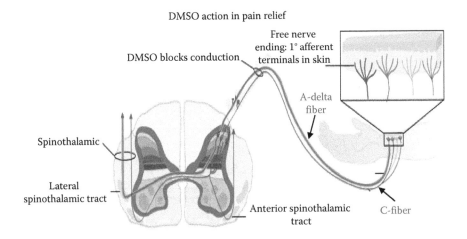

FIGURE 2.1 DMSO topical application can block both myelinated A-delta and unmyelinated C-pain fibers.[30,34] See text for details.

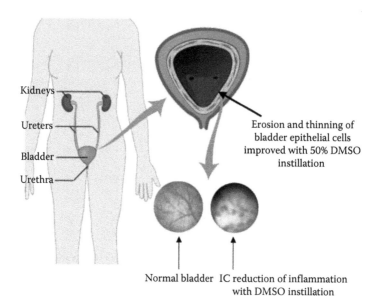

FIGURE 3.2 DMSO instillation for interstitial cystitis.

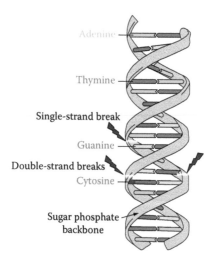

FIGURE 4.1 Ionizing radiation can create two types of breaks: (1) in the sugar phosphate backbone or (2) in the base pairs of the DNA. These breaks can cause either single (bolt) or double DNA strand breaks (bolts). Direct DNA damage occurs when radiation particles physically break one or two strands of DNA. Indirect damage occurs when radiation creates harmful free radicals to damage DNA, a target of repair for DMSO. Single-strand DNA breaks are more prone to damage than double-strand DNA breaks because more toxic agents target single-strand DNA regions than double strands. Ionizing radiation of single-strand breaks are more easily repaired by the cell than cell repair of similar ionizing radiation of double-strand breaks, which can be lethal to the cell. Double-strand DNA breaks are the cause of many diseases, including various types of cancer.

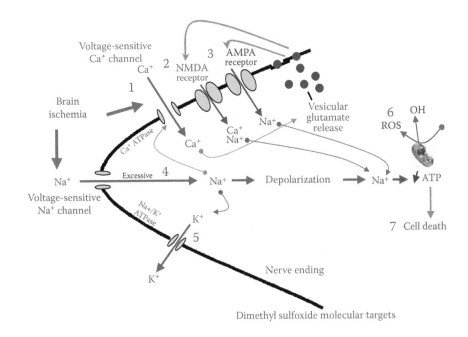

FIGURE 8.3 Scheme detailing some of the molecular mechanisms DMSO is reported to target at the subcellular level in nerve endings or neuronal cytoplasm following cerebral ischemia. DMSO is reported to suppress excessive calcium (Ca^{2+}) influx into cells (1) and block NMDA (*N*-methyl-D-aspartate) and AMPA (alpha-amino-3-hydroxy-5-methyl-4-isoxazolepropionate) receptor channels (2, 3) following ischemic events that can promote excitotoxic death of neurons. DMSO is reported to block sodium (Na^{2+}) channel activation (4), a reaction seen after CNS injury that may activate a biochemical cascade resulting in intra-cytoplasmic potassium efflux (5), cell energy hypometabolism (↓ATP), ROS and hydroxyl radical (OH) production (6), and cell death (7). DMSO has been shown to increase the levels of high-energy phosphates cAMP and creatine phosphate and inhibit tissue factor expression, thrombus formation, and vascular smooth muscle cell (VSMC) proliferation and migration (not shown), elements known to participate in cerebral ischemia and stroke. See text for details.

Rat brain MCA occlusion

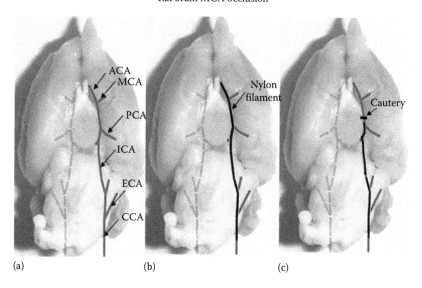

(a)   (b)   (c)

FIGURE 8.5   Normal arteries at the base of rat brain (a) and approaches to induce focal brain ischemia using a nylon filament introduced from the ICA to the MCA (b) and thrombotic occlusion (c) of the MCA. Such occlusions of the MCA will mimic human ischemic stroke and involve parts of the cortex and subcortical regions, producing sensory loss and paresis or paralysis of the contralateral side of the body. See text for details. *Key*: ACA (anterior cerebral artery), MCA (middle cerebral artery), PCA (posterior cerebral artery), ICA (internal carotid artery), ECA (external carotid artery), CCA (common carotid artery).

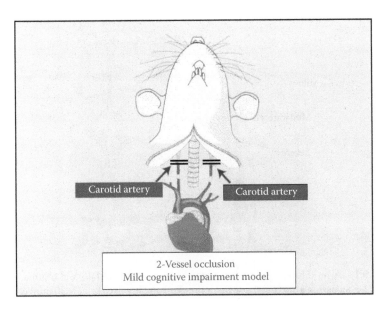

FIGURE 8.11 Permanent occlusion of both common carotid arteries (2-VO) in 14-month old rats. After 1 to 14 weeks, mild cognitive impairment characterized by progressive visuo-spatial memory impairment is observed that can be quantified and reversed with DMSO treatment. (From de la Torre, J.C. et al., *Brain Res.*, 779, 285, 1998.)

Severity of traumatic brain injury

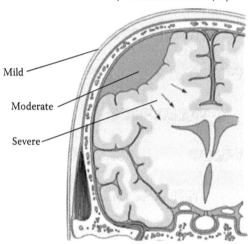

Mild

Moderate

Severe

FIGURE 9.1 A mild head injury refers to the severity of the initial blunt trauma, which is generally a concussion affecting the scalp or the skull and involves brief (<30 min) changes in mental status or consciousness as opposed to a severe injury, which results in extended LOC, coma, and posttraumatic amnesia (PTA) lasting days, months, or years. This injury has a high risk of mortality. LOC after a moderate head injury does not generally exceed 6 h, and PTA may extend up to 24 h. A moderate-to-severe head injury can result in an epidural hematoma, subdural hematoma, or intraparenchymal hemorrhage.

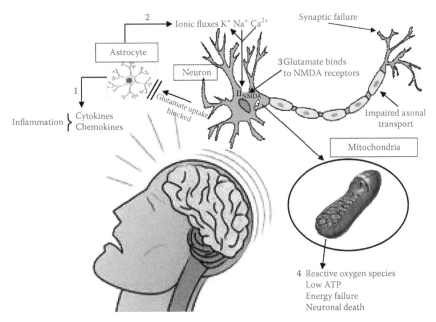

Subcellular secondary damage after TBI

FIGURE 9.4 Diagram shows major subcellular events associated with secondary damage following primary traumatic brain injury (TBI). Astrocytes undergo reactive transformation in response to blunt head injury and release cytotoxic cytokines and chemokines. Reactive astrocytes are unable to uptake glutamate. Glutamate, in turn, accumulates in the extracellular space at the site of injury where it binds and opens NMDA channel receptors in neurons. This action by glutamate creates an influx of calcium (Ca^{2+}) and sodium (Na^+) ions and an efflux of potassium (K^+) ions, which result in the depolarization of neurons. This ionic dyshomeostasis damages intracytoplasmic mitochondria that promote the formation of reactive oxygen species, reduced ATP synthesis, energy failure, and cell death. Studies have shown that DMSO is able to antagonize the formation of cytokines and chemokines in injured tissue (1), block Na^+ and Ca^{2+} influx into cells after neuronal depolarization (2), prevent glutamate induction of excitotoxic neuronal death by suppressing NMDA ion currents (3), and scavenge the formation of reactive oxygen species and the damage caused by these oxidizing radicals. See text for details.

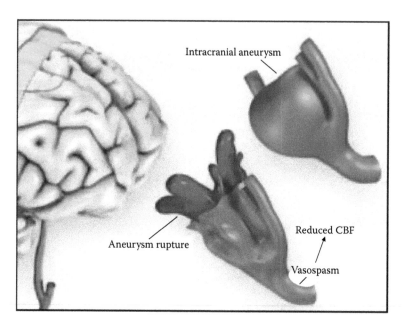

FIGURE 9.5 Rupture of intracranial aneurysm, depending on type, size, and location, can cause subarachnoid hemorrhage leading to serious and life-threatening complications including hemiplegia, aphasia, regional vasospasm, and brain swelling. Treatment with IV DMSO of ruptured aneurysms in people is reported to reverse progressive hemiplegia, vasospasm and resulting low CBF, and improve hemispheric motor deficits. (From Mullan, S. et al., Dimethyl sulfoxide in the management of postoperative hemiplegia, in: Wikins, R.H., ed., *Arterial Spasm*, William & Wilkins, Baltimore, MD, 1980, pp. 646–653.)

6 DMSO in Clinical Microbiology

HOW BACTERIA ACHIEVE ANTIBIOTIC RESISTANCE

Antibiotic resistance by bacteria refers to the ability of microorganisms to withstand the killing effects of one or more antibiotics. Bacteria resistant to multiple antibiotics are considered *multidrug resistant* and are referred more colloquially to as superbugs.

Drug-resistant bacteria pose one of the most serious and growing threats to health and survival and continue to be a global concern to contemporary medicine. Antibiotic-resistant infections complicate treatment and increase morbidity and mortality.

These infections are difficult to treat with first-line antibiotics and often result in second-line treatments that promote longer-lasting illnesses, more doctor visits or extended hospital stays, and the need for more expensive and less safe medications. Some resistant infections can cause death.

The first step in the emergence of antibiotic-resistant microorganisms is a genetic change in a bacterium. There are two ways that can happen. The first way is by spontaneous mutation of the bacterium's DNA. Many antibiotics kill bacteria by inactivating an essential bacterial protein, but either the microbe can develop a mutation that either prevents binding of the antibiotic to the protein or can increase the production of that protein. Last, the bacterium can shed the targeted protein from its cell body. The second way bacteria can become resistant to antibiotics is by the transfer of immunity from a resistant bacterium to its progeny, thus ensuring the survival of subsequent generations of microbes.

In some countries, including the United States, patients expect and demand antibiotics from doctors, even in situations where they are inappropriate or ineffective. Because microbes are always mutating, some random mutation eventually will protect against the drug. As people enjoy the benefits of antibiotics, their misuse promotes antibiotic resistance in many lethal bacteria.

Resensitizing resistant bacteria to antibiotics so that they can be killed is the objective of searching for antimicrobial adjuvants that can be used to boost otherwise ineffective antibiotics against methicillin-resistant *Staphylococcus aureus* (MRSA), *Mycobacterium tuberculosis*, *Pseudomonas*, *Clostridium difficile*, and other *superbugs* that can cause lethal infectious outbreaks in hospitals and institutions.

The emergence of multidrug-resistant and extensively drug-resistant *M. tuberculosis* throughout the developing world is very disturbing in the present scenario of tuberculosis (TB) management. TB has always been a global problem and tends to increase in prevalence mainly in the developing countries of the world,

115

many of the affected living in poverty, but is also problematic in industrialized nations, including the United States.[1]

M. tuberculosis usually attacks the lungs, but these pathogens can invade any part of the body such as the kidney, spine, and brain.

There are 10 drugs currently approved by the *U.S. Food and Drug Administration* (FDA) for treating TB. Before the emergence of antibiotic-resistant pathogens, potent drugs such as rifampicin, isoniazid, ethambutol, and pyrazinamide were effective as first-line anti-TB agents in the treatment of latent or active TB. Most patients with active TB can be treated initially with isoniazid, rifampicin, pyrazinamide, and ethambutol for 8 weeks, and this regimen can be followed by 18 weeks of treatment with isoniazid and rifampicin if needed.[2]

Multidrug-resistant TB is defined by resistance to the two most effective first-line drugs, isoniazid and rifampicin.[3] When bacteria are resistant to second-line antibiotics, they are called extremely resistant. Recently, strains of *M. tuberculosis* have been identified that are resistant to all available antibiotics. This is representative of the major problems of antibiotic resistance and the cause of considerable concern if these strains begin to quickly spread worldwide.

Approximately 3.3% of new multidrug-resistant TB cases are resistant to both isoniazid and rifampicin, so other drug combinations must be tried. However, most of these drug combinations fail to fully treat TB, and consequently, there is a fundamental need to explore alternative anti-TB agents that can kill strains resistant to currently available drugs and for new drugs that are better tolerated and will shorten treatment regimens.[4]

DMSO IN MULTIDRUG-RESISTANT TUBERCULOSIS

Researchers working in the field of microbiological pharmacotherapeutics are looking to find ways to make first-line antibiotic therapies effective against drug-resistant strains of bacteria. Part of this research is to search for adjunct therapy that can help current antibiotics function more effectively, even in multidrug-resistant strains of bacteria.

Many of the early studies examining the role of dimethyl sulfoxide (DMSO) in multidrug-resistant TB were done by Russian and Polish scientists in the mid-1970s. A summary of these studies details some of the properties observed when DMSO was used as an adjuvant to antibiotics against *M. tuberculosis*–resistant bacteria or by itself.

In one study, isoniazid-resistant TB bacteria were inoculated into guinea pigs that were divided into two groups: the first group underwent isoniazid treatment and the second group was treated with a single oral dose of DMSO and isoniazid 2 weeks after inoculation with tubercle bacilli.[5] When control animals died from the bacterial inoculation after 80 days, samples were taken from liver, lymph nodes, and spleens and cultured and examined microscopically.[5] All guinea pigs treated with DMSO–isoniazid survived beyond 1 year. Animals not treated with DMSO yielded tubercle bacilli in the collected organs and appeared entirely resistant to isoniazid treatment.[5] These findings indicated that reversion to isoniazid sensitivity by DMSO permitted a cure of the experimental animals infected with an isoniazid-resistant strain of tubercle bacilli, suggesting that this drug combination may be clinically useful in human antibiotic-resistant TB.

A subsequent study by the same investigators replicated the guinea pig experiments *in vitro* by isolating isoniazid- and rifampicin-resistant tubercle bacilli from patients with pulmonary TB.[6] This study consisted of cultures containing 61 strains of tubercle bacilli resistant to isoniazid and 19 strains resistant to rifampicin. DMSO 5% exposure to the cultures resulted in 19/61 strains resistant to isoniazid reverted to sensitivity and all 19 rifampicin-resistant strains also reverted to sensitivity after DMSO exposure.[6] The data showed that complete reversion of rifampicin sensitivity was obtained with DMSO and also reversion of some isoniazid-resistant tubercle bacilli strains indicated support for a clinical study of these antibiotics combined with DMSO.

DMSO has also been reported to stabilize rifampicin for periods of up to 8 months without any loss of potency by this antibiotic.[7]

Aside from its preservation of shelf life for many drugs, a common use for DMSO is its ability to act as a safe and versatile solvent for many antibiotics, including those used for antibiotic-resistant pathogens.[8] There are no data relating to DMSO acting or synergistically potentiating antibiotic activity when it is used as a solvent.

These findings created much interest among investigators interested in antibiotic-resistant therapy, and reports soon appeared in an attempt to further examine the ability of DMSO to restore sensitivity to chemotherapeutic compounds used to treat various resistant bacterial strains.

M. tuberculosis strains, resistant to rifampicin, isoniazid, streptomycin, paraminosalicylic acid, thioacetozone, ethambutol, and prothionamide, were grown in cultures containing these drugs together with either DMSO at 1% concentration or water.[9] Susceptibility tests were applied, and resensibilization to ethambutol, streptomycin, and thioacetozone was observed only in DMSO-treated cultures.[9] Moreover, many *M. tuberculosis* strains resistant to isoniazid and all strains resistant to rifampicin became sensitive to these drugs when combined with DMSO.[9]

The ribosomal subunit 30S is considered a site of action of streptomycin and also a site of resistance for certain strains of *Escherichia coli* and other bacteria. These mutant germs show an impermeability to streptomycin that confers resistance to the drug's action.[10] The use of DMSO to overcome *E. coli* resistance to streptomycin in cultures showed that these pathogens decreased their permeability and increased their uptake of streptomycin, thus allowing a killing effect by the antiobiotic.[11] The effect by DMSO on *E. coli* cell membrane with respect to enhanced streptomycin uptake can only be surmised.

Drug-resistant isolates of *Mycobacterium avium–Mycobacterium intracellulare* were isolated and cultured *in vitro* in the presence of single drugs alone and in combination with 2.5% DMSO.[12] Bacterial isolates were exposed to first-line antibiotics used for the treatment of TB, including rifampicin, streptomycin, ethambutol, and isoniazid, and all showed a degree of killing power that varied depending on the bacterial strain, but with the addition of DMSO to each antibiotic, a 26%–30% increase in inhibitory bacterial growth was observed.[12]

Thirty-two patients with destructive pulmonary TB were treated with streptomycin and penicillin in 10% and 25% solutions of DMSO through inhalation. Fourteen of the patients showed absence of *M. tuberculosis* excretion. Most patients showed improvement, including a decrease in manifestations of nonspecific endobronchitis, a decrease in the perifocal infiltration, and a decrease in the dimensions of destruction of lung tissue.[12a]

These early studies showing the ability of DMSO to enhance antibiotic sensitivity to drug-resistant pathogens partly introduced the concept of the adjuvant, defined as a compound that could increase the killing power of antibiotics and antimicrobial agents that by themselves failed to inhibit bacterial growth *in vitro* or *in vivo*.

No longer would *magic bullets* like penicillin create their magic on a growing list of superbugs; now, it was necessary to find adjuvants that could help fight germs and postpone the day when first-line antibiotics would no longer work.

The superbugs have devised ingenious ways to evolve in order to elude and weaken antibiotic drug action. The first of these is to increase their membrane permeability, thus preventing the antibiotic to accumulate to toxic levels within the cell. A second way is to make the drug less toxic using special enzymes to reduce drug potency, a third way is to inactivate the antibiotic, and a final way is to follow a metabolic pathway not targeted by the drug.[13]

One strategy to get around these neutralizing techniques used by superbugs on antibiotic monotherapy is to combine two to four different antibiotics in a shot-gun approach. However, the downside produced by multiple drug administration is serious side-effects and the real possibility of generating more multidrug-resistant bacteria, a therapeutic nightmare. Adjuvants to antibiotics may be more useful if they can either target the bacterial membrane, be more toxic to pathogens, or follow bacterial pathways that can be targeted by antibiotics. This is where DMSO, and similar agents, may become clinically useful, especially in drug-resistant TB.

Globally, an estimated 8.7 million people fell ill due to TB and 1.4 million died from this disease in 2011.

TB is second only to HIV/AIDS as the greatest killer worldwide due to a single infectious agent.

Thus, with almost 9 million new cases each year, TB remains one of the most feared diseases on the planet. Despite a reduction in incidence in several regions of the planet, TB remains a pandemic of worldwide significance. There is an urgent need to develop better regimens, to shorten treatment, and to effectively manage both drug-sensitive and drug-resistant disease.

TB is spread from person to person through the air. When people with lung TB cough, sneeze, or spit, they propel the TB germs into the air. A person needs to inhale only a few of these germs to become infected.

When a person develops active TB, the symptoms they develop, cough, fever, night sweats, weight loss, chest pains, and weakness, may be mild for many months. This mild beginning of TB can lead to delays in seeking care and may result in transmission of the *Mycobacterium* to others.

It is estimated that persons who develop TB can infect a dozen other people through close contact over the course of a year. Without proper and immediate treatment, up to two-thirds of people with TB will die.

Multidrug-resistant TB is a type of TB caused by mycobacteria that do not respond to isoniazid and rifampicin, the two most powerful, first-line, and standard anti-TB drugs. When bacteria fail to respond to first-line antibiotics, second-line antibiotics must be used.

As bad as drug-resistant TB can become, an even worse outcome is the development of more severe drug resistance. This severe resistance to antibiotics can lead to extensively drug-resistant TB, a form of multidrug-resistant TB that responds to even fewer available medicines, including the most effective second-line anti-TB drugs.

Multidrug-resistant *M. tuberculosis* and extensively drug-resistant TB are recognized as a one of the most serious challenges to global health with no solution in sight.[14]

A strategy to alter cell membrane permeability to overcome multidrug-resistant *M. tuberculosis* strains, using a combination of DMSO at varying doses and ethambutol, was explored by Jagannath and colleagues.[15] This group was aware of DMSO's ability to penetrate biological membranes[16] and reasoned that combining it with a membrane detergent such as ethambutol would enhance the delivery of antituberculous drugs at subminimum inhibitory concentrations.[15]

M. tuberculosis bacilli contain an unusual lipid-rich outer layer, containing mycolic acids, a unique characteristic among Gram-positive bacteria. This powerful lipid-rich layer in *M. tuberculosis* represents a formidable barrier for antibiotics to penetrate, although the synthesis of its cell wall has become a target of several effective antibiotics, such as isoniazid and ethambutol, which presumably interfere with cell wall mycolic acid production.

In the study by Jagannath and colleagues,[15] three multiple-drug resistant strains of *M. tuberculosis* resistant to rifampicin and ethambutol were suspended in broth and incubated for 3–5 days with ethambutol and 2%–5% DMSO. Bacilli were then washed to remove traces of ethambutol–DMSO and exposed to isoniazid, streptomycin, and rifampicin. It was found that combining DMSO with ethambutol resulted in a 16–64-fold increase for rifampicin inhibition of *M. tuberculosis* strains, a 16–33-fold increase for streptomycin, and a 4–16-fold increase of sensitivity to isoniazid.[15]

Moreover, DMSO or streptomycin by themselves showed no significant inhibition of *M. tuberculosis* when compared to untreated controls, but combining DMSO–streptomycin completely eliminated the cultured bacilli after 7 days incubation.[15] This study indicated that the action of DMSO combined with the effect of ethambutol had a significant effect on delaying or damaging cell wall biosynthesis in *M. tuberculosis* multidrug-resistant strains and may have altered the bacilli's cell wall permeability to enhance antibiotic delivery of rifampicin, isoniazid, and streptomycin (Figure 6.1).[15]

This experiment has not been replicated *in vivo*, despite its clinical implications in drug-resistant TB.

Another possibility that has not been explored as far as DMSO's adjuvant action to make multidrug-resistant *M. tuberculosis* bacilli more sensitive to antibiotic treatment is the effect of DMSO on inner membrane protein synthesis associated with energy metabolism in microorganisms such as *E. coli*.[17] All cells are bound by a cell membrane, but although the membranes of all cells are quite similar, those of bacteria differ from eukaryotic cells.

Bacterial cell membrane differences allow for selective action of antimicrobial agents. This little-studied action by DMSO on bacterial proteins responsible for energy metabolism would classify DMSO as a bacterial antimetabolite. Bacterial antimetabolites function in two ways: (1) by competitive inhibition of bacterial

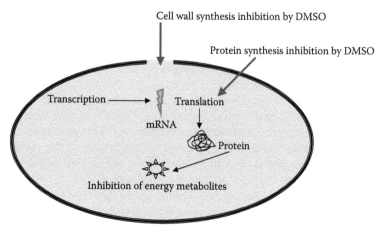

FIGURE 6.1 Potential inhibition of multidrug-resistant *M. tuberculosis* by DMSO when used as an antibiotic adjuvant, based on *in vitro* studies.[15,17,19–21] DMSO may act to delay or damage cell wall biosynthesis, thus allowing a greater concentration of first-line antibiotic penetration of the bacterial cell, a process that can alter transcription of mRNA and translation of protein synthesis impacting on the production of energy metabolites needed for bacterial cell growth and function.

enzymes and (2) by erroneous incorporation into bacterial nucleic acids, where proper base pairs cannot form during DNA replication and transcription (Figure 6.1).

It is unlikely that DMSO would act on bacterial cell membrane disruption by the second mechanism, since it lacks a structure and stereochemistry similar to purine and pyrimidine nucleotides found in microorganisms. This is not the case with some antibiotics, which have a similar chemical structure to purine and pyrimidines in microorganisms and are able to disrupt these pathogens by damaging their DNA.[18]

The mechanism involved in DMSO's ability to alter bacterial cell membrane remains unclear although it is known that DMSO affects the permeability of the cell membrane to facilitate the uptake of DNA,[19] and it is reported that DMSO destroys or damages the integrity of various bacterial membranes by increasing their membrane permeability.[20]

DMSO may act to facilitate transfection by opening transient pores or holes in the bacterial cell membrane, a process that would allow increased uptake of antibiotics or nucleic acids isolated from bacteriophages, which are harmful to bacteria and are otherwise kept from entry into the inner bacterial membrane (Figure 6.1).[19]

Recently, some support for this concept gained ground when it was reported that DMSO can induce water pores in biological membranes, which in turn promotes the transport of active substances across cell membranes.

Such membrane transport activity by DMSO could have significant clinical application for multidrug-resistant microorganisms.

In summary, DMSO has two important qualities that can facilitate the treatment of antibiotic-resistant organisms. The first is its potential as a powerful and safe solvent of many water-insoluble compounds; this ability can diminish the final toxicity of the antibiotic or antibacterial compound.

The second quality shown by DMSO is its ability to potentiate the penetration of antibacterial drugs by either *punching a hole* in the invading organism's cell wall,[15,21] for example, that of *M. tuberculosis*, to enhance antibiotic drug penetration and concentration inside the cell wall. Another mechanism by DMSO may involve its inhibition of the growth of pathogens through its blocking action on the pathogen's protein synthesis, which can impact the reduced production of the organism's energy metabolites and pathogenic potential.

REFERENCES

1. Hyman CL. Tuberculosis: A survey and review of current literature. *Curr Opin Pulm Med.* 1995;1(3):234–242.
2. Potter B, Rindfleisch K, Kraus CK. Management of active tuberculosis. *Am Fam Phys.* 2005;72(11):2225–2232.
3. Kalokhe AS, Shafiq M, Lee JC, Ray SM, Wang YF, Metchock B, Anderson AM, Nguyen ML. Multidrug-resistant tuberculosis drug susceptibility and molecular diagnostic testing. *Am J Med Sci.* 2013;345(2):143–148.
4. Field SK, Fisher D, Jarand JM, Cowie RL. New treatment options for multidrug-resistant tuberculosis. *Ther Adv Respir Dis.* 2012;6(5):255–268.
5. Szydlowska T, Pawlowska I. In vivo studies on reversion to sensitivity of INH-resistant tubercle bacilli under the influence of dimethylsulfoxide (DMSO). *Arch Immunol Ther Exp (Warsz).* 1974;22(4):559–561.
6. Szydłowska T, Pawłowska I. Comparative studies on the influence of dimethylsulfoxide (DMSO) on reversion to sensitivity to isonicotinic acid hydrazide (INH) and rifampicin (RMP) in resistant strains of tubercle bacilli. *Arch Immunol Ther Exp (Warsz).* 1976;24(4):575–577.
7. Karlson AG, Ulrich JA. Stability of rifampin in dimethylsulfoxide. *Appl Microbiol.* 1969;18(4):692–693.
8. Ollinger J, Bailey MA, Moraski GC, Casey A, Florio S, Alling T, Miller MJ, Parish T. A dual read-out assay to evaluate the potency of compounds active against *Mycobacterium tuberculosis*. *PLoS ONE.* 2013;8(4):e60531.
9. Müller U, Urbanczik R. Influence of dimethyl sulfoxide (DMSO) on restoring sensitivity of mycobacterial strains resistant to chemotherapeutic compounds. *J Antimicrob Chemother.* 1979;5(3):326–327.
10. Anand N, Davis BD, Armitage A. Uptake of streptomycin by *Escherichia coli. Nature.* 1960;185:23–24.
11. Feldman WE, Punch JD, Holden PC. In vivo and in vitro effects of dimethyl sulfoxide on streptomycin-sensitive and -resistant *Escherichia coli. Ann NY Acad Sci.* 1975;243:269–277.
12. Nash DR, Steingrube VA. In vitro drug sensitivity of *M. avium–intracellulare* complex in the presence and absence of dimethyl sulphoxide. *Microbios.* 1982;35(140):71–78.
12a. Liubinets VI, Kruk MV. Dimexide in the treatment of endobronchitis in patients with destructive forms of pulmonary tuberculosis. *Zh Ushn Nos Gorl Bolezn.* November–December 1969;29(6):68–71.
13. Alekshun MN, Levy SB. Molecular mechanisms of antibacterial multidrug resistance. *Cell.* 2007;128:1037–1050.

14. Worthington RJ, Melander C. Combination approaches to combat multidrug-resistant bacteria. *Trends Biotechnol.* 2013;31(3):177–184.

15. Jagannath C, Reddy VM, Gangadharam PR. Enhancement of drug susceptibility of multi-drug resistant strains of *Mycobacterium tuberculosis* by ethambutol and dimethyl sulphoxide. *J Antimicrob Chemother.* 1995;35(3):381–390.

16. Narula PN. The comparative penetrant-carrier action of dimethyl sulfoxide and ethyl alcohol in vivo. *Ann NY Acad Sci.* 1967;141(1):277–278.

17. Skotnicki ML, Rolfe BG. Effect of dimethyl sulphoxide on the expression of nitrogen fixation in bacteria. *Aust J Biol Sci.* 1977;30(1–2):141–153.

18. Harley JP, Klein DA, Prescott LM. *Microbiology.* Wm. C. Brown Publishers, Dubuque, IA, 1993.

19. Melkonyan H, Sorg C, Klempt M. Electroporation efficiency in mammalian cells is increased by dimethyl sulfoxide (DMSO). *Nucleic Acids Res.* 1996;24(21):4356–4357.

20. Murata Y, Watanabe T, Sato M, Momose Y, Nakahara T, Oka S, Iwahashi H. Dimethyl sulfoxide exposure facilitates phospholipid biosynthesis and cellular membrane proliferation in yeast cells. *J Biol Chem.* 2003;278(35):33185–33193.

21. He F, Liu W, Zheng S, Zhou L, Ye B, Qi Z. Ion transport through dimethyl sulfoxide (DMSO) induced transient water pores in cell membranes. *Mol Membr Biol.* 2012;29(3–4):107–113.

7 DMSO in Malignancy

DMSO AND NEOPLASIA

Neoplasia is characterized by populations of precursor cells with the ability to grow or divide abnormally. This growth of cells exceeds those of normal tissues around them, which can persist in the same excessive manner even when the promoting stimulus has dissipated.

The abnormal growth can create a tumor or mass that can be classified as benign, premalignant, or malignant and can also result in nontumor proliferation of immature white blood cells, such as those seen in leukemia.[1] The malignant neoplastic cells primarily result from damage to DNA, which originates from a variety of endogenous (inside the body) or exogenous (outside the body) sources.[2] Whatever the source of the stimulus, there is a deficiency in the DNA repair damage system, which can lead to DNA mutations and progression of the malignant process.[2]

An enhanced understanding in the past 50 years of the transformational process of neoplastic cancer cells that mediate their recruitment, activation, programming, and persistence presents interesting new targets for anticancer therapy.[3]

Cryoprotective agents such as dimethyl sulfoxide (DMSO) are reported to induce erythroid differentiation of Friend leukemic cells *in vitro*.[4] Since malignant cells are known to have the capacity to differentiate into benign cells,[5] chemical agents that can induce such cell differentiation are frequently associated with a decrease in, or a loss of, malignant potential.[5]

The complex problem of how to study and treat leukemia took a giant step forward in 1966, when a virus-induced erythroleukemic cell line in mice was established in the laboratories of Friend et al.[6]

FRIEND LEUKEMIA CELLS

The observations obtained in subsequent years using Friend's virus-induced erythroleukemic cell line suggested that it might be possible to treat malignant disease by inducing the differentiation of the tumor cells *in vivo*.[7,8] The Friend leukemia system provided an excellent model for testing this concept. Friend leukemia cells (FLCs) can be induced to differentiate *in vitro* by any one of a number of cryoprotective compounds, DMSO being the most notable of that group.[9] This FLC line has provided a mother lode of experiments in cancer research and is still in use today in the study of cell differentiation.

When FLC is inoculated intravenously into mice, malignant disease develops from leukemic cell infiltration of the spleen, bone marrow, liver, and lymph nodes.[9] When FLC is inoculated locally by subcutaneous injection into mice, the cells develop into a tumor, which has the characteristics of a myeloblastoma.[9] Most of these animals die when the tumor spreads to other tissues.

DMSO has been shown to be a prototypical inducer of cell differentiation whose activity was discovered when it was exposed to FLC, providing the first example of a nonphysiological chemical agent that possessed the ability to reverse the differentiation block imposed in leukemia cells.[10]

Studies in the 1970s indicated that induced differentiation found *in vitro* by differentiating agents, including DMSO, could have potential to the treatment of malignant disease *in vivo*. To prove this point, many experiments were designed using FLC and DMSO to see whether malignant cell differentiation could be turned benign.[10] Thus, the hunt for an ideal chemotherapeutic agent was begun. Such an agent would act selectively to program cancer cells to die, as suggested by Basile et al.[11]

For example, those early studies were able to show that both neuroblastoma cells and cells of certain leukemias can be made to mature when treated with certain naturally occurring substances, such as nerve growth factor.[12,13]

An experiment using undifferentiated proerythroblasts obtained from a Friend erythroleukemic cell line was used to test DMSO differentiation *in vitro*. In the absence of DMSO, some gigantic cells and many undifferentiated cells were seen in culture. With the addition of DMSO to cultures, striking morphological changes were seen as cells became smaller, the nuclear–cytoplasmic ratio and nucleoli reduced, and the formation of many differentiated normoblasts occurred, suggesting that cell maturation from DMSO exposure took place, preventing these cells from turning malignant.[14]

In another example, the potential by DMSO to modify leukemia cells was shown. Human acute myelogenous leukemia originates from pluripotent stem cells that transform into neoplastic cells.[15] A major phenotypic abnormality of these leukemia cells is their inability to differentiate to functional mature cells at the myeloblast or promyelocyte stage. Leukemic patients with this condition usually die of infection, because blast cells do not mature to their functional end cells, allowing the leukemic cells to rapidly multiply in the system.[15]

More recently, it has been found that Friend erythroleukemia cells differentiated by DMSO can produce hemoglobin and turn malignant cells into orthochromatic normoblasts.[16] DMSO-treated erythroleukemic cells undergo an irreversible pathway to maturation and, in that process, produce globin mRNA and accumulate large amounts of hemoglobin. Hemoglobin accumulation in FLC differentiated by DMSO is needed for the erythroid maturation process.[17] What is of interest is that treated cells express an ability to remember (so-called memory response) the previous exposure to the DMSO treatment and to continue their differentiation even after the DMSO exposure has been discontinued.[17] This terminal maturation process of erythroid cells and hemoglobin synthesis remains to be further clarified.

HL-60 HUMAN CELL LINE

The HL-60 human cell line was developed from peripheral blood leukocytes in a patient with acute promyelocytic leukemia, which are generally promyelocytes.[15] This HL-60 cell line has been used as a unique *in vitro* model to study various important aspects of human myeloid cell differentiation. When these promyelocyte cells underwent incubation for 5–7 days with appropriate concentrations of DMSO, they were able to differentiate into normal, peripheral blood granulocytes, despite their karyotypic abnormalities and leukemic origin.[18] These peripheral blood granulocytes were capable of phagocytizing bacteria and develop complement receptors.[18]

Treatment of human promyelocytic leukemia HL-60 cells with DMSO is known to cause differentiation of these cells along granulocytic/neutrophilic differentiation lineage.[19] After DMSO treatment, cultured granulocytic HL-60 cells eventually die via apoptosis. This process of apoptotic death induced by DMSO involves enhancing the effector proteins Bak and Bad while downregulating the apoptosis suppressor protein Bcl-2.[19] The ectopic expression of Bcl-2 protein is able to prevent apoptosis of granulocytes and monocyte/macrophage by a number of differentiating agents, but not when DMSO is used. Induction of apoptosis in macrophages by DMSO is of pathological significance because macrophages produce free radicals and inflammatory cytokines that can damage normal cells.[20]

METASTATIC LIVER DISEASE

Metastasis is the principal cause of death in cancer patients, and once it occurs, the most fundamental aspect of these cancer cells is their ability to sustain chronic growth. This growth differs from normal cell proliferation because the latter takes care to control the production and release of growth-promoting signals that ensure a homeostatic cell growth and cell number. Cancer cells have the unique ability to deregulate or block these signals and, essentially, do what they please.[21]

Metastatic liver disease is a malignant tumor in the liver that has spread from another organ affected by cancer. Metastatic liver disease is frequently associated with primary colorectal carcinoma. More than half of the patients with metastatic liver disease will die from metastatic complications. This disease is the sixth leading cancer worldwide and the second in cancer-related deaths in the United States.[22]

The reason the liver is a common site for cancer metastasis is due to its abundant blood supply, and since both lobes of the liver are usually involved, surgical removal is not possible unless a liver transplant is performed. Chemotherapy and a partial hepatectomy, especially in young people, is a possible treatment when only one lobe becomes cancerous. However, due to the rich blood supply of the liver, bleeding after surgery is a concern.

One procedure by Salim[23] using electrocoagulation + DMSO (500 mg by mouth) attempted to destroy liver metastases spread from advanced colonic cancer and reduce bleeding complications. A significant 5-year survival rate was obtained without major adverse effects in 78 patients with liver malignancy randomized to be treated with electrocoagulation + DMSO as compared to electrocoagulation only consisting of 77 patients.[23] The possible free radical–scavenging ability by DMSO was considered as a major factor in the long-term efficacy of this procedure.[23]

Tumor cells in culture generally grow in disarray unlike normal nontumorigenic cells. When these tumor cells are exposed to 1%–2% DMSO in culture, the random piling is drastically abolished within 3 days and instead, monolayers of cells in regular parallel orientation are seen to form, a pattern typical of nonmalignant fibroblasts.[24]

These morphological changes from malignant to nonmalignant cells in culture could be maintained indefinitely as long as DMSO was kept present.[24]

When DMSO was removed from the medium, reversal of these cells to the malignant state occurred after 3 days. It is possible that some of the properties that characterize DMSO, such as its stability, highly polar nature, and dielectric constant greater than that of water, which can enhance cell permeability, may explain, at least in part, the influence of this solvent on the hydration and solvation shells that surround the membrane components of the malignant cells and their ultimate transformation to benign cells.[24] These findings suggest that cancer is a disease of differentiation.[25]

However, the precise identification of the defect involved in the transformation of normal cells to malignant cells as well as the induction of cell differentiation by DMSO to normal cells is not easy to explain. Cancer cells seem to lose important control systems due to damage or loss of the genes inside the cells, a process known as mutation. Damage to the cell can be caused by a variety of factors. These factors, or carcinogens, can make the cell multiply abnormally to create cancer.

DMSO COMBINED WITH ANTICANCER AGENTS

A strain of spontaneous rat lymphatic leukemia inbred rats exposed continually to low, oral concentrations of DMSO produces physiological changes in these animals that increases the therapeutic effects of the antineoplastic agent cyclophosphamide while also lowering its threshold toxicity.[26] This positive finding in a rat leukemic model suggests that combining DMSO with antitumor compounds in the treatment of human cancer is both feasible and attarctive.[26] It also implies that the toxicity seen with many of the anticancer agents may be lowered when DMSO is used as an adjuvant.

This suggestion was supported by a sequential study from the same authors who reported that 12 anticancer agents administered to tumor-bearing rodents in the presence of 0.25% DMSO in the drinking water for over 1 year increased antineoplastic potency in 8 of these agents.[27] Several of the parenterally administered antineoplastic agents given with oral DMSO also inhibited the tumor growth rate in these animals.[27]

Several trials of DMSO on human cancer were attempted using cyclophosphamide and large doses of DMSO. It was reported that when 5%–10% DMSO was given orally or intravenously to 10 patients with a brain tumor, no enhancing effects of cyclophosphamide in the cerebrospinal fluid or plasma was seen.[28] In another study, 15 renal carcinoma patients were given thioTEPA combined with 2% DMSO by mouth, but no beneficial effects on the tumors were noted.[29]

Basal cell carcinoma is a malignant epithelial neoplasm of the skin, which remains the most common cutaneous malignancy in Caucasians, with as many as 2 million new cases expected to occur yearly in the United States.[30] Although basal cell carcinoma is not usually life threatening, it has become a medical problem of growing concern. Surgical excision remains the standard treatment for this tumor, but nonsurgical modalities also have achieved acceptable cure rates, particularly photodynamic therapy.[31]

The basis of photodynamic therapy depends on a photoactivating light, a photosensitizer such as 5-delta-aminolevulinic acid (5-ALA), tissue oxygen, and a target cell. Photosensitizers can be given systemically or locally on the skin. Topical photodynamic therapy is generally well tolerated, and the usual side effects following 5-ALA include erythema and pain at the site of application.[32]

Recent studies have shown that topical photodynamic therapy with 5-ALA and DMSO in 60 patients with basal cell carcinoma resulted in 81% of the patients becoming symptom-free 72 months after treatment with good cosmetic outcome.[33]

A similar study evaluated the long-term follow-up for 19 cases of Bowen's disease, a neoplastic skin disease also called squamous cell carcinoma, in which patients were given topical 5-ALA plus DMSO with a single exposure to 630 nm diode laser at differing energy dosages. At 60 months, 57% of the patients treated with 5-ALA and DMSO showed no histologic signs of squamous cell carcinoma.[34]

DMSO IN EXTRAVASATION

DMSO has been used to reduce extravasation of cytotoxic agents given to cancer patients intravenously. Extravasation or leakage of these agents from the blood into the surrounding tissue is a true medical emergency that requires immediate action to prevent ulceration or tissue necrosis. The only licensed drug that is used in the United States to reverse extravasation of anthracycline, a class of drugs used in cancer chemotherapy, is dexrazoxane,[35] a derivative of ethylene diamine tetraacetic acid (EDTA), which chelates iron ions bound to anthracycline to lower the formation of superoxide radicals.[36]

However, cooling and topical DMSO over the area of extravasation has been shown effective in reducing small anthracycline extravasations.[37] The advantages of DMSO and cooling are the rapidity of preventing anthracycline tissue damage following extravasation and substantial cost-effective savings due to the high cost of a hospital maintaining dexrazoxane in its inventory.

Although brief reports and anecdotal accounts have often appeared attesting the usefulness of DMSO in the treatment of lymphomas, glioblastomas, melanomas, and prostate and colon cancer when combined with conventional cancer therapeutic agents, these reports remain unconfirmed by strategic clinical trials.

These controversial findings suggest that DMSO may have an important role in anticancer chemotherapy. However, further research is needed, particularly on a larger population that can be studied long term in randomized clinical trials.

CANCER AND RADIATION

Radiation is a well-established carcinogen for humans, and 10% of invasive cancers are related to radiation.[38] A study was conducted that used high linear energy transfer (LET) particles, which are used in radiotherapy, to irradiate a human bronchial epithelial cell line, a stepwise neoplastic transformation of these cells occurred together with a consistent downregulation of the tumor suppressor *TGFBI* gene.[39] The irradiated cells showed fragmentation and shrinking of mitochondria, which was accompanied by a reduction in mitochondrial functions. The reduced mitochondrial functions were seen as lowering of cytochrome C oxidase, superoxide production, as well as succinate dehydrogenase activities when compared with nonirradiated controls.[39] Cytochrome C oxidase is the last enzyme in the respiratory electron transport chain of mitochondria and is important in the synthesis of ATP, the energy fuel of all cells. DMSO exposure of the irradiated epithelial cells was able to control the superoxide production and the damage caused by this free radical on the mitochondria.[39]

A combination therapy using DMSO and sodium bicarbonate infusion was recently reported by Hoang et al.[40] DMSO and sodium bicarbonate were mixed into a solution and infused daily for 6 months to nine patients suffering from advanced nonresectable biliary tract carcinoma, a highly fatal malignancy that has a poor response to chemotherapy. The treatment was well tolerated by all nine patients, and no adverse effects were recorded. The combined DMSO–sodium bicarbonate infusion was seen to improve pain control, blood biochemical parameters, and quality of life for the patients. Notably, this treatment led to a 6-month period of progression-free survival for all nine patients in the trial, a finding the authors consider significant of further research and application for the palliative care and extended survival of patients with this condition.[40]

These findings add fuel to the conclusion that more research is needed using anti-cancer agents mixed in DMSO solutions, a procedure that may provide more effective penetration and effectiveness in cancer suppression.

Onyx EMBOLIZATION

Onyx is composed of ethylene–vinyl alcohol copolymer dissolved in DMSO and suspended with micronized tantalum powder to provide contrast for visualization under fluoroscopy. Onyx liquid is used as an embolic agent of the vascular endothelium for a number of conditions including carotid-cavernous fistulas and cerebral arteriovenous malformations (AVMs).[41,42] Embolization of these vessels may prevent lethal hemorrhage if they burst or leak.

Onyx has been used with radiosurgery where it was successful in 80% of patients treated for AVMs, showing safety and efficacy with a low rate (5%) of clinical complications.[43] Radiosurgery involves a single high-dose fraction of radiation directed stereotactically at an intracranial region of interest.[44,45]

The flow-directed catheters used to deliver the liquid Onyx for embolization can be coated with DMSO to prevent early polymerization within the catheter.[46] It has not been investigated whether DMSO may serve a secondary role during embolization in stabilizing AVMs besides being a vehicle for the copolymer.

Transvenous occlusion of the cavernous sinus in dural arteriovenous fistula using Onyx embolization has been tried on patients with some success.[47] Onyx with DMSO was injected through the superior pharyngeal branch of the ascending pharyngeal artery where small transosseous feeders to the fistula arising from the superior pharyngeal branch of the ascending pharyngeal artery could be occluded by this approach especially when more conventional approaches cannot be achieved.[47]

Arteriovenous fistula of the spinal dura is also amenable to Onyx endovascular embolization. Although vascular malformation of the spinal cord is a rare clinical condition, it occurs when congestion of the spinal cord medullary venous plexus results in edema and neuronal ischemia.[48] If not repaired, axial or radicular back pain, gait imbalance, weakness of the lower limbs, numbness, and bladder or bowel disturbances can result.[49]

Onyx has been tried in order to embolize the abnormal communicating vessels to the dura and relieve the spinal cord edema that develops with a relatively high rate of success.[49] However, the neuronal ischemia is not always improved in arteriovenous fistula of the spinal dura using Onyx, and it is only speculative whether a high dose of intravenous DMSO separate from the Onyx embolization could aid control of this condition.[50]

REFERENCES

1. Almeida CA, Barry SA. *Cancer: Basic Sciences and Clinical Aspects*. Wiley-Blackwell, Hoboken, NJ, 2010.
2. Hanahan D, Weinberg RA. The hallmarks of cancer. *Cell*. 2000;100(1):57–70.
3. Hanahan D, Coussens LM. Accessories to the crime: Functions of cells recruited to the tumor microenvironment. *Cancer Cell*. March 20, 2012;21(3):309–322.
4. Lyman GH, Papahadjopoulos D, Preisler HD. Phospholipid membrane stabilization by dimethylsulfoxide and other inducers of Friend leukemic cell differentiation. *Biochim Biophys Acta*. October 19, 1976;448(3):460–473.
5. Lehman JM, Speers WC, Swartzendruber DE, Pierce GB. Neoplastic differentiation: Characteristics of cell lines derived from a murine teratocarcinoma. *J Cell Physiol*. August 1974;84(1):13–27.
6. Friend C, Patuleia MC, De Harven E. Erythrocytic maturation in vitro of murine (Friend) virus-induced leukemic cells. *Natl Cancer Inst Monogr*. September 1966;22:505–522.
7. Pierce GB, Wallace C. Differentiation of malignant to benign cells. *Cancer Res*. 1971;31:127–136.
8. Affabris E, Jemma C, Federico M, Rossi GB. Modulation of erythroid differentiation induced by DMSO in Friend cells treated with interferon: A clonal analysis. *Ann Ist Super Sanita*. 1982;18(3):425–427.
9. Preisler HD, Lyman G. Differentiation of erythroleukemia cells in vitro: Properties of chemical inducers. *Cell Differ*. 1975;4:179–187.
10. Friend C, Scher W, Holland JG, Sato T. Hemoglobin synthesis in murine virus-induced leukemic cells in vitro: Stimulation of erythroid differentiation by dimethylsulfoxide. *Proc Natl Acad Sci USA*. 1971;68:378–382.
11. Basile DV, Wood HN, Braun AC. Programming of cells for death under experimental conditions: Relevance to the tumor problem. *Proc Natl Acad Sci USA*. November 1973;70(11):3055–3059.
12. Paran M, Sachs L, Barak Y, Resnitzky P. In vitro induction of granulocyte differentiation in hematopoietic cells from leukemic and non-leukemic patients. *Proc Natl Acad Sci USA*. 1970;67:1542–1549.
13. Goldstein MN, Plurad S. Drug-induced differentiation of human neuroblastoma: Transformation into ganglion cells with mitomycin-C. *Results Probl Cell Differ*. 1980;11:259–264.
14. Fioritoni G, Bertolini L, Torlontano G, Revoltella R. Cytochemical characteristics of leukopoietic differentiation in murine erythroleukemic (Friend) cells. *Cancer Res*. March 1980;40(3):866–872.
15. Collins SJ, Ruscetti FW, Gallagher RE, Gallo RC. Terminal differentiation of human promyelocytic leukemia cells induced by dimethyl sulfoxide and other polar compounds. *Proc Natl Acad Sci USA*. May 1978;75(5):2458–2462.

16. Tsiftsoglou AS, Wong W. Molecular and cellular mechanisms of leukemic hemopoietic cell differentiation: An analysis of the Friend system. *Anticancer Res.* January–February 1985;5(1):81–99.

17. Tsiftsoglou AS, Sartorelli AC. Dimethyl sulfoxide-induced differentiation of Friend erythroleukemia cells in the absence of cytokinesis. *Cancer Res.* 1979;39(10):4058–4063.

18. Collins SJ, Ruscetti FW, Gallagher RE, Gallo RC. Normal functional characteristics of cultured human promyelocytic leukemia cells (HL-60) after induction of differentiation by dimethylsulfoxide. *J Exp Med.* April 1, 1979;149(4):969–974.

19. Naumovski L, Cleary ML. Bcl2 inhibits apoptosis associated with terminal differentiation of HL-60 myeloid leukemia cells. *Blood.* April 15, 1994;83(8):2261–2267.

20. Nathan CF, Murray HW, Cohn ZA. The macrophage as an effector cell. *N Engl J Med.* September 11, 1980;303(11):622–626.

21. Witsch E, Sela M, Yarden Y. Roles for growth factors in cancer progression. *Physiology (Bethesda)* 2012;25:85–101.

22. Ananthakrishnan A, Gogineni V, Saeian K. Epidemiology of primary and secondary liver cancers. *Semin Intervent Radiol.* March 2006;23(1):47–63.

23. Salim AS. Scavengers of oxygen-derived free radicals prolong survival in advanced colonic cancer. A new approach. *Tumour Biol.* 1993;14(1):9–17.

24. Borenfreund E, Steinglass M, Korngold G, Bendich A. Effect of dimethyl sulfoxide and dimethylformamide on the growth and morphology of tumor cells. *Ann NY Acad Sci.* January 27, 1975;243:164–167.

25. Markert CL. Neoplasia: A disease of cell differentiation. *Cancer Res.* September 1968;28(9):1908–1914.

26. Warren J, Sacksteder MR, Jarosz H, Wasserman B, Andreotti PE. Potentiation of antineoplastic compounds by oral dimethyl sulfoxide in tumor-bearing rats. *Ann NY Acad Sci.* January 27, 1975;243:194–208.

27. Thuning CA, Fanshaw MS, Warren J. Mechanisms of the synergistic effect of oral dimethyl sulfoxide on antineoplastic therapy. *Ann NY Acad Sci.* 1983;411:150–160.

28. Egorin MJ, Kaplan RS, Salcman M, Aisner J, Colvin M, Wiernik PH, Bachur NR. Cyclophosphamide plasma and cerebrospinal fluid kinetics with and without dimethyl sulfoxide. *Clin Pharmacol Ther.* July 1982;32(1):122–128.

29. Aisner J, Wiernik PH. Thiotepa and dimethyl sulfoxide in the treatment of renal cell carcinoma. *Cancer Clin Trials.* 1978;Spring:23–25.

30. Roewert-Huber J, Lange-Asschenfeldt B, Stockfleth E, Kerl H. Epidemiology and aetiology of basal cell carcinoma. *Br J Dermatol.* December 2007;157(Suppl 2):47–51.

31. Kalka K, Merk H, Mukhtar H. Photodynamic therapy in dermatology. *J Am Acad Dermatol.* March 2000;42(3):389–413 (quiz 414–416).

32. Itkin AM, Gilchrest B. δ-Aminolevulinic acid and blue light photodynamic therapy for treatment of multiple basal cell carcinomas in two patients with nevoid basal cell carcinoma syndrome. *Dermatol Surg.* 2004;30:1054–1061.

33. Christensen E, Skogvoll E, Viset T, Waroe T, Sundstrom S. Photodynamic therapy with 5-aminolaevulinic acid, dimethylsulfoxide and curettage in basal cell carcinoma: A 6-year clinical and histological follow-up. *J Eur Acad Dermatol Venereol.* 2009;23(1):58–66.

34. Souza CS, Felicio LV, Ferreira J et al. Long-term follow-up of topical 5-aminolaevulinic acid photodynamic therapy diode laser single session for non-melanoma skin cancer. *Photodiagn Photodyn Ther.* 2009;6(3–4):207–213.

35. Jones RL. Utility of dexrazoxane for the reduction of anthracycline-induced cardiotoxicity. *Exp Rev Cardiovasc Ther.* 2008;6(10):1311–1317.

36. Kane RC, McGuinn WD, Dagher R, Justice R, Pazdur R. Dexrazoxane (Totect™): FDA review and approval for the treatment of accidental extravasation following intravenous anthracycline chemotherapy. *Oncologist.* 2008;13:445–450.

37. Conde-Estévez D, Mateu-de Antonio J. Treatment of anthracycline extravasations using dexrazoxane. *Clin Transl Oncol*. January 2014;16(1):11–17.
38. Anand P, Kunnumakkara AB, Kunnumakara AB, Sundaram C, Harikumar KB, Tharakan ST, Lai OS, Sung B, Aggarwal BB. Cancer is a preventable disease that requires major lifestyle changes. *Pharm Res*. 2008;25(9):2097–2116.
39. Hei TK. Mitochondrial damage and radiation carcinogenesis. *J Radiat Res*. March 1, 2014;55(Suppl 1):i17.
40. Hoang BX, Tran HQ, Vu UV, Pham QT, Shaw DG. Palliative treatment for advanced biliary adenocarcinomas with combination dimethyl sulfoxide-sodium bicarbonate infusion and S-adenosyl-L-methionine. *J Pain Palliat Care Pharmacother*. 2014;28(3):206–211.
41. Ayad M, Eskioglu E, Mericle RA. Onyx: A unique neuroembolic agent. *Exp Rev Med Dev*. November 2006;3(6):705–715.
42. Jalaly J, Dalfino J, Mousa SA. Onyx® in the management of cranial arteriovenous malformations. *Exp Rev Med Dev*. July 2013;10(4):453–459.
43. Pierot L, Kadziolka K, Litré F, Rousseaux P. Combined treatment of brain AVMs with use of Onyx embolization followed by radiosurgery. *AJNR Am J Neuroradiol*. July 2013;34(7):1395–1400.
44. Leksell L. The stereotaxic method and radiosurgery of the brain. *Acta Chir Scand*. 1951;102(4):316.
45. Dalyai R, Theofanis T, Starke RM et al. Stereotactic radiosurgery with neoadjuvant embolization of larger arteriovenous malformations: An institutional experience. *Biomed Res Int*. 2014;2014:306518.
46. Tapping CR, Dixon S, Little MW, Boardman P, Sharma RA, Anthony S. Liquid embolization of the gastroduodenal artery before selective internal radiotherapy (SIRT). *Clin Radiol*. August 2012;67(8):789–792.
47. Pero G, Quilici L, Piano M, Valvassori L, Boccardi E. Onyx embolization of dural arteriovenous fistulas of the cavernous sinus through the superior pharyngeal branch of the ascending pharyngeal artery. *BMJ Case Rep*. April 23, 2014;2014. pii: bcr2013011067.
48. Bradac GB, Daniele D, Riva A et al. Spinal dural arteriovenous fistulas: An underestimated cause of myelopathy. *Eur Neurol*. 1994;34:87–94.
49. Lv X, Li Y, Yang X, Jiang C, Wu Z. Endovascular embolization for symptomatic perimedullary AVF and intramedullary AVM: A series and a literature review. *Neuroradiology*. April 2012;54(4):349–359.
50. Jacob SW, de la Torre JC. Pharmacology of dimethyl sulfoxide in cardiac and CNS damage. *Pharmacol Rep*. 2009;61(2):225–235.

8 DMSO in Basic Neuroprotection

BRAIN TRAUMA OVERVIEW

The search for safe and effective agents that can reverse, arrest, or minimize the consequences of traumatic brain injury has become a major research priority throughout the world due to the rising epidemic of brain trauma resulting from falls, motor vehicle accidents, sports injuries, physical assaults, and military and civilian gunshot wounds.

The leading causes of traumatic brain injury in the general population are falls (35.2%), motor vehicle crashes (17.3%), blunt impact (i.e., being struck by or against a moving or stationary object) (16.5%), and assaults (10%).[1] These numbers do not include U.S. military personnel sustaining war-related injuries in Iraq and Afghanistan in the past decade. There were a total 253,330 traumatic brain injuries sustained by military personnel between 2000 and 2012; 194,561 were described as mild, 42,063 were moderate, and 6,476 were severe or penetrating wounds.[2]

The Centers for Disease Control (CDC) estimated that at least 2.4 million emergency department visits, hospitalizations, or deaths were related to traumatic brain injury in 2012.[3] Of these brain trauma cases, 52,000 deaths are recorded annually in the United States.[4] Approximately 75% of these brain injuries are mild concussions and do not require hospitalization or aggressive treatment.

Children and older adults are more likely to sustain a traumatic brain injury, usually from a fall, and it is one of the leading causes of death and disability for this segment of the population.[5] The direct and indirect economic costs of brain injury in the United States in 2010 were estimated at $76.5 billion annually.[6]

Moreover, about 3.2 million persons in the United States are living with life-long neurologic deficits from the consequences of head injuries.[7]

A workshop on head injuries, sponsored by the National Institute of Neurological Disorders and Stroke (NINDS), was held in May 2000 gathering some of the leading experts in the field of traumatic brain injury.[8] The group concluded that severe, closed head injury was a life-threatening injury that required immediate medical help and the need for the development of more effective drugs for the treatment of this trauma. It was also the consensus that none of previously tested treatments had been shown to significantly modify the outcome of severe, closed head injury. Additionally, the workshop members estimated that of the 50,000 patients hospitalized in the United States annually with severe, closed head injuries, 17,500 (33%) died irrespective of any treatment given even when the best head injury centers were considered.[8] This was a call to action to urgently find more effective and safe interventions that could reduce the high fatality and permanent disabling deficits following moderate-to-severe head injuries.

There was also a stark warning by the workshop members that current therapies exhibit side effects that are potentially harmful to the patient and that can compromise the outcome of trauma to the brain.[8]

Management of such head injuries presently relies on controlling high intracranial pressure (ICP) from progressive edema buildup, increasing a low cerebral perfusion, and correcting pulmonary distress. Unfortunately, no single agent is capable of addressing these consequences after a head injury and multiple therapy is generally required to control each problem separately.

The drugs given to treat and manage the head injury complications in the past have been osmotic agents such as urea, glycerol, and mannitol; for control of ICP, steroids, which was given for many years until it was found to be contraindicated for head injuries due to inefficacy and increased mortality; and furosemide, a powerful diuretic also later abandoned as a treatment to lower brain pressure and swelling. If the head-injured patient survived, phenobarbital was given for the first 3 weeks after admission,[9] a protocol that has been modified in favor of outcome. Presently, barbiturate coma is used temporarily, until the patient regains consciousness after which the barbiturate dose is reduced and withdrawn. The basis for using induced barbiturate coma is to reduce cerebral blood flow and metabolic demand in a patient who has sustained severe trauma to the brain.

However, controversy has risen as to the benefits of barbiturate coma since it does not appear to prevent brain damage, possibly because the reduction of high ICP is not sustained.[10]

A Cochrane analysis of randomized controlled trials of using barbiturate therapy and barbiturate coma showed no evidence that barbiturate therapy improved outcome in patients with acute severe head injury.[11] This analysis concluded that barbiturate therapy results in a fall in blood pressure in 25% of patients treated and this drug's promotion of a hypotensive response will impair lowering ICP and improving cerebral perfusion pressure (CPP).[11] When trying to monitor the level of consciousness in patients with head injury, barbiturates are impractical because they interfere with this clinical exam.

Although initially effective in the control of elevated ICP, the main problems with the hypertonic diuretics (urea, mannitol, and furosemide) were their side effects when used repeatedly and also their tendency to rebound after multiple doses and create higher ICP than that on admission.[12,13]

Hypothermia was suggested as a treatment for head injury. Although hypothermia can reduce ICP, the analysis of high-quality studies using therapeutic hypothermia showed an association with cerebrovascular disturbances, including a lowering of cerebral perfusion and an increase in morbidity on rewarming and possibly a risk of pneumonia in adult patients.[14]

It was said that although these drugs saved a few from head injury death, they could not be correlated with the recovery outcome.[9] Osmotic agents such as magnesium sulfate to lower ICP have had a time-honored place in neurosurgical practice and with the introduction, by Javid[15] in 1958, of intravenous urea to treat brain swelling, a new era of drugs with similar action leading to the search for better osmotics.

Although intravenous urea still remains one of the most reliable drugs to combat acute brain swelling, another osmotic agent, mannitol, has shown fewer complications in treating head injuries, but its action is half as short as that of urea (about 4 h), plus, it does not improve brain tissue pO_2 in severe head trauma and its effect is not as dramatic when brain edema is severe.[16]

A fundamental guide to emergency department management of head injuries at the present time is ensuring adequate CPP, adequate oxygenation, and preventing even brief or transient episodes of hypotension, hypoxia, seizures, hyperglycemia, and hypocapnia.[17] CPP is a function of ICP and systemic blood pressure, and it must be monitored and controlled at all times to prevent further structural and functional damager.[18]

When severe brain tissue trauma occurs, brain cells respond in a typical manner. The anatomic changes in neurons following trauma have been well studied and consist of cytoplasmic swelling secondary to sodium influx and a failure of membrane ionic regulation.[19] This is followed by characteristic cytoplasmic eosinophilia, severe loss of the endoplasmic reticulum and Golgi apparatus, chromatin clumping, nuclear shrinking with peripheral shifting, and disappearance of axonal microtubules, a phenomenon that generates synaptic dystrophy and neurotransmission failure.[20]

Depending on its location and severity, a focal traumatic insult to the brain will destabilize ischemic neurons by inducing hypoxia, glutamate excitotoxicity, a massive and disruptive intracellular rise of Ca^{2+}, mitochondrial energy loss, formation of oxygen and nitric oxide free radicals, and impairment of adenosine triphosphate (ATP) synthesis that fuel all brain cells.[21] The biochemical cascade associated with anatomic changes in ischemic neurons and glial cells is the product of irreversible oxygen–glucose deprivation resulting from the severe ischemia that leads inevitably to ATP production failure and a neurometabolic energy crisis in adjacent hypoperfused neurons.[22]

The reduced mortality and improved outcome for patients with severe traumatic brain injury in the past decade has been attributed to the partial but incomplete success of trying to squeeze oxygenated blood through a swollen brain. This plan requires quantification of cerebral perfusion by monitoring the ICP and by prompt treatment of cerebral hypoperfusion to lower the incidence of secondary injury. Although this plan can mean the difference between good survival and incapacitating, long-lasting neurological deficits or death, it is not generally possible to achieve due to the absence of an agent that can address cerebral hypoperfusion, brain edema, excitotoxic release of glutamate and cytokines, ionic cell shifts involving sodium and calcium channels, and free radical formation. Is there an agent that can address the pathologic cascade results from severe brain trauma and clinical success and improved outcome?

SECONDARY INJURY AND ISCHEMIC PENUMBRAL NEURONS

Because traumatic brain injury results in irreversible damage at the site of impact and initiates cellular and molecular processes that lead to secondary neural injury in the surrounding tissue, understanding the pathophysiology of secondary injury is critical to the development of effective therapy.

Adding to the cell and molecular damage that results at the site of trauma, secondary injury to the surrounding brain tissue also undergoes electrophysiological changes without the anatomic necrotic changes to neurons described earlier. The secondary injury primarily involves reduced cerebral perfusion, inflammatory changes, lowered cerebral energy metabolism, release of excitatory neurotransmitters, and oxidative stress.[23,24] These cell and molecular abnormalities not only add to the site of injury but can give rise to an area of the brain adjacent to the injury core where a ring of ischemic penumbral neurons can accumulate.

In his classic paper describing the ischemic penumbra, Astrup and his colleagues[25] recognized two stages where penumbral neurons can experience functional impairment and irreparable degenerative damage, usually with the former stage preceding the latter.

However, irreparable damage to penumbral neurons is not always the case, and it is clear that at least in embolic stroke, these neurons are salvageable with the proper treatment, for example, using intra-arterial thrombolytics such as the tissue plasminogen activators (tPAs). These clot busters are able to accelerate the conversion of plasminogen to plasmin, the major enzyme that catalyzes clot breakdown. However, one of the disadvantages of tPA therapy is that it must be administered as quickly as possible after an embolic stroke, usually within 4½ h after the onset of symptoms.[26]

This brief window of opportunity means that only 3% of persons qualify for treatment with tPAs since most patients wait beyond this time period to seek medical help after the onset of stroke symptoms. Another serious disadvantage of tPAs is the risk of hemorrhage in the brain, worsening the primary damage from stroke.

Due to the bleeding risk, tPA has not been assessed in any traumatic brain injury trial that we are aware of, but other agents have been used to address the disturbances in platelet function, leading to hyperaggregability following a head injury.[27]

Neurons in the ischemic penumbra are formed as a result of critically subnormal brain perfusion following a stroke or local brain trauma.[28] These penumbral neurons retain their structural integrity but lose their ability to function normally.[29] The penumbral region is therefore a functionally silent population of neurons. The physiological outcome of penumbral neurons is dependent on the degree and duration of ischemia.[30] It has been reported that an ischemic penumbra results when local cerebral blood flow dips between 12 and 18 mL/100 g tissue/min following brain trauma.[31] Below 12 mL/100 g tissue/min, the tissue becomes irreparably damaged and soon dies.[31]

The penumbral neurons are more likely to die if normal cerebral perfusion is not quickly restored. However, the grace time before penumbral neurons die from a reduced blood flow supply is variable and dependent on the type of injury involved, severity, region of the brain affected, and many other constants. Penumbral neurons have been reported to remain viable for hours, days, months, or even beyond a year.[32-34] During traumatic brain injury, astrocytes become reactive and increase proinflammatory cytokines to produce excitotoxicity through the excessive production of extracellular glutamate. This results in brain edema and swelling, a condition that promotes neuronal death from the activation of reactive oxygen species (ROS).

In addition, the activation of Janus kinase and signal transducer and activator of transcription signaling pathway is one of the most important transducing signals from the cell surface to the nucleus in response to cytokines, particularly the proinflammatory interleukin-6 cytokine family.[36] The Janus kinase and signal transducer and activator of transcription signaling pathway is the focus of much research at the present time due to studies that report its involvement in the pathophysiology of brain ischemia, brain trauma, and spinal cord injury (SCI).[37,38]

It is clear that when primary traumatic brain injury cannot be prevented, secondary injuries must be restricted with appropriate neuroprotective strategies. Nevertheless, present drug treatments are not generally able to do that consistently or effectively.[39]

The ability of pharmacotherapy to rescue salvageable neurons following brain trauma and arrest or reverse the quantitative damage that results from such injuries remains elusive and extremely limited.

A recent Cochrane Database of Systematic Reviews and Cochrane Injury Group conducted an exhaustive literature search analyzing drug therapeutic studies dealing with severe head injury. The search concluded that the majority of these clinical trials included only a modest number of patients largely due to the uncertainty of the trial drug effectiveness.[40] Not surprisingly, the quality of these clinical trials was regarded as more or less questionable.[40]

For this reason, there is an expectation that the effects of dimethyl sulfoxide (DMSO) used to treat brain trauma, stroke, Alzheimer's disease (AD), and SCI will receive the clinical attention it deserves. When brain trauma, stroke, AD, and SCI are combined, it represents a block of disorders that affect 8% of adult Americans and more than 15 million individuals in the United States alone.

The findings reviewed here will reveal a very unusual molecule possessing a number of biological actions that may explain its positive therapeutic effects following traumatic brain injury and other CNS insults.

DMSO IN EXPERIMENTAL BRAIN TRAUMA

The mechanism of action(s) exerted by DMSO in improving the outcome of traumatic brain injury remains unknown. However, DMSO is reported to display a range of properties, which may be useful in the management of the brain trauma patient or animal. For example, DMSO increases cerebral blood flow without altering blood pressure,[41,42] quickly reduces ICP without a rebound effect,[43–46] is a potent diuretic that does not affect cardiac rate or output,[41,42] reduces cerebral edema,[47,49] blocks sodium channel activation,[50,51] is a powerful free radical scavenger,[52,53] prevents glutamate excitotoxic neuron death,[54] lowers the Janus kinase and signal transducer and activator of transcription signaling pathway whose activation can be induced by excitotoxic cytokines following brain ischemia,[55] and suppresses NMDA-AMPA-induced (NMDA, N-methyl-D-aspartate; AMPA, alpha-amino-3-hydroxy-5-methylisoxazole-4-propionate) ion currents and excessive calcium influx known to damage and kill neurons.[54,56] Additionally, DMSO has a protective and stabilizing effect on cell membranes[57–59] and inhibits inflammation, a reaction that is generally associated with moderate-to-severe brain trauma.[60–62]

The biological actions listed earlier for DMSO are relevant to our review of the cellular–molecular events associated with CNS injuries, including traumatic brain injury, stroke, or neurodegenerative conditions since these events generally involve sodium channel activation, free radical formation, and glutamate excitotoxic signaling involving NMDA-AMPA-induced ion channels and calcium signaling in cell-death pathways.[50,51,54,61] These cellular–molecular actions modified by DMSO will be discussed in detail in this section of this chapter and in Chapter 9. Chapter 9 will detail how the basic research of DMSO in cerebral injuries has been successfully applied to humans sustaining a variety of insults to the brain.

It is important to point out that the cellular and subcellular biological actions of DMSO with the exception of lowering brain edema after trauma are not known to be associated with mannitol, a drug commonly used for the treatment of head injuries. This may be one of the reasons why DMSO treatment has been shown to be more efficacious and safer than mannitol when used in patients with severe, closed head injury.[44,45]

The initial studies with DMSO were first carried out in the early 1970s in the Division of Neurosurgery at the University of Chicago by a team of investigators led by Jack de la Torre. How DMSO was chosen to be tested against an animal model of head injury among dozens of other potentially worthwhile compounds is worth mentioning. At that time, about a dozen head injury centers existed in the United States funded by National Institutes of Health (NIH) and most of these centers were desperately seeking new agents to address the rising incidence of head trauma. These head injuries occurred from falls, especially in children and the elderly. Automobile accidents were also high on the list of head injury causes.

Since seat belts were not in common use in the early 1970s, about 50,000 deaths per year were recorded from acceleration–deceleration brain damage sustained mainly from falls and car crashes. Most of these deaths occurred from complications that followed injury to the brain, generally within several days following hospital admissions. The frustration from personnel working in the emergency room after admitting a severe head-injured patient, whether the patient was a child or an adult, was the inconsistent nature of the treatments available at the time, which sometimes worked, but more often did not.

At the research meeting held every other week in the Division of Neurosurgery at the University of Chicago in 1971, DMSO was proposed by a member of the faculty, Jack de la Torre, as a possible agent to test in experimental animals with severe head injury. When asked by Sean Mullan, the chief of neurosurgery, for the reasons behind this proposal, de la Torre pointed out DMSO's unique properties as a powerful diuretic, anti-inflammatory, and cell membrane protector from freezing damage, properties, he argued, that could potentially counteract cerebral edema buildup resulting from increased ICP. The research project was approved, and the first series of experiments involving traumatic brain injury in rhesus monkeys were scheduled soon after. A vivid account by Jack C. de la Torre described in a letter to a colleague what happened on the very first experiment following a single, acute intravenous dose of DMSO as the primary treatment for this devastating brain insult.

> I remember it was bitterly cold in Chicago that March morning of 1971. Tucked away
> on a corner of the second floor of Billings Hospital, my technician and I struggled with
> a myriad of cables, wires and electronic monitors set up to detect any changes in the
> brain-injured animal lying on the table. I was a newly appointed Assistant Professor

to the Division of Neurosurgery at the University of Chicago, the lowest professorial rank in academics. The academic trail for me would involve over 200 publications, six books and hundreds of international medical conferences where I would lecture about injuries to the brain and spinal cord. However, that March morning, only one thing was on my mind. Will the animal survive a severe brain trauma when given a new drug that had never been tested for that purpose? The odds were not good. We had already tested, as part of being one of seven Head Injury Centers in the U.S., dozens of worthless treatments reputed to benefit this usually lethal injury. And, we would keep on searching when…on that morning, "looky, looky, look at that…!!" my technician's eyes rolled off the animal as I finished my intravenous administration of the drug. She pointed excitedly to the monitoring charts. The charts were going crazy, instead of cardiac collapse, respiratory arrest, a flat EEG and sure death, the heart rhythm stabilized, breathing returned, at first in gigantic gasps, then in steady, normal, breathing pattern. The electroencephalogram, monitoring brain cell activity, returned in full force and blood flow to the brain, which had ceased in the final stages of the injury, began flowing again and reviving the almost dead brain. It was, as if the hand of God had somehow touched the animal's forehead. 'I don't believe it', I stammered. But it was true. I felt a tingling in my spine because this reawakening of a virtually dead animal had all the markings of a medical breakthrough. The drug was dimethyl sulfoxide or DMSO for short, used years earlier as a pain lotion and anti-inflammatory agent. What we would discover and publish about dimethyl sulfoxide in the next 8 years at the University of Chicago laboratories, would be pharmacological actions of a simple molecule that should have sent shock waves through the field of medical therapeutics as one of the most important drug discoveries of the century in treating devastating brain and spinal cord trauma. Instead, the discovery, the potential for saving lives and the continued research that should have uncovered other uses for dimethyl sulfoxide and similar agents was quietly laid to rest in the coffers of forgotten medicine. It was a baffling paradox that defied a reasonable explanation and to this day still remains unclear to me.

The model to simulate acute head injury in rhesus monkeys employed in some head injury centers' research laboratories consisted of an intracranial balloon inserted extradurally via a cranial burr hole near the temporoparietal region to produce progressive brain compression and increased ICP. This injury simulated and provoked an expanding extradural hematoma, common in acceleration–deceleration injuries from motor vehicle accidents. This type of slowly evolving compression injury is also commonly seen in falls, aggression, and sports-related injuries. Once in place in the extradural space, the balloon was slowly expanded with 0.2 mL water increments until respiratory arrest was reached. One hour after each treatment, the balloon was decompressed.[63] We were later to discover that no other agent tested for head injury in our laboratories, including steroids and hypertonic agents, had the dramatic effects seen with DMSO.

In the initial experiments by de la Torre,[63] 40 monkeys were divided into three groups (Table 8.1). Group 1 received 1 g/kg urea; Group 2, 2 g/kg of a 50% solution of DMSO; and Group 3, physiological saline at the same volume as the experimental treatments. All drugs were given intravenously at the rate of 8 mL/min. The results using DMSO or urea on brain swelling secondary to an expanding extradural lesion in monkeys showed that 10 of 15 urea-treated monkeys survived (33% mortality) and

TABLE 8.1
Epidural Balloon Compression

Treatment	Number	Survivors	% Mortality
Saline (untreated)	10	0	100
Urea	15	10	33
DMSO	15	14	7

Notes: Results of epidural balloon compression simulating epidural hematoma in rhesus monkeys. Mortality refers to 24 h survival following trauma. Neurological deficits were seen in 4 urea- and 1 DMSO-treated animals. All other survivors recovered uneventfully over 2 weeks.

the examination of these animals after 24 h showed 4 survivors sustained right-side paresis, uneven pupil size (anisocoria), and general lethargy.[63]

Of the DMSO-treated monkeys, 14 survived (7% mortality) and 1 survivor showed mild right-side paresis only. In the monkeys treated with physiological saline, 9 of 10 animals died (90% mortality) within an hour after injury, a lone survivor showed severe neurological deficits and died the following day.[63] Respiration, systemic arterial pressure, carotid blood flow, ICP, and electroencephalography (EEG) were significantly modified during the endpoint of brain compression but returned to baseline in the surviving monkeys that were free of deficits upon recovery.

On the average, cortical blood flow as seen through a pial window installed in every animal prior to injury returned to normal 23 min after treatment with DMSO and 45 min after treatment with urea. Copious volumes of urine were excreted from all monkeys within an hour after the administration of DMSO and considerably less after urea treatment.[63,64]

All neurological deficits seen in the monkeys treated with either urea or DMSO were temporary, and recovery was observed from 1 to 5 weeks after the experimental head injury.[63,64] This remarkable study in nonhuman primates indicated that a promising new agent, DMSO, was significantly more effective than urea, the clinical drug of choice at the time and used even presently for severe, closed head injury[63,64] (Table 8.1). As we will review here and in Chapter 9, further studies confirmed and extended these observations that a safe and highly effective agent called DMSO had been found that could manage severe trauma to the brain.

As will be seen and fully discussed in Chapter 9, the results obtained with DMSO treatment of rhesus monkeys subjected to brain compression resulting in consequential brain swelling were applied to humans sustaining severe closed head injuries in several pilot clinical trials. The findings from these trials clearly indicated that lives were spared and neurological deficits minimized in many patients treated with DMSO. Thus, the experimental brain trauma induced in the experimental rhesus monkeys was a price that paid off, considering the potential of saving millions of lives who have sustained a severe, closed head injury, using DMSO when it is finally approved by regulatory agencies.

Handling and storing pure 100% DMSO for intravenous injections in animals allowed some useful recommendations to be made for future work.[64] One observation related to the arterial catheters and needle hubs used during experiments. For example, arterial catheters made of polycarbonate, polystyrene, styrene acrylonitrile, and polyvinyl chloride needed to be avoided because DMSO could partly dissolve or chemically leach these products. DMSO was found safe to use with conventional or linear polyethylene, polyvinyl chloride, polypropylene, and teflon catheter systems, and products using these plastics also included hubs in disposable needles.[64]

One question that arose from these studies was the potential of DMSO to dissolve plastics. In the United States, most intravenous bags and tubing are made of polyvinyl chloride, which are safe to use with DMSO.

The preliminary reports on DMSO published by the University of Chicago neurosurgery team were sufficiently encouraging to try and extend those findings by targeting other CNS injuries, as we will review in the following text. These findings also stimulated other independent groups to examine DMSO's potential in other types of brain injuries.

When a drug is experimentally tested against a primary drug of choice for a clinical condition and is reported to be superior to the standard treatment, a flurry of extracurricular research studies by other investigators usually follows to verify or not the merits of the new drug. This flurry of research studies on DMSO by other investigators exploring possible brain injury treatments was soon to confirm the data reported by the University of Chicago investigators.

First, a three-part study from the University of California, San Diego, was undertaken to test DMSO in the control of lethal ICP and subsequent brain edema using a rabbit model.[42] All animals were monitored for EEG activity, ICP, central venous pressure, systemic arterial pressure, and water content in the brain. Animals had a circular area of the parietal cortex frozen for 90 s using a probe previously cooled in liquid nitrogen. This lesion created a consistent localized area of necrosis and hemorrhage at the surface and a few millimeters below the cortex. When animals were injected with Evans blue, the dye was observed to extravasate just beyond the lesion site only in the cryogenic lesion group, allowing the measurement of the dye spread following each treatemnt.[42] It was found that EEG activity after 1 g/kg of a 10% DMSO solution given 24 h after the cryogenic lesion was an improvement in voltage and the number of slow waves in five of six animals treated when compared to untreated lesioned animals. The ICP was significantly reduced 5 min after DMSO with no changes seen in the central venous pressure, which rose only in untreated lesioned animals.

Brain water content was significantly reduced after DMSO in both hemispheres and was increased in the untreated cryogenic lesioned animals. This study concluded that after a cryogenic lesion, DMSO stabilized systemic arterial pressure, water content in the brain, and central venous pressure and reduced ICP while increasing cerebral perfusion when compared to untreated lesioned animals.[42]

This cryogenic lesion study led to a second study by the San Diego group to further test DMSO's ability to reduce toxic brain edema at variable doses that could be effective and still maintain electrolyte and fluid balance.[43] The same rabbit model and cryogenic lesion was used, and the findings suggested that a 1 g/kg of a 10% DMSO solution given as an intravenous bolus was as effective in reducing ICP as a 1 g/kg of a 40% DMSO bolus, a dose that could create red cell hemolysis.

These authors found that while a 10% bolus of DMSO would avoid red cell hemolysis, it would also increase central venous pressure and fluid overload. However, the authors determined that the central venous pressure increase and fluid overload side effects could be avoided when 2 g/kg of a 20% DMSO infusion was given continuously over 1 h. This approach seemed as effective as the 10% bolus injections.[65] Since the study was terminated 1 h following all treatments, it was hard to judge whether a bolus or slow infusion was preferable in a patient developing a rapid extradural hematoma following a closed head injury. The optimal dose of intravenous DMSO would become a controversial stumbling block in subsequent clinical and experimental studies examining how best to treat a brain injury.

This impasse of DMSO given as an intravenous bolus or slow infusion at an optimal concentration was finally solved, as we shall see in Chapter 9, in two clinical pilot trials performed by Turkish investigators treating severe closed head injury patients using DMSO.[44,45]

The third study carried out by the San Diego researchers[43a] focused on extending the infusion time of DMSO from 1 to 3 h using the same cryogenic rabbit model of brain trauma used in the first and second studies. This experiment examined whether DMSO at a dose of 2 g/kg 20% infusion given 24 h after the cryogenic lesion would still lower ICP and reduce other side effects when given over 3 h rather than over 1 h as reported in the second study.

The results indicated that DMSO infused over a 3 h period was able to significantly lower ICP from 12.9 mm Hg before treatment to 7.3 torr after 45 min and to 4.2 mm torr after 3 h. Systemic arterial pressure and central venous pressure remained unchanged during and after DMSO infusion.[43a] Brain electrolyte levels of sodium and potassium showed a reduction of both electrolytes in the lesioned and unlesioned contralateral hemispheres after DMSO, which was similar to untreated lesioned animals.

The San Diego studies determined an optimal dose of DMSO at 2 g/kg in a 20% solution given over a 3 h period as a more optimal approach to lessen DMSO possible side effects such as volume overload, a rise of potassium brain levels, and red blood cell (RBC) hemolysis.[65] We now know that the optimal dose of DMSO in human patients sustaining a severe brain injury is 1 g/kg in a 28% solution mixed with 5% dextrose in water.[44,45]

It should be noted that the gross hemolysis seen after a rapid bolus of high intravenous doses of DMSO has been shown to be a temporary side effect that does not involve organ damage, including the kidneys.[66] Hemolysis after high rapid doses of DMSO is a reaction that always disappears after the discontinuation of DMSO. This was demonstrated in a pathologic study monitoring blood chemistry, hematology, urine, ocular, neurological, and cardiovascular systems in rhesus monkeys given repeated high doses of DMSO (3 g/kg 40%) for 9 days and followed up for 4 months.[65] At the end of the study, all animals underwent postmortem analyses, including gross and microscopic pathological examinations. No significant or long-lasting changes were recorded in any of the parameters studied when these data were compared to those animals given normal physiological saline.[66]

The three-part study by the San Diego group also showed that DMSO could significantly reduce ICP without affecting systemic pressure or central venous pressure while lowering the brain water content in the lesioned hemisphere. These studies also suggested that a slow infusion of 2 g/kg of a 20% DMSO solution over a 3 h period was as effective in lowering ICP as 1 g/kg of a 40% DMSO bolus but without the gross hemolysis seen with the bolus.[42–44]

Although this conclusion is probably valid for a cryogenic lesion to the cortex where local necrosis and hemorrhage develop immediately and remain a nonexpanding lesion, it is not the case with a relatively slower-building epidural hematoma (also called extradural hematoma) that results from a severe blunt trauma to the skull and meninges.

An extradural or epidural hematoma occurs when there is a rupture of a blood vessel, usually an artery, which then bleeds into the space between the dura mater and the inner wall of the skull. An epidural hematoma is generally an expanding mass that can reach its peak in 4–8 h and is capable of stripping the dura apart and threatening the formation of malignant brain edema as well as inducing secondary brain damage and death. This means that a brain trauma patient's arrival at the emergency department requires quick evaluation and if an expanding epidural hematoma is found, rapid surgical evacuation is mandatory. Drug treatment is generally considered when refractory high ICP is detected.[67,68] This implies that if brain edema is present, rapid intravenous infusion (fast drip or bolus) of an antiedema drug should be applied to control rising ICP and failing cerebral perfusion.[66]

The cryogenic lesion model was again used by the San Diego researchers who compared the acute effect of DMSO and pentobarbital, an agent used to reduce ICP often by creating barbiturate coma, on the development of brain edema in rabbits.[69] DMSO was given 1 mg/kg in a 10% solution by intravenous bolus, and pentobarbital 40 mg/kg was administered intravenously over a 30 min infusion. It was found that ICP was quickly lowered by both DMSO and pentobarbital with central venous pressure unchanged. Where the treatments differed was in the lowering of systemic arterial pressure by pentobarbital but not by DMSO, an outcome that likely produced a reduction in cerebral perfusion by the barbiturate.

To find out the physiological basis for DMSO's action in lowering ICP and reducing brain edema after a cryogenic brain lesion in rabbits, cerebral blood flow was measured before and after 2 g/kg/h DMSO 20% infusion for 2 h or 1 g/kg DMSO bolus preceded by a bolus of 20 mg indomethacin, a nonsteroidal anti-inflammatory agent and a nonselective prostaglandin (PG) synthase inhibitor.[65] The results suggested that following the DMSO bolus injection in the absence of indomethacin, an immediate rise in cerebral blood flow was followed by a decrease 30 min later despite a constant lowering of ICP.

These rise and paradoxical dip of cerebral blood flow were not observed when DMSO was given as a slow infusion where ICP was still lowered and cerebral blood flow rose consistently throughout the infusion. When indomethacin was given 15 min prior to DMSO infusion, cerebral blood flow failed to increase immediately until 60 min had passed, an action not observed when indomethacin was administered prior to a DMSO bolus. The authors concluded that the action of reducing ICP immediately after DMSO infusion may have been due to the ability of DMSO to extract brain water rapidly and lower brain pressure as a result. The decrease of cerebral blood flow induced for 60 min after a bolus of DMSO in animals receiving indomethacin implies that ICP was quickly lowered and was not a factor in affecting cerebral blood flow levels.[65]

Another explanation may be that indomethacin's inhibition of the cyclooxygenase pathway controlling the synthesis of PGs may temporarily block DMSO's ability to stimulate vasodilating PGs such as E_1 or I_2 in cerebral endothelium to increase cerebral blood flow.[70–72]

FREE RADICALS IN BRAIN INJURY

Free radicals are atoms that contain an unpaired electron in its outer orbit, making this atom highly reactive. This highly reactive process takes place by the free radical stealing an electron from a healthy molecule in order to become stable. Instead, the original free radical creates a new free radical, resulting in a chain reaction. Free radicals can be generated during the production of ATP in mitochondria.

These free radicals can leak from the mitochondria to form ROS, including superoxide anion and hydroxyl radicals. Free radicals use a number of biological pathways to create cell damage. For example, free radicals attack the nitrogenous bases of cell DNA where damaging mutations can result and prevent oxidative phosphorylation from taking place, thus preventing the cell from making ATP to fuel its energy metabolism. Repair of cell damage in the presence of decreased energy production following mitochondrial dysfunction aggravates cell damage by promoting secondary damage.[73]

Secondary injury following brain trauma in humans or animals is characterized by a host of pathophysiologic cascades that can additionally damage more brain mitochondria. These cascades include exposure of neurons to excitotoxic levels of excitatory neurotransmitters with intracellular calcium influx, of ROS, of production of peptides that participate in programmed cell death, and of glutamate generation, a primary mediator of mitochondrial dysfunction and delayed cell death.[74]

Excessive free radical production can react with cell membranes, lipids, and nucleic acids in a site-specific fashion that can unbalance proteins in healthy cells and damage or kill these cells.[75]

Although free radicals perform some beneficial functions within the immune system, their excess production can cause deadly toxic oxidation or oxidative stress.[76] The brain is particularly vulnerable to oxidative stress following injury, and hence, much research has focused on how to arrest or reverse free radical damage after a traumatic or ischemic brain injury.

The role of toxic free radicals in brain trauma has been the object of numerous studies in the past, and many therapies have been introduced to neutralize the primary and potentially secondary damages caused by these reactive molecules.

However, therapeutic strategies to control secondary damage following brain trauma by focusing on blocking free radicals that generate oxidative products remain ineffective.[77]

Nevertheless, studies carried out in preclinical models of traumatic brain injury using DMSO have provided information that free radicals appear to be intimately involved in producing devastating brain cell damage and that blocking their chain reaction can lead to preserving brain tissue from further damage.

A study by Ikeda and his colleagues,[78] from Johns Hopkins University, looked at the effects of DMSO and other hydroxyl radical scavengers such as dimethylthiourea and deferoxamine in cats. A standard cortical freezing method for the production of vasogenic brain edema was used. The authors found a reduction in brain edema that correlated with a reduction of hydroxyl radicals in animals treated with DMSO and deferoxamine.[78] These results could be partly explained by the scavenging properties of DMSO when chain-propagating radicals such as the superoxide anion are present.[79,80]

Superoxide has been associated with the pathophysiology of many diseases including brain trauma, ischemic stroke, and inflammatory reactions. Superoxide can leak from brain mitochondria in response to these insults and convert to hydrogen peroxide, an important contributor to oxidative damage.[81]

One study that supports the antioxidant-related damage caused by some molecules that commonly develop after trauma showed DMSO's antioxidant properties in rat brain homogenates were able to reduce both lipid peroxidation and protein oxidation induced by ferrous chloride/hydrogen peroxide.[82]

The brain is a highly oxidative organ consuming 20% of the body's oxygen even though it represents only 2% of the total body weight. For this reason, oxidative activity involving ischemia or hypoxia resulting from brain trauma or stroke can have devastating effects on brain function if left untreated or when treatment is delayed.[83]

Oxidative stress from ROS has been associated with the onset of systemic inflammation, and it is generally recognized that increased oxidative activity plays an important role in pathological process of brain tissue damage.[84] The presence of excessive free radical, ROS, and release of local inflammatory mediators, including cytokines, are a characteristic initiator of pathologic cascades following traumatic brain injury.[85,86]

These inflammatory cytokines include inteleukin-6, nitric oxide, and tumor necrosis factor-alpha (TNF-α). They can lead to complications affecting immunosuppression and result in severe, systemic inflammation with multiorgan damage, causing death within days or weeks following brain trauma.[87] As discussed earlier (see also the section "Anti-Inflammatory" in Chapter 2), DMSO has been shown to significantly decrease the levels of nitric oxide, IL-6, and TNF-α in injured rats,[21,36,37] as well as the oxidative stress marker malondialdehyde (see the section "Burns and Scar Tissue" in Chapter 2).

NF-κB (nuclear factor kappa-light-chain-enhancer of activated B cells) is a protein complex that controls the transcription of DNA and modulates a large number of normal cellular processes, such as immune and inflammatory responses. DMSO has been shown to suppress the effects of activation of NF-κB immunoreactivity, which is exerted by decreasing the production of TNF-α and inhibiting the transcription of critically important adhesion molecules.[88]

Finding a neuroprotective agent that can reduce systemic inflammation, cytokine production, and free radicals after brain trauma is the task of numerous investigators that continues to this day. As seen in Figure 8.1, DMSO appears to fulfill many of these requirements based on the research of this compound dating back to the early 1970s. This research will be reviewed in this section of this chapter beginning with a primary understanding of how cerebral blood flow behaves normally and after brain injury.

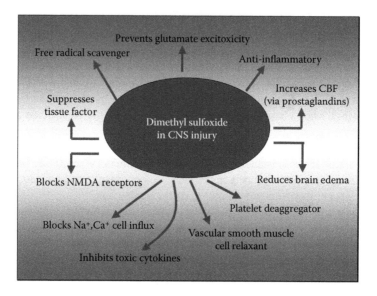

FIGURE 8.1 Reported biological activity for DMSO in central nervous system (CNS) injury. Most of the molecular and cellular factors listed in this illustration result from physical and physiological damages to the brain, but the chronology and the degree of damage caused by many of these factors on brain tissue have not been entirely clarified.

CEREBRAL HEMODYNAMIC FUNCTION IN BRAIN INJURY

Traumatic brain injury is known to impair cerebral hemodynamic function. Understanding and correcting hemodynamic dysfunction has become an important therapeutic target in arresting or reversing secondary damage resulting from brain trauma, ischemia, or neurodegeneration from worsening or becoming irreversible.

Restoring cerebral blood flow to preinjury levels is a key therapeutic target because when trauma, ischemia, or neurodegenerative damage occurs, nutrient delivery to the brain, including oxygen and glucose, which are the primordial molecules necessary for maintaining normal brain cell function, can fail to supply brain cells' demand to carry out all metabolic activities.

Since most of the energy of the brain is obtained exclusively from aerobic metabolic processes, a low supply and high demand for energy nutrients in the presence of trauma not only creates a neuronal energy crisis from the uncoupling of cerebral blood flow and metabolism but also ensures that motor, sensory, and cognitive functions remain impaired or irreparably lost. Additionally, the brain cannot manufacture or store energy nutrients such as glucose and is totally dependent on optimal cerebral blood flow to deliver these nutrients as needed.[89]

We will see in the "Cerebral Hemodynamic Function in Brain Injury" section how brain trauma, stroke, and neurodegenerative conditions such as vascular dementia (VaD) and AD can uncouple cerebral blood flow and metabolism by involving a multifactorial list of factors that can act independently or as part of a pathologic cascade to alter blood flow to the brain. This analysis is relevant in attempting to understand the role DMSO plays when used therapeutically to improve the outcome following primary injuries to the brain.

The most common factors influencing cerebral blood flow are cardiac output, ICP, blood gas content, vasoactive substances such as nitric oxide, PGs, endothelins, thromboxane A_2, endothelial-derived hyperpolarizing factor, adenosine, and K^+ and H^+ ions.[34]

CPP is the driving force that moves blood flow through the cerebral circulation. Cerebral blood flow varies directly with CPP, which is defined as the difference between mean arterial and ICPs, and inversely with cerebrovascular resistance. Normally, CPP can be considered equal to mean arterial pressure. Immediately after brain trauma, two compensatory mechanisms to counter low CPP from vessel stenosis or vessel obstruction (e.g., platelet aggregation) can be activated: autoregulation and oxygen extraction fraction. As CPP falls, cerebral blood flow is maintained by the vasodilation of resistance arterioles, which results as reflex autoregulation is activated.[90]

If cerebral blood flow falls beyond the autoregulatory range, oxygen extraction factor kicks in order to maintain oxygen levels, but up to a point.[91] When CPP falls beyond the point where neither autoregulation nor increases in oxygen extraction fraction can compensate, ischemia resulting in an insufficient delivery of oxygen to meet metabolic demands will occur with almost certain damage to the brain.

Cerebrovascular resistance results mainly from the friction caused by the blood on the vessel wall and is determined by vessel length, which does not vary significantly in normal physiology, and from blood viscosity, which normally stays within a small physiological range. Blood viscosity, however, increases with blood loss, with ischemia, or with the inability of RBCs to deform in tight capillary lumens.[92] Blood viscosity depends on hematocrit, erythrocyte aggregability, and plasma viscosity.[93]

When blood viscosity increases, cerebral blood flow decreases, and inversely, when viscosity is decreased, cerebral blood flow is increased. This occurs because resistance arterioles regulate the amount of blood flow supplied to the brain by changing their diameters according to the viscosity and the friction of the blood passing through them, also called shear stress. Consequently, the most important aspect of resistance in regulating brain blood flow is the diameter of the vessel whose main activator is blood viscosity.

There is a vasomotor duality that produces either vasodilation or vasoconstriction of brain arterioles, and it is thought to occur from endogenous arteriolar generation of either 20-hydroxyeicosatetraenoic acid (20-HETE), a powerful vasoconstrictor derived from arachidonic acid (AA), or from conversion of AA to the vasodilator epoxyeicosatrienoic acid.[94]

Normal cerebral blood flow (CBF) in adult humans is about 60 mL/100 g/min and, regionally, about 70 mL/100 g/min in gray matter, and 20 mL/100 g/min in white matter. These values normally decrease with age by an estimated rate of about 0.5% per year.[95] Between the ages of 20 and 65, normal CBF generally declines about 15%–20%.[96,97] It is generally accepted that when CBF reaches 30 mL/100 g/min following brain injury, neurologic symptoms can appear and when CBF falls to 15–20 mL/100 g/min, electrical failure or irreversible neuronal damage can occur even within minutes. Neuroimaging studies show that vascular disturbances can reduce CBF to generate ischemic neurons (<12 mL/100 g/min), penumbral neurons

(12–22 mL/100 g/min), or hypoperfused neurons (>22 mL/100 g/min. Penumbral neurons are functionally impaired but, unlike ischemic neurons, lack structural damage, while hypoperfused neurons may or may not exhibit anatomic or functional changes for extended time periods.[27]

Penumbral and hypoperfused neurons therefore may be salvageable following brain trauma in the event that a safe and effective treatment can be found. Such a treatment should address most if not all of the following factors:

1. Inhibit platelet aggregability
2. Block vasoconstrictors including thromboxane A_2, endothelins, and 20-HETE
3. Suppress tissue factor to prevent thrombosis
4. Block Na^+ and Ca^+ intracellular influx
5. Block glutamatergic activation of NMDA receptors
6. Suppress inflammation
7. Inhibit free radical formation
8. Reduce brain edema
9. Stabilize cell membrane from damage
10. Inhibit toxic cytokines
11. Increase cerebral blood flow by (a) activating smooth muscle cells in brain to increase vasodilation and (b) stimulating PG vasodilators E_1 and prostacyclin (I_2)

These factors that are linked with injuries to the brain and spinal cord and the role DMSO plays in their activation are reviewed in the following text and summarized in Figure 8.1.

PROSTAGLANDINS

Concussive brain trauma is often accompanied by platelet aggregation and platelet dysfunction in pial arteries and arterioles due to coagulation disturbances caused by the sudden trauma.[98,99]

In brain trauma patients, early coagulopathy is common on hospital admission with the coagulopathic process beginning at the time of trauma.[100] As discussed in the section "Cerebral Hemodynamic Function in Brain Injury" in this chapter, blocked or obstructed brain vessels can lead to energy delivery disturbances whose outcome is brain cell damage or death. For this reason, traumatic brain injury has been the focus of treatments using antiplatelet aggregators such as prostacyclin, a potent vasodilator and platelet deaggregant.[101]

Although microcirculation is improved following prostacyclin treatment after brain trauma,[101] brain edema and cerebral blood flow do not necessarily improve. This limitation has discouraged the use of prostacyclin and other antiplatelet aggregators both in basic and clinical strategies in treating brain trauma.

DMSO has been consistently reported to be an effective platelet deaggregator[101–103] and, unlike prostacyclin, has been shown, following severe insults to the brain, to improve brain edema, cerebral blood flow, CPP, and electroencephalographic activity, while lowering ICP and cerebrovascular reactivity.[41,47–49]

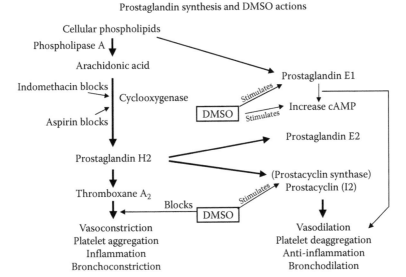

FIGURE 8.2 Simplified scheme of prostaglandin biosynthesis, which starts with the release of a fatty acid nearly always derived from the 2-position of a membrane phospholipid. Arachidonic acid is created from diacylglycerol via phospholipase-A_2 and then brought to either the cyclooxygenase pathway where either prostaglandins (PG) or thromboxane is formed from synthases that yield prostacyclin (PGI_2) and PGE_2 or thromboxane A_2. PGE_1, PGI_2, and PGE_2 are vasodilators, anti-inflammatory, and platelet deaggregators, while thromboxane A_2 acts as a vasoconstrictor, inflammatory, and platelet aggregator. DMSO is reported to block thromboxane A_2 and promote the release of PGE_1, PGI_2, and PGE_2 either directly or via cAMP stimulation.

The first step in the PG synthesis is the release of the substrate fatty acid, such as AA, from the cellular phospholipids, by the action of the enzyme phospholipase A_2 (Figure 8.2).

There are three ways that PGs can be elaborated. The monoenoic pathway forms PGE_1 and PGE_1-alpha from dihomo-y-linolenate. The bisenoic pathway leads to the elaboration of the endoperoxides (PGG_2 and PGH_2), which are the controlling agents in subpathways producing PGI_2, TXA_2, and PGE_2 (Figure 8.2).

It can be seen from Figure 8.2 that PGI_2 (prostacyclin) and, to a lesser extent, PGE_2 act as potent vasodilators and platelet deaggregators. The antiplatelet aggregation activity by PGI_2 and PGE_2 results from their increase of cAMP levels in platelets.[104,105]

The other products, especially TXA_2, are potent vasoconstrictors and platelet aggregators. There is a third trienoic pathway derived from eicosapentaenoic acid, which originates with the membrane-bound fatty acid, but little is known about the effects of its products, TXA_3 and PGI_3, on the cerebral vascular bed. However, it is of interest that the vascular-platelet action of PGI_3 is similar to that of PGI_2, although TXA_3 is not considered analogous to TXA_2 since the latter can increase the levels of cAMP in platelets, thus neutralizing their potential aggregation.[105,106]

Formation of PGE_2 can stimulate the levels of adenosine diphosphate (ADP), a mild platelet deaggregator and antagonist of noradrenaline production.[107]

Vascular constriction and platelet clumping can be antagonized if PGI_2 is released from endothelial cells in the microvasculature.[104] PGI_2 appears to do this by stimulating adenylate cyclase in many tissues including peripheral neurons leading to an increase in cAMP levels in platelets.[108] An increase in PGI_2 synthesis or availability at the vessel wall would have a positive and direct effect on cerebral ischemia by antagonizing platelet aggregation and vasoconstriction (Figure 8.2).

DMSO also has a direct effect on platelet cAMP levels. It increases cAMP presumably by inhibiting phosphodiesterase (PDE)[109,110] although an indirect action on PGI_2-induced elevation of platelet cAMP by DMSO can also be considered (Figure 8.2).

Research has revealed that of the 60 different platelet isoforms described in mammalian species, platelets are known to possess three main PDEs: PDE_2, PDE_3, and PDE_5.[111,112] Despite many studies carried out in the past four decades examining the role of DMSO in PG synthesis, it remains unclear which of these PDEs DMSO targets for the inhibition of platelet aggregation.

Platelet function is, to a large extent, dependent on PDEs, because their inhibition can dampen platelet aggregation by increasing cAMP and cyclic guanosine 3'-5'-monophosphate (cGMP), an action that limits the levels of intracellular nucleotides.[113]

Cyclic AMP is a powerful platelet deaggregator, and evidence exists that DMSO can increase the levels of cAMP in various biological preparations[108,110,114,115] (Figure 8.2). Increasing cAMP by DMSO suggests that DMSO can manipulate signal transduction mediated by second messenger molecules such as cAMP. This means that the inhibition of platelet signal transduction by DMSO may suppress platelet activation regardless of the initial stimulus.[111] The mechanism that would allow DMSO to directly inhibit platelet-activating second messengers like cAMP and cGMP is by its interaction with intracellular signalling pathways. In addition, DMSO may act indirectly on the endothelium-derived PGI_2, which is known to increase intracellular platelet cAMP through the activation of membrane adenylate cyclase, a process that begins when PGI_2 binds to its receptors on the platelet surface.[111]

The role of DMSO in PG biosynthesis remains partly theoretical, but a number of intriguing findings lend support to the specific actions by DMSO on PG activity and their catalytic enzymes (Figure 8.2).

For example, studies have reported that DMSO can increase the synthesis of PGE_1, a moderate vasodilator.[71,72] PGE_1 can reduce platelet aggregation by increasing cAMP levels, and this PG also inhibits the calcium-induced release of noradrenaline in nerve terminals, an effect that may antagonize vasoconstriction following tissue trauma with the resulting effect of preventing cerebral blood flow reduction[116] (Figure 8.2).

An increase in PGI_2 synthesis or availability at the vessel wall would have a positive and direct effect on ischemia by antagonizing platelet aggregation and vasoconstriction, thus allowing better blood perfusion of the hypoperfused tissue. DMSO may also antagonize the synthesis or release of thromboxane A_2[117,118] and thus indirectly counteract platelet aggregation and vasoconstriction. The presence of TXA_2 in blood will cause strong vasoconstriction and platelet aggregation, while the formation of $PGF_2\alpha$ will also negatively affect the vascular and platelet systems, but the reaction will not be as severe as with TXA_2.[111]

Theoretical considerations to explain platelet deaggregation by DMSO appears to involve PG synthesis in the brain.[118]

One study has shown evidence that DMSO has the ability to block the prothrombotic release of thromboxane A_2.[119] Although the exact mechanism of DMSO's blocking action of thromboxane A_2 to reduce platelet aggregation has not been clarified, it may possibly mimic the action of hydralazine or dipyridamole since DMSO shares a number of similar biochemical features with these agents, specifically their increase of cAMP levels.[109,118]

The actions of DMSO on the biosynthesis of specific PGs may have clinical application not only in cerebral ischemia secondary to traumatic brain injury but also in conditions where inflammatory cytokines, platelet aggregation, and vasospasm result from disease and physical injury.

DMSO is reported to exert protective activity on tissue factor expression in human endothelial cells in response to TNF-α or thrombin exposure.[120]

TISSUE FACTOR

The best known function of tissue factor is its role in blood coagulation. Blood coagulation begins when a mammalian vessel is damaged and reacts by beginning a coagulation cascade in order to preserve the integrity of the circulation. The coagulation cascade can also be generated by mediators of the inflammatory response, such as cytokines. The coagulation cascade leads to clot formation by the exposure of flowing blood to tissue factor, also called thromboplastin.

Tissue factor is the cellular receptor and cofactor for factor VII and binds both factors VII and VIIa.[121] Tissue factor is the high-affinity receptor for factors VII and VIIa and is the cofactor necessary for the catalytic function of factor VIIa.[122] As tissue factor binds factor VIIa, it activates Factor IX and Factor X, a process that results in fibrin formation.[123] Fibrin is the building block of the hemostatic plug.

Tissue factor is a key protein in the activation of coagulation and thrombus formation and has been associated with acute coronary syndromes, myocardial infarction,[121] and microvascular perfusion defects following focal cerebral ischemia.[124]

Electron microscopic evidence has shown the presence of fibrin in cerebral microvessels derived from degranulated platelets and poymorphonuclear leukocytes following middle cerebral artery (MCA) occlusion.[125] The use of antitissue factor agents as DMSO has been shown to be, may be, clinically useful to prevent the coagulation cascade that can lead to microvascular plugs following the ischemic process related to blunt brain trauma and stroke.

It has been reported that DMSO has an effect on blocking Na^+ and Ca^+ entry into cells.[50,51] Since substantial Na^+ and Ca^+ entry into myocytes typically occurs after cardiac arrhythmias and myocardial infarction, DMSO administration may prevent this inward cellular ion flux while preserving K^+ outflux from cardiac tissue. The mechanisms exerted by DMSO on Na^+ and Ca^+ channels need to be further investigated in mammalian models of ischemic injury since the results of such studies could produce an extremely useful and relatively safe agent for a variety of cardiac and brain ischemic disorders affected by changes involving these cations (Figure 8.3). Drugs that block abnormal Na^+ entry into damaged brain cells have

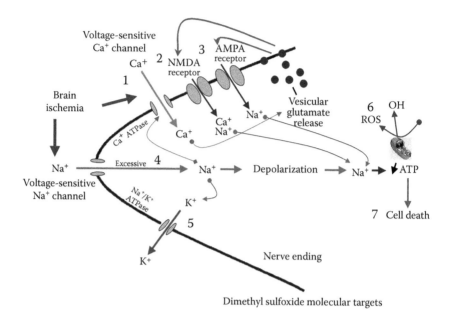

Dimethyl sulfoxide molecular targets

FIGURE 8.3   (See color insert.) Scheme detailing some of the molecular mechanisms DMSO is reported to target at the subcellular level in nerve endings or neuronal cytoplasm following cerebral ischemia. DMSO is reported to suppress excessive calcium (Ca^{2+}) influx into cells (1) and block NMDA (*N*-methyl-D-aspartate) and AMPA (alpha-amino-3-hydroxy-5-methyl-4-isoxazolepropionate) receptor channels (2, 3) following ischemic events that can promote excitotoxic death of neurons. DMSO is reported to block sodium (Na^{2+}) channel activation (4), a reaction seen after CNS injury that may activate a biochemical cascade resulting in intracytoplasmic potassium efflux (5), cell energy hypometabolism ($\downarrow$ATP), ROS and hydroxyl radical (OH) production (6), and cell death (7). DMSO has been shown to increase the levels of high-energy phosphates cAMP and creatine phosphate and inhibit tissue factor expression, thrombus formation, and vascular smooth muscle cell (VSMC) proliferation and migration (not shown), elements known to participate in cerebral ischemia and stroke. See text for details.

been shown to exhibit strong, neuroprotective activity in animal models of brain ischemia and hypoxia, and a number of clinical trials are now in progress to test these class of drugs when cerebral ischemia is present.[126–128] Curiously, DMSO is not presently being considered for such clinical testing.

DMSO given at clinical doses has been reported to suppress in a relatively rapid (minutes) and reversible manner excessive calcium influx into cells and channel-opening of the ionotropic receptor channels NMDA and AMPA, which are known to be activated by glutamate during oxidative or metabolic stress[54] (Figure 8.3). This *excitotoxic* process by glutamate leads to electrophysiological responses and calcium influx into brain cells when activated is blocked by DMSO, resulting in the protection of hippocampal neurons against deadly pathway.[54]

Although the mechanism responsible for DMSO's suppression of glutamate excitotoxicity is unclear, two possible explanations may be relevant. First, it is shown that

DMSO can reduce Na^+ and K^+ currents in cardiac myocytes,[129] and as we have seen, this can also occur in neurons to reduce their excitability and action potentials,[51] a process that could quell Na^+ channel current into excitable glutamatergic neurons.

A second possibility is the anti-inflammatory modulation that DMSO demonstrates against proinflammatory trauma-induced cytokines such as TNF-α, interleukin-6, and NF-kB, a transcription factor involved in inflammatory responses. Part of the modulation of toxic cytokines by DMSO is the scavenging of free radicals and antioxidant activity, an effect that can help resolve the development of brain edema after trauma.[47,52,130,131]

Stabilization of cellular membranes by DMSO is yet another property that helps tissue resist the effects of physical and physiological damage that can occur after blunt or ischemic injury to the brain. This cell membrane protection has been shown consistently by DMSO on cells exposed to freezing, radiation, and sonic stress.[59,132,133]

Lim and Mullan[59] showed that astrocytoma cells subjected to sonic stress from a sonicator in an *in vitro* system was capable of tearing apart the cell membrane by violent collapse. They found that exposure of 10% DMSO to the astrocytoma cells afforded significant protection and survival rate of these cells when compared to no treatment controls or to glycerol, an agent used in brain edema to control brain swelling.[59]

Another important property that may explain DMSO's ability to reduce brain injury and edema is its effect on the Janus kinase/signal transducer and activator of transcription (JAK2/STAT) signaling pathway. JAK2/STAT signaling pathway appears to be one of the main pathways downstream of cytokine receptors and growth factor receptors that functions by transducing signals from cell surface to the nucleus.[55] This activity has been shown to have a protective role in brain trauma.[55] Recent studies have revealed that the JAK2/STAT signaling pathway plays a role in the regulation of gene expression during inflammation.[134,135] Immunohistochemical examination following contusive brain injury in rats showed that JAK2/STAT pathway was mildly activated in both neurons and astrocytes 3 h after injury.[55] Conversely, the inhibition of the JAK2/STAT pathway was associated with the promotion of brain edema, which can increase the severity of neurological deficits.[55]

When DMSO was given intraperitoneally 20 min prior to pericontusional trauma to rat cortex, increased levels of the JAK2/STAT pathway were observed together with improved neurological recovery and reduced neurological deficits when compared to animals not given DMSO.[55] It is unclear how DMSO is able to accelerate or increase the activation of this neuroprotective pathway.

CEREBRAL BLOOD FLOW

DMSO has been reported to increase cerebral blood flow in many preparations involving brain trauma, stroke, and SCI,[35,41,61,118,136,137] but the mechanisms DMSO may tap to increase cerebral perfusion have not been investigated.

VSMCs in brain vessels are known to modulate vessel tone when they receive signals to vasodilate,[35] but the only evidence that DMSO contacts VSMC is in preventing their migration,[119] a process known to be associated with vascular damage and atherogenesis.[138]

One curious observation that should be mentioned is the fact that injected DMSO, even at extremely high doses, does not appear to increase cerebral blood flow appreciably in normal dog brain.[139] Normal dogs given mega intravenous doses of 2–8 g/kg of 100% DMSO showed an increase in cerebral blood flow in the cortex and caudate nucleus but only after 6 g/kg, a dose that produced hemolysis and increased intravascular volume.[139] Cerebellar tissue and brain stem blood flow remained unchanged.[139]

This study indicates that DMSO's effect in increasing cerebral blood occurs when a neuropathological state is present, for example, brain trauma or stroke, where doses ranging from 0.5 to 1 g/kg of a 25%–40% DMSO solution are sufficient to significantly increase cerebral perfusion.[41,61,137] This outcome implies that DMSO's ability to increase cerebral blood flow from an ischemic state may occur by affecting associated factors: (1) reducing cerebrovascular resistance and blood viscosity, (2) reducing tissue edema and ICP, (3) modulating PG control of vessel tone and/or platelet aggregability, (4) relief of inflammatory, and (5) a combination of these factors.

The physiopathological responses to pressure-induced cerebral ischemia in canine brain were investigated using DMSO. Brain retraction pressure over the somatosensory cortex was achieved using a De Martel retractor fixed into place by a micromanipulator, and DMSO was administered at 250 mg/kg prior to retraction and every 12 h after retraction for 3 days. DMSO treatment was significantly better than untreated controls with respect to CPP, focal swelling, and tissue necrosis.[140]

These gross findings were reinforced by neurobehavioral scoring following brain retraction, with DMSO treatment showing improved recovery when compared to mannitol, methyl prednisolone, furosemide, and barbiturate treatments.[140] This study received support by the same group of investigators who used the brain retraction model to induce focal ischemia in animals pretreated with ethanol to simulate a drunken vehicle driver sustaining a head injury.[140] In the untreated animals, ethanol intoxication led to a large source of hydroxyl-free radicals and neuronal tissue damage, an effect that was markedly attenuated by DMSO treatment.[140]

CONCUSSIONS

Concussions are a type of mild traumatic brain injury (MTBI) that occurs when the brain is jarred or shaken and are most commonly seen in sports-related brain injury.

Sports-related concussion has been described as an *epidemic* by the Centers for Disease Control and Prevention. Visits to the emergency department for concussions have increased 62% between 2001 and 2009, and it is estimated that up to 250,000 concussions occur each year from high school sports alone.[141] Given the millions of young persons involved in high school, collegiate, and older professional players participating in contact sports that involve repetitive brain trauma, as well as military personnel exposed to repeated brain trauma from combat-related injuries, concussions have become an important public health concern requiring public education and immediate clinical attention to find an effective treatment.

Concussions are the result of biomechanical forces that result in rapid acceleration and decelerations of the brain often involving helmet-to-helmet contact, leading to functional impairments even in the absence of overt structural damage.[142] Football is not the only sport that produces concussions; every contact sport such as hockey,

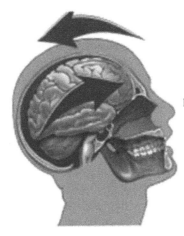

The head strikes a
hard object creating
a concussion-type
injury

FIGURE 8.4 Rapid acceleration–deceleration of the brain created by the head striking a moving or fixed object.

soccer, boxing, and rugby also contributes to this epidemic.[143] Concussions can also result from the head striking a hard object, as it often occurs in falls, creating a more severe acceleration–deceleration of the brain than those seen after football helmet-to-helmet contact (Figure 8.4).

Concussions have been defined as a temporary impairment of neurological function that heals by itself over time, and unlike severe, closed head injury, neuroimaging normally shows no gross structural changes to the brain as a result of the trauma.[144]

Loss of consciousness is a clinical hallmark of concussion but is not required to make the diagnosis. Symptoms include confusion, disorientation, unsteadiness, dizziness, headache, and visual disturbances.[145] These postconcussive deficits occur with minimal detectable anatomic pathology and often resolve completely over time, suggesting that they are based on temporary neuronal dysfunction rather than cell death as seen in severe head injury.[146] Neuronal dysfunction can occur due to cerebral hypoperfusion, ionic shifts, altered metabolism, impaired synaptic connectivity, or changes in neurotransmission, such as diffuse axonal injury.[147]

Repeated concussions, even mild blows to the head from contact sports, can cause cumulative pathology on the brain. Successive concussions can have devastating consequences, including progressive decline of memory and cognition, depression, suicidal behavior, poor impulse control, aggressiveness, and, eventually, dementia.[148]

The metabolic cascade following this type of injury has been studied extensively in animal models as well as humans.[149,150] This metabolic cascade immediately after biomechanical injury to the brain consists of a sudden release of neurotransmitters involving excitatory transmitters, such as glutamate binding to the NMDA receptor, which leads to further neuronal depolarization with the efflux of potassium and influx of calcium into brain cells.[149] These ionic shifts lead to acute and subacute changes in cellular physiology, diminished cerebral perfusion, and brain cell energy crisis, making the brain less able to respond adequately to a second injury and potentially leading to longer-lasting deficits.[151]

Unfortunately, studies targeting the metabolic cascade have failed to produce a viable treatment option given the rapid onset and the shortness of time in reaching a concussed patient in time to stop this cascade from progressing. One option, however, is to keep a rapidly acting, safe, and effective agent such as DMSO near the sports facility where games are played and have someone present when players come in contact with each other that can administer DMSO intravenously if it is determined that intervention must be applied.

Some studies have shown that cerebral blood flow decreases immediately following a concussion, and the amount of time cerebral perfusion remains lowered and seems to depend on the severity of the injury.[152,153]

Due to their lack of structural damage, concussion is also referred to as MTBI. More commonly, athletes present with a brief alteration of consciousness, headache, and amnesia and require careful examination and observation before returning to competition. The diagnosis of MTBI is based on subjective findings and subtle changes in mental status.

The diagnosis of concussion is largely based on the clinical examination, yet certain imaging studies can be considered, such as computed tomography or magnetic resonance imaging (MRI), especially if an intracranial bleed is suspected. MTBI requiring urgent neurosurgical attention should be ruled out.[154] Computerized tomography and MRI are usually without significant findings in MTBI. Functional magnetic resonance images and positron emission tomographic scans have been used in concussion research but serve no role in the clinical management of concussion.[155]

Neuropsychological testing is also very sensitive in the evaluation of brain injuries in athletes and may become more clinically useful in the future. Neuropsychological testing may demonstrate areas of deficiencies; however, results may be difficult to interpret because of confounding factors.[156]

Complications following MTBI such as second impact syndrome have been known to occur, and when this happens, a rapid and effective treatment to counteract the pathologic cascade that follows may be life-saving.

The search for an effective and safe concussion treatment when prevention does not work is higher now than ever before. A better understanding of the detrimental molecular and cellular effects caused by a concussion is the key to finding a safe, rapid, and effective treatment.

Treatment strategies focused on managing the neuroinflammatory responses to concussion may prove to be a tangible target since it is hypothesized that these delayed secondary effects may be the primary source of long-term neural damage. There are promising data to suggest that manipulating neuroinflammation can be used as treatment strategies to manage the long-term deficits produced by a concussive injury.[151]

However, the use of general anti-inflammatory drugs such as nonsteroidal anti-inflammatory drugs (NSAIDs) will not serve as a *magic bullet* for this type of injury. Rather, it seems that a tailored drug that targets the inflammatory and metabolic cascade that follow a concussion when given at particular temporal intervals will likely be useful in treating the complexity of the inflammatory response to concussion. DMSO, an efficient anti-inflammatory agent, may become a treatment of choice as we review its properties as a neuroprotective agent.

DMSO IN BRAIN CONCUSSION

DMSO may be an ideal agent to treat the metabolic cascade that often follows the rapid acceleration and decelerations of brain concussions.[155,157]

Glutamate is an amino acid in the brain that functions as the principal excitatory neuronal neurotransmitter. During excitatory neurotransmission in the healthy CNS, glutamate is briefly released at a high concentration into the extracellular space where normal physiological levels are maintained by reuptake of glutamate through neuronal or glial glutamate transporters. When excessive extracellular glutamate accumulates in the synaptic space after brain traumatic injury, it causes a significant increase in Ca^{2+} influx into neurons, thereby triggering neuronal excitotoxicity. Traumatic injury to neurons can produce disastrous results when the excessive release of intracellular glutamate concentrations overload the synaptic space. Moreover, many glutamate-induced CNS neurodegenerative disorders, such as epilepsy, amyotrophic lateral sclerosis, brain ischemia, and traumatic brain injury, are reported to associate with the dysfunction of glutamate excitotoxicity.[158]

When primary hippocampal cell cultures exposed to toxic levels of glutamate were given DMSO at clinical doses, DMSO was shown to suppress, in a rapidly reversible manner, excessive calcium influx into cells and channel opening of the ionotropic receptor channels NMDA and AMPA, which are known to be activated by glutamate during brain trauma due to the development of oxidative or metabolic stress.[54]

Moreover, this *excitotoxic* process by glutamate, which can damage or kill neurons, was reported to be blocked by DMSO.[54]

The treatment of brain edema after MTBI presents a challenge for many clinicians. Considerable animal and human studies indicate that DMSO can quickly lower edema and ICP following severe traumatic brain injury (TBI).[41,43–45,61,159] These findings suggest that DMSO at lower doses than those used for severe, closed head injury may be beneficial to mild brain trauma. This concussion is supported by the pharmacological properties of DMSO, including its antiedema property, its potent diuretic action, its anti-inflammatory properties, and its increase of cerebral perfusion.[160] Moreover, its blocking effect of excessive glutamate, sodium, and calcium entry into cells and efflux of potassium (Figure 8.3) that can induce cytotoxicity demonstrates that it can therapeutically target the biochemical cascade described in animal and human brain following MTBI (concussions) or severe TBI.[50,51,54]

DMSO COMBINED WITH A GLYCOLYTIC INTERMEDIATE

Because brain injury can lead to an impairment of mitochondrial oxidative phosphorylation with the loss of ATP as well as to the formation of cytotoxic free radicals, de la Torre and his group[161] reasoned that combining DMSO with an intermediate of anaerobic glycolytic metabolism, fructose-1,6-diphosphate (FDP), might be more useful than either drug alone in preventing the neuronal and energy compromise associated with severe, concussive head injury in a mouse model.[161]

FDP is an intermediate of anaerobic glycolytic metabolism and has been shown to restore the activity of the Embden–Meyerhof pathway and oxidative phosphorylation when administered during prolonged hypoperfusion states.[162] FDP can also inhibit oxygen-free radicals, stimulate anaerobic glycolysis, and increase the production of ATP, the main energy fuel of cells and neurons.[163] It is this activity by FDP that may limit injury to the brain following ischemic-hypoxia.[163]

The combination of DMSO + FDP to treat experimental brain trauma in mice thus seemed a reasonable option to prevent or restore loss of ATP in ischemic brain cells while simultaneously reducing progressive cerebral edema caused by the injury.[161] Consequently, it was believed that FDP would act primarily to protect ischemic brain tissue from energy substrate depletion, a common outcome after brain trauma, while DMSO would act to stabilize cell membranes from excess free radical formation and abnormal Ca$^+$ entry into cells while reducing ICP by improving CBF.[160]

Both FDP and DMSO have each been reported to be useful when administered in head injury[42,45,61,163] or cerebral ischemia in animals or humans.[159,163,164] FDP has been shown to prevent the ischemic-induced loss of ATP and accumulation of intracellular Ca^{2+} in experimental animals, possibly by stimulating glycolysis while inhibiting gluconeogenesis.[164] The administration of FDP during brain ischemia would seem to be particularly useful, since it readily crosses the blood–brain barrier (arias) and can yield twice as many moles of ATP as glucose.[164]

In an experiment to test the synergic potential of DMSO-FDP, male CD-1 mice weighing 22–29 g were restrained and their heads were positioned under a head injury apparatus with their chins resting firmly on a flat surface at the base of the apparatus.[161] The apparatus consisted of a 25 cm long glass shaft held vertically by clamps on a ring stand and a tubular lead cylinder weighing 40 g. The lead cylinder was allowed to drop 20 cm through the shaft onto the center of the head, thus creating an 800 g/cm force injury on the cranium. An imaginary line was drawn coronally just anterior to the ears in order to create a consistent injury site on the animal's head. This head injury mouse model has been previously used to screen the effectiveness of potential pharmacological agents after brain trauma.[165] FDP was given at 350 mg/kg in a 10% solution, and DMSO was administered at 1 gm/kg in a 28% solution both intravenously via the tail vein.

All mice were pretested on the string prior to head injury, and all were able to grasp the string for a 90 s trial period and to travel with ease from the center of the string to one of the end posts holding the string. The extent of motor deficits reflected by the reduced ability of mice to remain gripping the string after head injury has been shown to be inversely proportional to the increased severity of the head injury.[166]

Following head injury, testing showed 11 of 12 mice treated with DMSO + FDP retained sensory function as compared with 5 of 12 mice given vehicle nontreatment, 8 of 12 treated with DMSO only, and 7 of 12 treated with FDP.[161] The combination of DMSO + FDP significantly protected mice from motor deficits after injury, as indicated by an increased mean grip test score (76.3 s) shown by this group 1 h after head injury when compared with vehicle nontreated (5.8 s), FDP-treated (20.8 s), and DMSO-treated (19.6 s) animals. The pattern of protection continued for at least 2 h after head injury, with all groups showing some improvement over the previous grip test taken 1 h after head injury. However, the only significant protection after 2 h was observed in the group treated with DMSO + FDP. Two hours after head injury, the group treated with FDP + DMSO had a mean grip test score of 86.5 s, which was significantly better than from the DMSO only (33.8 s), FDP only (24.3 s), and vehicle nontreated groups (12.2 s).[161]

This study was the first to demonstrate that the combined DMSO + FDP given to mice that have been subjected to a moderate 800 g/cm concussion head injury can significantly modify the sensory-motor deficits, incidence of mortality, and histopathology. DMSO + FDP appeared superior in reducing the trauma-induced neurological deficits and fatal outcome to either agent administered alone.

Besides the improved sensory-motor function of mice treated with DMSO + FDP, a possible fringe benefit to the head-injured brain where inadequate cerebral perfusion is expected may be the ability of these two agents to allow RBCs to maintain their deformability. Under normal conditions, RBCs with a 7–8 μm diameter can squeeze through capillaries that are 3–5 μm wide so that oxygen can be delivered to the tissues. The RBCs' viscoelastic properties or deformability is possible because of a spectrin–actin interaction within the erythrocyte, which is dependent on its stored ATP and 2,3-diphosphoglycerate.[167]

If 2,3-diphosphoglycerate or ATP stores are reduced by ischemia-hypoxia (Heckler/Marcel), it is probable that RBCs will increase their rigidity and thus be unable to adequately perfuse the cerebral microvasculature. This consequence would prevent optimal oxygen delivery to neurons and glial cells, which rely on oxygen for the elaboration of ATP and for their basic cellular metabolism. Infusion of FDP raises the concentration of 2,3-diphosphoglycerate,[168] allowing RBCs to deform normally and consequently facilitating oxygen disassociation from hemoglobin[169,170] during ischemia-hypoxia. Conversely, DMSO is reported to promote the formation of ATP from ADP[171] and to preserve mitochondrial oxidative phosphorylation, thus ensuring ATP production after experimental brain ischemia[172] This activity could additionally favor RBC deformability during tissue ischemic-hypoxia, a process that would improve cerebral perfusion.

Another advantage of adding FDP to DMSO is that FDP has been shown to be effective in preventing ischemic injury to the brain[173,174] ostensibly by inhibiting excitotoxic release of glutamate during brain ischemia and preventing ATP loss by substantially increasing the flux of glucose into the pentose-phosphate pathway. Indirect evidence has also demonstrated that FDP can modulate cerebrovascular tone, myocardial contractility, platelet aggregation, and blood pressure,[175,176] systems that can exacerbate damage following severe head injury.

The exact mechanism for the synergic activity demonstrated by the DMSO + FDP combination has not yet been determined, but it may be assumed to lie in the ability of each agent to address different but critical metabolic abnormalities associated with the pathogenesis of severe head injury, for example, ATP loss leading to neuronal energy deficiency, metabolic acidosis, abnormal Ca^{2+} entry into cells, platelet aggregation, inflammatory response, cerebrovascular constriction, glutamate toxicity, ICP increase, and free radical formation. This drug combination is reviewed in terms of its effect on experimental and clinical ischemic strokes in this chapter and Chapter 9.

DMSO IN EXPERIMENTAL BRAIN ISCHEMIA

Stroke is the most fatal neurological disease and the leading cause of long-term disability in the United States. Stroke is divided into two major types, ischemic stroke, where an artery is blocked by a blood clot, and hemorrhagic stroke, where a brain vessel bleeds from a leak or a rupture. About 80% of strokes involve blockage of a major cerebral artery, and this is where most of the animal research in brain ischemia is presently focused.[177] A considerable amount in our current understanding of cerebral ischemia induced by stroke comes from animal models representing global and focal ischemia.

Global ischemia is the product of mild-to-severe reduction of cerebral blood flow throughout the brain, which can occur either acutely or chronically. If blood flow cannot be restored within a given time limit (depending on the cause), permanent damage to the brain or death can result. When blood flow is restored after global ischemia, reperfusion injury can occur if the brain is not pharmacologically protected. Many diseases can cause chronic global brain ischemia such as atherosclerosis of a conducting brain artery, diabetes, sickle cell anemia, cardiac arrhythmias, and congenital heart defects, while acute global ischemia can result from a heart attack.[178]

Focal brain ischemia is less severe than global ischemia but more common both clinically and using experimental animal models. Focal ischemia reduces blood supply to a specific region of the brain where immediate or delayed neuronal death can occur.[178] This area of injury is called the core ischemic zone and is marked by severe ischemia with blood flow falling 75%–90% of baseline in conjunction with the loss of oxygen and glucose, leading to rapid depletion energy stores.[178] Severe focal ischemia can result in brain cell death from the necrosis of neurons within the ischemic core, but glial cells may be spared if blood flow is restored early.[179]

Surrounding the ischemic core is an area called the penumbra where brain cells are kept structurally intact by collateral blood vessels but functionally impaired.[180]

The ischemic penumbra is a zone where pharmacological therapy could most likely be effective in restoring the loss of function.[181] The ischemic penumbra blood perfusion is generally too low to maintain electric activity but sufficient to preserve ion channels and structural integrity.[180]

Research using animal models of stroke have provided not only an understanding of the pathomechanisms underlying injury after ischemic stroke but also a way to develop effective therapy to prevent and arrest the severe pathology associated with stroke. Although many models of stroke have been used for research, including nonhuman primates, rodents are more often used due to their lower cost, ease of handling, substantial characterization of pathological expectations from previous research, and potential for genetic engineering.

Experimental focal ischemia has generally involved occlusion of the MCA that is the largest branch of the internal carotid artery (ICA) that supplies a portion of the anterolateral frontal lobe, most of the temporal lobe, and the parietal lobe. Occlusion of the MCA in animals can result in sensory-motor deficits and cognitive impairment

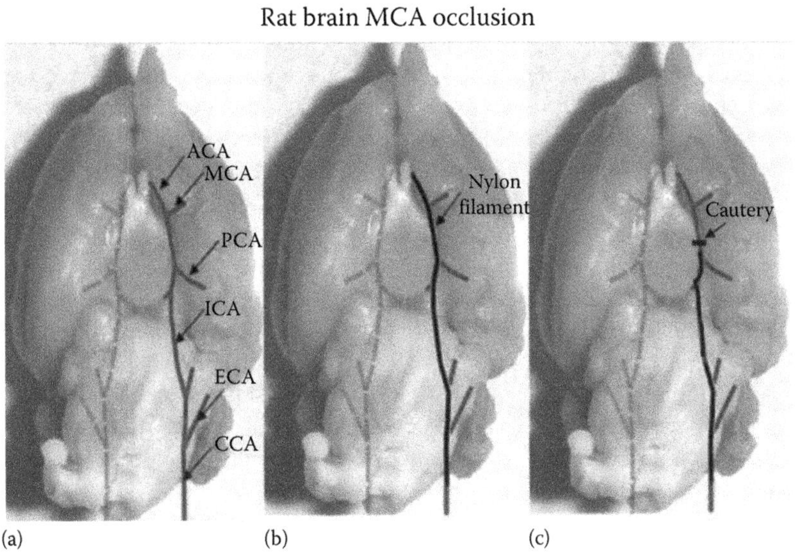

Rat brain MCA occlusion

(a)   (b)   (c)

FIGURE 8.5 **(See color insert.)** Normal arteries at the base of rat brain (a) and approaches to induce focal brain ischemia using a nylon filament introduced from the ICA to the MCA (b) and thrombotic occlusion (c) of the MCA. Such occlusions of the MCA will mimic human ischemic stroke and involve parts of the cortex and subcortical regions, producing sensory loss and paresis or paralysis of the contralateral side of the body. See text for details. *Key*: ACA (anterior cerebral artery), MCA (middle cerebral artery), PCA (posterior cerebral artery), ICA (internal carotid artery), ECA (external carotid artery), CCA (common carotid artery).

that can be monitored, quantitated, and analyzed by a variety of research tools when drug evaluation is done.

In the context of cerebral arterial thrombosis, the effects of thrombotic stroke can be induced experimentally by cautery or by photothrombosis of the MCA, which differs from those due to cerebral ischemia induced by mechanical ligation or intraluminal occlusions of intracranial brain arteries[182] (Figure 8.5).

There are two commonly used surgical approaches to provide direct access to the arterial circulation of rodents. The first approach is to perform a craniotomy that allows access to distal branches of the ICA, such as the much used MCA, to be occluded either with a neurofilament or by cautery[183,184] (Figure 8.5). Other types of occlusion to achieve focal ischemia have been reported such as photothrombosis, ligation of an arterial vessel, or drugs to cause local embolism, such as endothelin, a powerful vasoconstrictor.[185]

One early report by Tamura and his colleagues[186] described an approach that occludes MCA proximal to the lenticulostriate arteries and produces the infarction of both the cortex and the lateral part of striatum. Figure 8.4 shows two popular techniques used in rats to induce focal brain ischemia.

The intraluminal suture model, developed by Koizumi and his team (koizumi) and Longa and his group,[187] is undoubtedly the most frequently used focal ischemia model in rats and mice but is also one of the trickiest to produce a consistent outcome. This may be due largely to the varied technical approaches, rodent model, rodent age, and gender of rodents.

The intraluminal monofilament model of MCA occlusion involves the insertion of a surgical filament into the external carotid artery and threading it forward into the ICA until the tip occludes the origin of the MCA, resulting in a cessation of blood flow and subsequent brain infarction in the MCA territory (Figure 8.5).

The technique has a number of variations and can be used to model permanent or transient occlusion. If the suture is removed after a certain interval, controlled reperfusion can be achieved accompanied by an ischemic penumbra, which drug therapy can target. If the filament is left in place over 24 h, the procedure is suitable as a model of permanent occlusion. One useful advantage of the filament occlusion is that it does not require craniectomy or craniotomy, which can result in undesired variables to the experiment.

Because the intraluminal suture model occludes the tip of the intracranial ICA, it cannot be considered a strict MCA occlusion.[188]

For this reason, the filament occlusion technique results in a wider ischemic zone and higher mortality than when MCA is directly occluded, especially when permanent occlusion is made.[188] Nevertheless, the filament occlusion technique has become the most frequently used method to mimic permanent and transient focal cerebral ischemia in rats and mice and has also been used in rabbits,[189] gerbils,[190] and marmosets.[191]

However, despite their usefulness in stroke research, these focal ischemic models are not consistently reliable. The surgery to access and manipulate the cerebral vessels requires skilled and experienced hands, and for practical purposes, the results can vary from one laboratory to another and even within the same laboratory where findings are often highly variable. But the variabilities in stroke research are not due to technical difficulties only.

It should be noted that the intensive research done in the past and up to the present, using experimental stroke models, displays many technical, behavioral, and pathological shortcomings to mimic human stroke, and a considerable number of pharmacological agents have been seen to dramatically improve or reverse tissue injury after brain ischemia in rodents have failed to be translated into effective clinical use. Many reasons have been suggested for this bench to bedside failure, and some of these reasons have been reviewed elsewhere.[192,193]

One obvious reason for the lack of success in translating rodent findings into effective human stroke therapy is the basic difference between the rodent cerebral circulation, hemodynamic responses to vascular stress, and relative tolerance to brain ischemia with those of humans.[194] In addition to anatomic and physiological differences between rodent and human cerebrovasculature, most human strokes occur in elderly individuals, while a majority of animal stroke models are performed on young rodents.[195]

This lack of correlation between small mammals such as rodents and human brain vasculature has produced no effective treatments in the last century for human ischemic stroke other than fibrinolytic therapy. The only FDA-approved therapy for the treatment of acute ischemic stroke is the thrombolytic tPA,[196] and curiously, the initial basic research was done in rabbits.[197,198]

As effective as they are when properly used, fibrinolytics have the limitation that makes them ineffective when administered after a 3 h time window following ischemic stroke onset.[199] This time window is often difficult to obtain due to the onset of symptoms in a stroke patient who often waits several or more hours before seeking medical help at the emergency department.

The effect of the large number of failed human trials that followed dramatic preclinical results obtained primarily from rodent research has visibly reduced current levels of funding among the pharmaceutical industry to invest in research of new neuroprotective agents.[194]

Another major reason for the lack of progress in ischemic stroke can be found in the difference between human brain and those of smaller mammals such as rodents, marmosets, and rabbits. Human brain is gyrencephalic (cortex convoluted with gyri and sulci), and those of many smaller animals are lissencephalic (smooth cortex). Human cortex and subcortex are highly organized, which includes deep gray nuclei and white matter tracts like the internal capsule.[200] Moreover, human brain lacks the collateralization of cerebral vessels found in rodents and other small mammals so their vulnerability to ischemia is enhanced.

These species differences between human and rodent brain can be diminished when nonhuman primates are used to mimic human ischemic stroke.

This has been the recommendation of the Stroke Therapy Academic Industry Roundtable (STAIR), a collaborative group made up of industry, academic members,[201] and other many independent investigators.[202–204]

This group met to discuss the failure of finding therapy that could be useful in human stroke and came to the conclusion that stroke research would benefit from using an animal with a gyrencephalic brain such as macaque rhesus monkeys.

ETHICAL CONSIDERATIONS FOR USING NONHUMAN PRIMATES IN CNS INJURIES

The use of macaques and other nonhuman primates comes with the ethical challenges that need to be considered and that involve the need to eradicate or diminish pain, emotional stress, and depression associated with the neurological deficits following experimental stroke. This can be achieved to some extent by providing proper analgesia during and after surgery and intensive veterinary care to compensate during the recovery period.[205] The trade-off in using a nonhuman primate model would likely accelerate the discovery of a treatment that could reverse or significantly limit the neurological and cognitive deficits seen after human or animal stroke.

Another more vulgar consideration in using nonhuman primates is their high cost, specialized technical personnel needed, veterinary care, and enriched living environment, factors that can discourage most researchers with limited research budgets.

It should be emphasized that the use of successful therapy against ischemic stroke induced in nonhuman primates does not guarantee success in human stroke.

A number of examples support this conclusion. For example, the experimental radical trapping agent NXY-059 was used in marmoset monkeys 4 h after permanent MCA occlusion and obtained excellent therapeutic results 10 weeks after treatment, resulting in fewer neurological deficits and better neurohistologic picture when compared to saline controls.[206,207] When NXY-059 was tried in human ischemic stroke in a randomized, double-blind study comprising 3190 patients, it was found ineffective when given 6 h after stroke onset.[208]

Thus, the recommendations of the STAIR group for using nonhuman primates for ischemic stroke research have not been validated except for one instance, the use of intravenous DMSO. After canvassing the extensive stroke literature, DMSO appears to be the only agent that has so far shown significant neuroprotection in nonhuman primate ischemic stroke and human ischemic stroke.[209,210] The medical use of DMSO in humans is discussed in Chapter 9.

ROLE OF DMSO IN EXPERIMENTAL STROKE

The brain region vascularized by the MCA is the most commonly affected area following a cerebral infarction. The reason is due to the large size of the MCA and the direct flow it receives at an angle from the internal carotid artery (ICA) providing the easiest path for thromboembolism. The MCA is therefore a vessel of choice in studying stroke in animals.

In nonhuman primates, DMSO was studied in 20 rhesus monkeys (*Macaca mulatta*) subjected to ischemic stroke via a transorbital approach to visualize and clip the MCA under direct microscopy.[209] The animals were divided into four groups: (1) shams, surgery but no occlusion (N = 2); (2) saline controls (N = 8), surgery and clipping of the MCA; (3) clipping of the MCA and treated with 3 mg/kg dexamethasone (N = 5); (4) clipping of the MCA and treated with DMSO (N = 5) 2.5 g/kg 50% solution mixed in saline. All treatments were given 4 h after occlusion of the MCA with the clip still in place. The occluding Mayfield clip was removed after 16½ h, and a second intravenous dose of dexamethasone or DMSO was repeated for all treated animals 30 min after clip removal. All animals were given a full neurological examination daily for 7 days following occlusion of the MCA. After 7 days, animals were killed for neurohistological analysis. Arteriograms were taken in all animals before occlusion, 30 min after occlusion, 16½ h after occlusion, and 30 min following unclipping of the MCA (Figure 8.6). Table 8.2 summarizes the results of the neurological examination after 16 h with the MCA clip in place and at 7 days following removal of the clip.[209]

Mean average cerebral blood flow measured near the MCA prior to clipping was 89 mL/100 g tissue/min. This value dropped by 43%–54% after MCA clipping and essentially returned to preocclusion values with no significant group differences following unclipping. Angiographic changes following MCA occlusion can be summarized as follows. ICA narrowing (arterial spasm?) was seen at the level of the siphon in all animals 30 min following MCA occlusion (Figure 8.6). All animals except two saline controls survived the initial MCA clipping. Postmortem tissue weights showed DMSO attenuated cerebral tissue edema when compared to dexamethasone or saline controls.

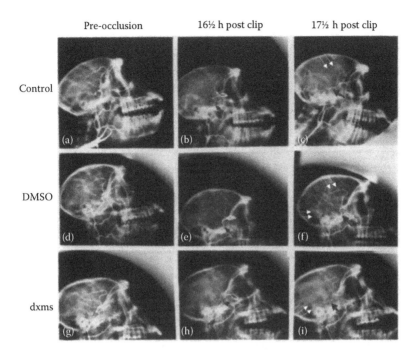

FIGURE 8.6 Arteriograms of rhesus monkeys showing characteristic vascular response after 16½ occlusion and 17 h when MCA was deoccluded. Saline-treated controls (a–c), DMSO treated (d–f), and dexamethasone treated (dxms g–i) before and after deocclusion of the MCAt. Panels (a), (d), (g) show preocclusion vascularization prior to MCA occlusion. Treatments were given 4 h after clipping the MCA and repeated shortly before clip was removed. At 16½ h after clipping, no blood flow is seen filling the MCA or its branches and severe narrowing of the ICA at the siphon is visible in saline controls and dexamethasone treated (b, h arrow), but no narrowing of ICA is observed in DMSO treated (e arrow). After unclipping at 17½ h, ICA narrowing continues in controls (c bold arrow) and dexamethasone (i bold arrow). No narrowing of ICA seen in DMSO (f bold arrow). In addition, the pericallosal artery and posterior parietal vasculature is absent in saline controls (c double arrows) and significantly reduced in dexamethasone (i double arrows). In DMSO (f double arrows), these vessels are seen to have recovered substantially as compared to preocclusion arteriogram. (Adapted from de la Torre, J.C. and Surgeon, J.W., *Stroke*, 7(6), 577, November–December, 1976.)

After 16½ h and with MCA still occluded, the narrowed carotid was still seen in the saline- and dexamethasone-treated groups. No narrowing of the ICA was observed in animals treated with DMSO[209] (Figure 8.6).

Reperfusion of the posterior parietal and temporal branches of the MCA was more pronounced 30 min following unclipping of the MCA in DMSO than in controls or dexamethasone-treated animals. Moreover, brain swelling of the traumatized hemisphere examined at postmortem was moderately high in controls, mildly reduced in the dexamethasone group, and absent in DMSO-treated monkeys.[209]

Postmortem histological examination of the ischemic hemisphere revealed less structural damage of the inferior thalamic peduncle and fewer chromatolytic and

TABLE 8.2

Neurological Deficits Observed in Rhesus Monkeys 24 h and 7 Days after Occlusion of the MCA and Following No Treatment (Saline Control, N = 8)), Dexamethasone (N = 5), and DMSO (N = 5)

| | Neurological Status in Monkeys after MCA Occlusion | | | | | | | |
| | 24 h Postocclusion | | | | 7 Days Postocclusion | | | |
Treatment	A	B	C	D	A	B	C	D
Saline (control)	—	—	—	8	—	—	—	6
Dexamethasone	—	—	—	5	—	—	1	4
DMSO	—	4	1	—	2	3	—	—

Source: Adapted from de la Torre, J.C. and Surgeon, J.W., *Stroke*, 7(6), 577, November–December 1976.

Notes: No neurological deficits (A); contralateral paresis, alert (B); contralateral paresis, lethargic (C); severe lethargy, or coma (D). Two saline-treated monkeys died on the fifth and sixth day postocclusion. Significant improvement was observed in the neurologic status of DMSO-treated animals after 24 h and after 7 days following MCA occlusion as compared to saline controls and dexamethasone-treated animals.

swollen neurons in the pre-occipital cortex and caudate-lentiform region in DMSO-treated animals when compared to dexamethasone or saline controls. The caudate-lentiform region is an area of the brain supplied by the lenticulostriate branches of the MCA, and because they are *end-arteries*, the brain regions they feed lack a significant collateral blood supply.[209] These end-arteries are not only a common site for thrombosis but are also difficult to treat with thrombolytics because they remain localized to the regions where they terminate, such as the caudate-putamen and internal capsule.

In addition to light microscopic analysis, selective brain tissue regions from all groups of macaque monkeys were examined by electron microscopy on a blind to treatment basis.[47] Brain sections from the nontraumatized hemisphere showed no evidence of swelling or distortion of cellular components, including well preservation of synapses, synaptic vesicles, and unremarkable neuropil. By contrast, the traumatized hemisphere in dexamethasone-treated macaques showed severe swelling of neurons near the MCA occlusion and loss of a clearly delineated plasmalemma.[47]

Ribosomes and osmophilic particles were seen to aggregate throughout the fluid-filled intracellular space, while mitochondria in neurons appeared to be undergoing degenerative changes. Loss of neuronal mitochondria in brain tissue is known to reduce energy synthesis of ATP and make neurons more vulnerable to ischemic damage. Dexamethasone-treated animals also showed swelling of boutons (axon terminals) and considerable loss of synaptic vesicles. These vesicles store neurotransmitters at the axon terminal that are released at the synapse and are essential in propagating nerve impulses between neurons.

Brain sections from DMSO-treated macaques showed swelling of some boutons and synaptic vesicles within neurons, but the axo-dendritic synaptic contacts revealed good preservation, and little to no damage of neuronal mitochondria or neuropil was appreciated.[47] There was no expansion of the extracellular space or compression of the neuropil as compared to dexamethasone or no treatment animals. These findings showed significantly less subcellular damage after DMSO treatment when compared to dexamethasone or no treatment macaques. In particular, loss of axon terminals and synaptic vesicles in dexamethasone- and saline-treated monkeys, but much lesser after DMSO treatment, implies impairment of synaptic neurotransmission and dysfunction of sensory-motor pathways.[47]

The reduction of focal brain edema by DMSO was confirmed in a study using ferric chloride injection into rat brain to induce the formation of lipid peroxidation and free radicals.[211] DMSO caused a decreased formation of brain peroxidation at the injection site, as measured by the formation of NMDA, and a significant reduction of tissue edema of the injured cortex when given with or without the antioxidant tocopherol.[211]

Another brief study using baboons subjected to MCA occlusion and treated with DMSO or DMSO-barbiturate coma appeared less definitive in its conclusions.[213] Four baboons underwent 6 h of temporary MCA occlusion and were treated with 1 g/kg of intravenous DMSO 30 min and 2 h after occlusion. DMSO infusion was continued for the next 8 h at 1 g/kg. The same DMSO-treated animals additionally received pentobarbital at a loading of 30 mg/kg begun 4 h after MCA occlusion.

Continuous barbiturate infusion was titrated to maintain an isoelectric EEG while not allowing a reduction of arterial blood pressure of more than 15%.[212] This group was compared to a DMSO-only group of four baboons given similar doses but without pentobarbital therapy. A neurological score was used to assess each of the four baboons treated with DMSO–pentobarbital and DMSO alone resulted in only one baboon of four (25%) with a good survival and few deficits for DMSO alone and with good survival with mild deficits for one of four baboons given DMSO–pentobarbital.[212] All untreated baboons had a high mortality rate and suffered severe neurologic deficits. This study pointed out the success of DMSO in only 1 of 4 baboons subjected to MCA occlusion. Possible reasons for the difference in findings from those using rhesus monkeys could be the DMSO dose that was twice that used (2.5 g/kg vs. 1 g/kg) and speed of delivery (bolus vs. slow infusion) in the baboon study, as well as the fact that primate brains differ considerably in their vascular anatomy and physiology, including those of baboons and rhesus monkeys.[213]

One major difference between baboons and macaque monkeys is the higher vulnerability of baboons to MCA occlusion producing marked edema and high risk of death.[214,215] This conclusion is supported by the fact that MCA occlusion in the baboon study using DMSO resulted in a 75% mortality regardless of treatment used and 100% mortality or severely poor outcome in 8 untreated baboons.[213] Consequently, it is likely that the MCA lesion in this baboon study may have been too severe to adequately test the value of DMSO or any other treatment.

A second ischemic stroke study using DMSO was carried out by the de la Torre team[48,64] in 45 squirrel monkeys. Survival rate was used to evaluate the efficacy of each treatment at the end of 7 days. The left MCA was accessed via a retro-orbital craniectomy and occluded with a small Mayfield clip for 4 h. The animals were randomly separated into groups of 5 animals/group except for DMSO only, which included 10 animals. The following treatments were given intravenously: Group 1, physiological saline; Group 2, DMSO 1 g/kg in a 50% solution; Group 3, hemodilution (HD) with 6% dextran solution to reduce hematocrit by 50%; Group 4, hyperbaric oxygen (HBO) 2 atmospheres pressure for 4 h; Group 5, DMSO + HBO; Group 6, DMSO + HD. This study was prompted by controversial findings regarding the usefulness of HBO and HD in experimental stroke.

The results of this study confirmed the potential value of DMSO in cerebral infarction. Seven days after each treatment, 8 of 10 DMSO-treated animals were alive with 2 animals showing mild paresis of the contralateral side. Animals treated with DMSO–HD or DMSO–HBO had 50% and 75% survival rate, respectively, with mild-to-moderate neurologic deficits. Animals given single HD or HBO showed 34% and 75% survival rates, respectively, but neurological outcome ranged from moderate-to-severe deficits involving hemiparesis, lethargy, and lack of alertness.[48,64] Combining DMSO to HBO or to HD did not appear to improve neurologic outcome after MCA occlusion when animals were examined at the end of 7 days.

Using only HBO or HD suggested that these therapies by themselves were not helpful in providing neuroprotection to the brain tissue after infarction of the MCA. The lack of neuroprotection by these agents was evident despite the increased oxygenation to brain cells presumably offered by HBO and increased cerebral blood flow ostensibly provided by HD as suggested by previous studies.

We now know that while HBO is known to increase oxygen to tissues including brain, it reduces CBF by reducing the powerful vasodilator nitric oxide while increasing hyperoxic vasoconstriction.[212a] Conversely, HD increases CBF but reduces oxygen delivery to brain by lowering the number of red cells in the circulation.[212b]

In the early 1980s when the squirrel monkey study was performed,[64] HD was a technique believed to be beneficial as a treatment for acute ischemic stroke. It consisted mainly of administering low-molecular-weight dextran in order to increase cerebral blood flow by dropping the hematocrit and hemoglobin levels by about 25%–50%. However, randomized multicenter studies have since shown no benefit for this approach when applied to ischemic stroke in humans.[216] Similarly, HBO was theorized to be beneficial in stroke patients whose blood supply was compromised by the ischemic injury, and consequently, it was thought that increasing the supply of oxygen would reduce the extent of damage to the brain by limiting hypoxia. However, a systematic review of the evidence now indicates that this technique has not been shown to be of benefit in stroke patients.[217]

These findings suggest that a subtle balance between CBF and oxygenation is critically required of an agent to counteract hypoxia and ischemia seen after a stroke. That balance between CBF and oxygenation of organ tissue may be provided by DMSO, which is reported to increase both CBF and oxygen delivery.[41,118,137,136,209]

It is not surprising therefore, in reviewing the use of DMSO combined with HD or with HBO, that no synergism was observed when these combined treatments were compared to DMSO given alone following MCA occlusion in squirrel monkeys.[64] In fact, the combination of DMSO with HBO or HD may have fallen a notch or two from the effectiveness of DMSO given alone. Nevertheless, the treated animals showed less paresis when compared to untreated animals and were more alert following recovery from stroke but only when DMSO was added to either HBO or HD.[63,64]

The Mongolian gerbil is an animal that has been used in ischemic stroke research because it lacks a posterior communicating artery in the brain that makes up part of the circle of Willis. The function of the circle of Willis is to supply blood to the brain and surrounding structures by creating a loose ring of collaterals to the cerebral circulation from the two ICAs and one basilar artery. Lacking the posterior communicating artery means that unilateral ligation of a common carotid artery can result in an ipsilateral brain infarction of the entire hemisphere. About 300 of these gerbils were used to test whether DMSO could protect the brain from ischemic infarction following ligation of the common carotid artery. The results of this immense study showed that DMSO treatment significantly lowered the cerebral infarct volume as revealed by trypan blue staining when compared to saline-treated controls.[218]

Reduction of infarct volume following DMSO treatment was confirmed by Shimizu and his colleagues[219] using the filament occlusion method for the MCA to create a focal brain infarction. DMSO at doses of 0.1–1.0 mL injected intraperitoneally 30 min prior to infarction resulted in significantly less infarction volume than untreated controls.[219]

Further support for reduced infarct volume with DMSO therapy was obtained in rats by Bardutzky and his colleagues.[220] This study was unique in the sense that it examined different time points of drug administration relative to ischemia onset in the same experimental setting. Moreover, it determined whether DMSO increases cerebral blood flow to prevent injury after focal brain ischemia. It then evaluated the surviving rats for neurological status in treated and untreated animals. Thus, the study evaluated the effects of DMSO on the spatiotemporal evolution of the ischemic lesion using quantitative perfusion and MRI and examined the influence of DMSO on cerebral blood flow mechanics.[220] Permanent focal cerebral ischemia was induced using the intraluminal filament occlusion technique in the MCA. It was found that DMSO given at 1.5 g/kg resulted in a highly significant reduction in infarct size when the infusion was initiated 20 h before or 1 h after MCA occlusion. DMSO was also effective in reducing infarct volume at a dose of 0.75 g/kg initiated 1 h after occlusion, but the extent of neuroprotection was less robust compared with the higher dose of 1.5 g/kg.[220]

The neurologic scores 24 h after MCA occlusion were significantly improved when DMSO was given 20 h prior to occlusion compared to the no treatment group.[220] As well, DMSO given 1 h after occlusion was also improved over the controls.[220]

Neuroprotection by DMSO was shown in the 3-day survival group with 50% of the untreated animals dying prematurely between 48 and 72 h after MCA occlusion, but no rat died prematurely in the DMSO group given 1.5 g/kg DMSO starting 1 h after MCA occlusion.[220]

Cerebral blood flow in the ischemic hemisphere that was reduced to an average of 36% compared to the nonischemic hemisphere in untreated control animals remained constant in both control and DMSO groups without significant differences.

This study demonstrated robust neuroprotection using DMSO in a widely used and established model of permanent focal cerebral ischemia.[220] The neuroprotective effect by DMSO was shown *in vivo* by diffusion and perfusion imaging, by neurologic scoring, and at postmortem by the volume of tetrazolium staining of the infarcted region. The effective doses of DMSO used were also clinically relevant and shown to be well tolerated in humans.[220]

The effect of DMSO on energy metabolism and function following brain ischemia and reoxygenation had not been studied in detail until 1991. A study by Gilboe and colleagues[221] reported the following metabolic and functional effects of DMSO during a 14 min period of reoxygenation of the ischemic isolated dog brain. Blood gases, EEG, auditory-evoked potentials, cerebral metabolic rate for glucose (CMRglc), and cerebral metabolic rate for oxygen (CMRO$_2$) were monitored throughout the experiment.[221]

At the end of the study, brain samples were taken to measure brain tissue high-energy phosphates, carbohydrate content, and thiobarbituric acid-reactive material (TBAR is an indicator of lipid peroxidation). After 60 min of reoxygenation in the nontreated 14 min ischemic brains, markers of tissue damage, including lactate, adenosine monophosphate (AMP), creatine (Cr), intracellular hydrogen ion concentration (iH+), and TBAR values were significantly higher than noninjured brain tissue, and energy molecules ATP, creatine phosphate (PCr), CMRglc, CMRO$_2$, and energy charge (EC) values were significantly lower when compared to normoxic brain control samples.[221] In brains exposed to DMSO, TBAR values were near control levels and marked recovery of EEG and auditory-evoked potentials was observed in brains treated with DMSO. Restoration of auditory-evoked potentials and EEG suggests that DMSO is involved in the preservation and restoration of cell membrane structures within the brain subjected to ischemia.[221]

DMSO treatment also significantly increased ATP, CMRO$_2$, and PCr values in postischemic brains and lowered noxious lactate and creatine when compared to untreated brains.[221] These findings indicated that DMSO appears to come close to normalizing brain metabolism during postischemic reperfusion, a finding that could explain in part its benefit following ischemic insults to the brain. This effect may not be totally due to free radical scavenging by DMSO since reducing free radicals alone following ischemia does not improve recovery of brain metabolism. A more likely explanation for the neuroprotective action of DMSO has to do with partially restoring to normal the energy metabolism of ischemic brain cells before bioenergetic failure has taken hold. This outcome is supported by evidence showing the effect of DMSO in increasing the energy fuels, ATP and PCr, while reducing cellular oxygen consumption and synaptic dysfunction.[222,223] In this regard, DMSO has been reported to increase the latency of anoxic depolarization of neurons after acute brain ischemia.[50]

It is important to point out that when bioenergetic failure occurs for any reason, brain cell metabolism rapidly diminishes and the cell begins a process leading to necrosis and death. When cerebral ischemia occurs, the source of ATP in the form of oxygen and glucose can rapidly diminish due to inadequate delivery of cerebral blood flow, a process that leads to depletion or cessation of oxidative phosphorylation.

Oxidative phosphorylation is the metabolic pathway within the mitochondria of brain cells where the citric acid cycle takes place, where enzymes and energy are released by the oxidation of nutrients to generate ATP. ATP is the molecule that supplies energy to cells and allows cellular metabolism to function. Once mitochondria, or the process that catalyzes the synthesis of ATP, is irreversibly damaged, as can happen in ischemic stroke, no neuroprotective agent or intervention can reverse neuronal death. What is important about this simple concept is the erroneous impression some investigators distill from studies where a lethal lesion is created, irreversible neuronal death is obtained, treatment of a putative neuroprotective agent is given, failure to reverse neuronal death and its consequences is observed, and a conclusion about the *failure* of the agent is assumed.

Under unusual circumstances, the neuroprotective agent can *only* prevent the brain injury from expanding or potentially protect the injured brain cells from dying, as may be the case in the ischemic penumbra where neurons may lie dormant until revival or death.

A case in point is the relative efficacy of tPA, the only FDA-approved treatment for ischemic stroke. Studies have shown that within the 3–4½ h time window where tPA is most effective, 26% of patients return to normal spontaneously without treatment, while an additional 13% return essentially to normal if treated with tPA.[224] This finding indicates that even when part of the brain is still thought to be salvageable after an acute ischemic stroke, the chances for returning the patient to a preocclusion status following the best treatment available at the present are limited indeed. The potential harm of tPA thrombolysis is another consideration of the potential for neuroprotection by a pharmacological agent. There has also been considerable concern that tPA can harm patients. For example, the percentage of patients who develop symptomatic intracerebral hemorrhages was reported increased by tPA therapy from 0.6% in placebo-treated patients to 6.4% in tPA-treated patients.[225]

This may be the reason why some experiments testing the potential ability of DMSO in a variety of injuries to the brain and spinal cord show partial or incomplete success in preventing or reversing the damage caused by the lesion. The success and benefit of such treatments should be weighed in terms of the gravity of the injury, the relative benefit of a placebo or no treatment, and the safety of the experimental agent.

In this sense, DMSO's ability to prevent lipid peroxidation and increase high-energy phosphates such as ATP and PCr in the face of brain tissue headed for bioenergetic failure constitutes a dramatic advancement in the field of neuroprotection that mandates further exploration.

Examination of another property by DMSO that seems to be the basis of action for thrombolytics like tPA, acetylsalicylic acid (aspirin), and clopidogrel bisulfate (Plavix) is their action in inhibiting platelet aggregation, a condition that generally results from prolonged cerebral ischemia. Multiple studies have shown that thrombosis of the brain is the major cause of stroke. Cerebral thrombosis can occur when a blood clot blocks a cerebral vessel either partially or completely. The clot or thrombus is caused by an inappropriate activation of the hemostatic process in an uninjured or mildly injured brain vessel through the aggregation of platelets that form a platelet plug. The clot can adhere to the vessel wall and impede or completely cut off cerebral blood flow, resulting in tissue death. Treatment of cerebral thrombosis can be started as soon as a diagnosis is confirmed. These consist of using thrombolytics such as tPA, clopidogrel, and aspirin to break up the clot made up of mostly aggregated platelets. Blood thinners (anticoagulants) can be given after platelet deaggregation to prevent further clots from forming.

DMSO when given at 2 g/kg 6 h after embolic stroke has been shown to prevent thrombosis in mongrel dogs subjected to an embolus introduced into the MCA. The DMSO-treated animals were observed to have normal behavior and no neurological deficits following the experimental embolization.[226] By contrast, three of nine control animals died after embolization, showing large infarctions of the thalamus, internal capsule, and piriform cortex, while the survivors showed contralateral hemiplegia and impaired consciousness.[226] The authors of this study postulated that the positive actions of DMSO in the postischemic period may have been due to its action as a platelet deaggregator. This idea seemed reasonable because previous studies had shown that agents, such as methohexital and in this case DMSO, are able to protect the endothelial membrane in ischemic blood vessels, thus allowing prostacyclin to be released from endothelial cells at the site of ischemia to inhibit platelet aggregation.[227] This notion is the basis for using tPA to dissolve cerebral vessel clots made up of platelet aggregates and clopidogrel or aspirin to prevent further platelet aggregation.

The assumption that DMSO could function as a platelet deaggregator was confirmed in a later study in rats whose carotid arteries had been occluded.

Dujovny and his team[102] examined experimentally occluded carotid arteries using scanning electron microscopy and observed within these vessels heavy deposits of platelets, red cell clumps, fibrin, and endothelial flattening in untreated control animals. Animals treated with intravenous DMSO (2 g/kg) given 1 h prior to carotid occlusion revealed endothelial preservation and no platelet or fibrin clumping.[102]

The findings by Dujovny et al.[102] suggested that the beneficial action of DMSO in experimental brain trauma and cerebral ischemia may have been due, at least partly, to the protective effects by DMSO on the microcirculation by decreasing platelet aggregation and thus preventing the evolution of postischemic reperfusion injury.

This effect on platelets by DMSO seems entirely plausible from the point of view that present therapy for arterial thrombosis consists of administering platelet deaggregators such as cyclooxygenase inhibitors (aspirin), ADP receptor inhibitors (clopidogrel, ticlopidine), adenosine reuptake inhibitors (dipyridamole), PDE inhibitors (cilostazol), and glycoprotein IIb/IIIa receptor antagonists that can block the platelet receptor for fibrinogen (tirofiban). What class of antiplatelet drugs DMSO might belong to remains to be determined.

DMSO IN MISSILE INJURY TO THE BRAIN

From a historical point of view, one of the more wretched creations introduced to society is the invention of the pistol not for hunting or target practice but for the sole purpose of killing people. Since the 1970s, firearm deaths have increased dramatically in the United States every year. Statistics indicate that in 1988, there were about 34,000 gunshot deaths in the United States and many of these were cranial gunshot wounds. By comparison, Canada statistics show 1450 deaths from gunshot wounds around the same period.[228] In large U.S. cities, gunshot wounds to the head rank as the highest cause of all trauma to the brain.

Bullets fired from guns tend to be classified as low velocity (<300 m/s) or high velocity (>300 m/s), but this is a technicality since both high- and low-velocity bullets act similarly in creating their pathology. High- or low-velocity missile wounds to the head tend to result in a temporary cavity that produces bursting injuries of the tissue, including blood vessels. When key blood vessels are damaged and bleeding cannot be stopped, the patient will likely die from blood loss at the site of injury.

The trajectory of a bullet through the brain carries a major significance as it crushes structures along its track. Bullets that cut through the brainstem, or the ventricular chambers where spinal fluid is located, are especially lethal.

Gunshot wounds to the head are nearly always fatal and, in the best-case scenario, can leave the victim with severe neurological disability. It is estimated that two-thirds of the victims shot in the head die before reaching a hospital.

When a high-velocity bullet hits the head, the skull often shatters into small pieces, creating not only more wounds to the brain but also a source for infection. Gunshot wounds to the head are classified as penetrating, where the bullet remains lodged in the brain, or perforating where the bullet exits the brain.

If the patient with a gunshot wound to the head reaches the hospital, a number of critical systems must be addressed if the patient is to survive. These include treating high ICP, low oxygenation, hypertension, falling cerebral blood flow, and rising cerebrovascular resistance that must be treated aggressively if the patient is to survive with a good outcome. If a hematoma is confirmed by CT scan, a craniotomy and surgical evacuation of the clot can be performed, but high ICP may persist. The high pressure in the brain may respond to mannitol treatment but a rebound effect where ICP returns higher than before mannitol treatment is often seen.[229]

When patients are deeply comatose and there is minimal evidence of brainstem function in the absence of an intracranial hematoma that might be causing the coma, a fatal outcome is nearly always certain.

An important study was reported by Brown and his colleagues[230] from the Division of Neurosurgery at the University of Chicago Hospitals and Clinics, a hospital which is classed as a level 1 trauma center. Trauma centers are designed to take care of severe injuries such as car crashes, explosions, and gunshot wounds.

The study by Brown and his associates[230] was prompted by the horrific rate of deaths due to gunshot wounds to the head seen in many large U.S. cities, including Chicago, in the 1980s, which dismally continue to this day. The objective of the study was to find out the clinical usefulness of mannitol as a treatment for a missile injury simulating a gunshot wound to the head and to compare this osmolyte to untreated controls. In addition, because DMSO had already been shown to be effective in nonhuman primate brain trauma, and in focal ischemic brain injury, the effects of mannitol were also compared to animals treated with DMSO.

Seven rhesus monkeys were included in the study. After anesthesia, monkeys were monitored for ICP, cerebral blood flow, CPP, cerebrovascular resistance, blood pressure, and cerebral metabolic rate of oxygen consumption, along with blood gases and hematocrit.[230] A BB pellet missile was fired at a velocity of 90 m/s through a burr hole made in the right cerebral hemisphere. Animals received either mannitol (0.05 g/kg, 25% solution) or DMSO (0.05 g/kg in a 50% solution) as an intravenous bolus 1 h following injury, and these doses were repeated when ICP rose above 20 mm Hg pressure. All parameters were measured 1 h after treatment was initiated. Animals were killed 6½ h after the last measurements were made. The results are summarized in Figures 8.7 through 8.10.

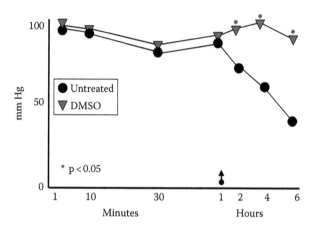

FIGURE 8.7 Mean arterial pressure plunged 1 h after injury in untreated controls (dark circles) but remained stable following DMSO (triangles) treatment beginning 1 h after treatment. (Adapted from Brown, F.D. et al., *Ann. NY Acad. Sci.*, 411, 245, 1983.)

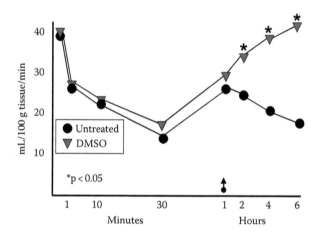

FIGURE 8.8 Cerebral blood flow dropped to more than 50% after injury and rose after 1 h following DMSO (triangles) as compared to untreated controls (dark circles). (Adapted from Brown, F.D. et al., *Ann. NY Acad. Sci.*, 411, 245, 1983.)

The ICP rose in all animals 1 min after injury and fell significantly 1 h following DMSO, reaching pre-injury levels after 6 h.[230]

In untreated animals, mean arterial pressure briefly rose after injury and decreased progressively for the next 6 h until death. DMSO dramatically reversed the hypotensive process 1 h following injury[230] (Figure 8.7).

DMSO was able to stabilize mean arterial pressure, cerebral blood flow, CPP, and metabolic rate of oxygen consumption 1 h following treatment. ICP was also reduced to preinjury levels by DMSO.[230]

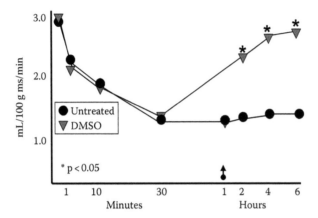

FIGURE 8.9 Cerebral metabolic rate of oxygen consumption decreased in all animals after injury and rose progressively after DMSO (triangles) as compared to untreated animals (dark circles). (Adapted from Brown, F.D. et al., *Ann. NY Acad. Sci.*, 411, 245, 1983.)

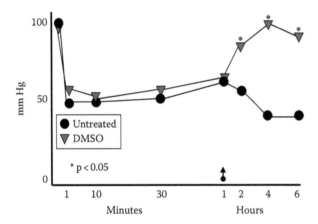

FIGURE 8.10 CPP fell to nearly half of the baseline values within 10 min of injury and was reversed by DMSO (triangles). Two hours after DMSO treatment, CPP values had returned to baseline unlike untreated animals (dark circles). (Adapted from Brown, F.D. et al., *Ann. NY Acad. Sci.*, 411, 245, 1983.)

Cerebral blood flow rose markedly after DMSO (Figure 8.8), and this rise paralleled those seen for the cerebral metabolic rate of oxygen (Figure 8.9) and CPP (Figure 8.10) as compared to significant decreases in untreated animals.[230]

CPP is the difference between the mean arterial pressure and the ICP and represents the vascular pressure gradient across the cerebral beds. It is now well recognized that low cerebral blood flow and CPP are potentially harmful in patients with brain trauma and are associated with poor outcome.[231,232]

What is less clear is whether altering cerebral blood flow or CPP will lead to clinical improvement. The study by Brown et al.[230] suggests that the traumatized brain requires a normal CPP before adequate cerebral blood flow and satisfactory oxygen metabolism can be attained. This thinking is now supported by the work of investigators from the Brain Trauma Foundation.

With respect to DMSO's ability to stabilize mean arterial pressure after a missile injury, several studies have documented that brain trauma worsens in patients who experience episodes of hypotension in the first few hours after injury.[233,234]

DMSO COMPARED TO MANNITOL IN MISSILE INJURY

Based on the positive results obtained with DMSO versus no treatment following missile injury to the brain, another study was designed by Brown and colleagues. This study[230] compared DMSO to the standard treatment for cerebral trauma, mannitol, following a missile injury to the brain. While mannitol was observed to be better than no treatment in each parameter measured, CPP, mean arterial pressure, cerebral blood flow, and oxygen metabolism indicated DMSO as vastly more superior to mannitol treatment.[230] The authors' conclusion from these findings was that this difference indicated that DMSO is a more effective agent than mannitol in their model injury and may have been due to many incalculable factors, including the ability of DMSO to maintain good cardiac output following cardiac ischemia, a problem often seen complicating brain trauma.[137,230,235]

DMSO IN AGING RESEARCH

The average human life expectancy in the United States has increased from 50 to 78 years in males and 80 years in females in the last 100 years. According to U.S. Census Bureau projections, by 2050, one in five Americans will be 65 or older, and at least 400,000 will be 100 or older.[236] Some futurists think even more radical changes are coming, including medical treatments that could slow, stop, or reverse the aging process and allow humans to remain healthy and productive to the age of 120 or more.

Aging research is now the focus of thousands of laboratories that include studies on telomerase to stem cells to energy and oxyradical formation. Moreover, the manipulation of the cellular and molecular mechanisms that extend lifespan such as prevention of DNA damage and genetic engineering as well as lifestyle objectives, involving a healthy diet, exercise, and risk assessment to disease, is an important mechanism in healthy life extension.

Aging is a complex process driven by diverse molecular pathways and biochemical events, and it is not too surprising that lifespan is the outcome of a variety of integrated processes. It is this diversity of processes that make it difficult to understand aging exclusively in terms of molecular or genetic mechanisms. While one research approach to study longevity can concentrate on identifying specific biochemical and biophysical processes of aging, another approach should promote the understanding of aging phenotypes and their variability.

Much of the research into life extension is presently done using a roundworm called *Caenorhabditis elegans*. This nematode is a noninfectious, nonpathogenic, and nonparasitic organism. *C. elegans* is a transparent nematode about 1 mm in length that has become a model organism for aging research. *C. elegans* is a very primitive organism that nonetheless shares many of the essential biological activities that are central to human biology. As simple as this organism is, it has a nervous system that contains only 302 neurons and about 7000 synapses, compared to human brain that has about 100 billion neurons and probably one thousand times more synapses.

Its simple nervous system seems to allow it to express behavior and even rudimentary learning.

Another useful feature of *C. elegans* is its transparency, a detail that allows the study of cell differentiation and cell development, characteristics that are important in the study of aging.

The Nobel Prize in Physiology or Medicine was awarded in 2002 to Sydney Brenner, Robert Horvitz, and John Sulston, for their discoveries concerning the genetic regulation of organ development and programmed cell death (apoptosis) and the significant contribution of experimental organisms such as *C. elegans*. This work has contributed much to our understanding of human physiology and pathophysiology.

C. elegans can be used for testing compounds that may have an effect on aging and longevity. The use of *C. elegans* is a useful approach to study longevity because the worm has a fast, conventional lifespan of about 14–20 days, is small in size, is easy to maintain in the laboratory, and has a powerful genetic toolkit where manipulation of the genome by adding, removing, or altering specific genes already known to lengthen lifespan can be made by relatively routine procedures. The worm can be frozen at −80°C and stored for 10 years with good survival.

DMSO has been used in several studies as a solvent for compounds reported to extend the lifespan of *C. elegans*, but it was not determined in those studies whether DMSO was inert or played a role in extending lifespan.[237,238]

Assuming that DMSO was inert, the authors failed to notice any positive function of DMSO on the lifespan of *C. elegans*.

However, it has been recently reported that DMSO itself extends the lifespan of *C. elegans*. A study by Wang et al.[239] revealed that DMSO extended the lifespan of *C. elegans* in a dose-dependent manner. The mean lifespan of *C. elegans* increased with DMSO treatment 15% and 23% at doses ranging from 0.01% to 2% and reached a maximum at 0.5%. At concentrations of 5% DMSO, it decreased the mean lifespan of *C. elegans*, indicating that DMSO is toxic at that concentration.[239]

This study also attempted to clarify the underlying mechanism of DMSO to increase lifespan extension in this model nematode. Several assays suggested that DMSO increased the lifespan of *C. elegans* by modulating gene expression. These assays pointed to DMSO modulating gene expression in the insulin/IGF-1 signaling pathway (Wang). Genetic analysis suggested that DMSO did not function through dietary restriction or thermal resistance since there was no life extension by DMSO during thermal stress of this organism.[239]

The discovery of antiaging properties of DMSO from Wang et al.'s[239] study provides a good basis to try DMSO in other aging models to detail more precisely its role in life extension.

DMSO IN EXPERIMENTAL DEMENTIA

Dementia is the term used for a group of symptoms caused by disorders that affect brain function and, as such, is not a specific disease. Dementia is characterized by progressive deterioration of intellectual functions and inability to carry out daily activities. Memory loss is an early and common symptom of dementia.

AD and VaD are the two most common forms of dementia among older people. Although traditionally considered separate pathologies, strong evidence indicates that AD and VaD share common pathogenic mechanisms including vascular risk factors, histopathology, clinical signs, diagnostic criteria, and treatment approach.[240]

AD is an insidious disorder that progressively ravages the brain, resulting in synaptic and neuronal loss, ultimately destroying memory, intellect, and human dignity in the process.

AD begins slowly involving parts of the brain that control thought, planning, language, and memory. People with AD may have trouble remembering things that happened recently or names of people they know.

Over time, AD symptoms get worse. People may not recognize family members or have trouble speaking, reading, or writing. They may forget how to brush their teeth or comb their hair. Later on, they may become anxious or aggressive.

AD usually begins after age 60. The risk goes up as you get older. No treatment can stop the disease. It is estimated that AD affects 5½ million people in the United States and will increase to 14 million by 2050.

Although it is not presently known what causes AD, it has been speculated for sometime that abnormal deposits of proteins called Abeta form amyloid plaques and produce neurofibrillary tangles throughout the brain. This accumulation of plaques and tangles is believed to cause massive neurodegeneration and stop communication between neurons, thus leading to loss of cognitive function and eventually dementia.

However, a conceptual shift in our understanding of dementia came in 1993, when de la Torre and Mussivand[241] proposed that sporadic AD was a vascular disorder with neurodegenerative consequences rather than vice versa. This proposal was supported by strong epidemiological, pathological, pharmacotherapeutic, and neuroimaging data, which collectively explained how the metabolic, inflammatory, hemodynamic, and physiopathological events seen to develop preclinically in AD are associated with a cerebral energy crisis secondary to chronic brain hypoperfusion. Reduced blood flow to the brain appears to result from the presence of one or more of two dozen described vascular risk factors.[242] These risk factors include atherosclerosis, dyslipidemia, cardiovascular disease, diabetes type 2, and stroke, all which commonly affect elderly individuals who are more vulnerable to damage from these conditions during advanced aging.[243]

It is not surprising that many animal models have been created to study this devastating dementia. The animal models used mainly serve to test and screen potential pharmacotherapy aimed at arresting, slowing, or reversing the pathological process that begins at the onset of AD. Models of AD include transgenic mice that produce abnormally high levels of Abeta peptides,[244,245] animals whose cholinergic system can be manipulated,[246] animals with induced amnesia,[247] knockout mice for tau,[248] animals with lesions of memory-related sites,[249] and even the nematode *C. elegans*.[250]

A rat dementia model was developed[251] and used to study the ability of DMSO combined with the glycolytic intermediate FDP.[252,253]

This model was unique in the sense that most other models previously used induced brain ischemia by lesioning the brain with cytotoxic chemicals, cold freezing, or focal electrolytic lesions. Temporary or transient bilateral common carotid artery occlusion (2-vessel occlusion [2-VO]) did not injure the brain directly, allowing instead a process where chronic brain hypoperfusion and its effect in producing memory impairment take place. This pathologic and progressive process mimics severe human carotid artery stenosis or occlusion, which over a protracted period of time results in mild cognitive impairment, the preclinical precursor of AD.[254,255]

Older rats 14 months old were subjected to either permanent bilateral carotid artery occlusion (2-VO) (Figure 8.11) or sham occlusion where surgery but no occlusion was done. Following surgical recovery, rats were tested for visuospatial memory function every 2 weeks for a total of six sessions.[252] On week 14 postsurgery, four rats from the 2-VO group who in previous trials had shown persistent and severe visuospatial memory impairment were selected to receive DMSO + FDP pharmacotherapy. These four rats were given once a day, a combined 28% solution composed of DMSO–FDP (250:130 mg/kg) intraperitoneally for seven consecutive days.

Nontreated 2-VO rats that only showed moderate visuospatial memory deficits and sham controls that had no memory impairment were given an equivalent volume of the vehicle, dextrose 5% intraperitoneally for 7 days. On week 15, after 7 days of daily treatments, all rats were retested in a water maze for memory function. Following this testing session, all treatments were discontinued for an additional

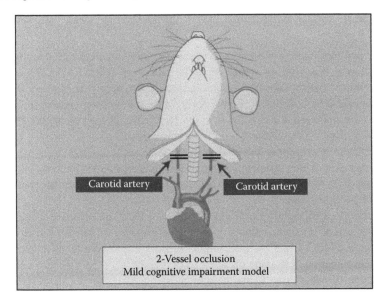

FIGURE 8.11 **(See color insert.)** Permanent occlusion of both common carotid arteries (2-VO) in 14-month old rats. After 1 to 14 weeks, mild cognitive impairment characterized by progressive visuospatial memory impairment is observed that can be quantified and reversed with DMSO treatment. (From de la Torre, J.C. et al., *Brain Res.*, 779, 285, 1998.)

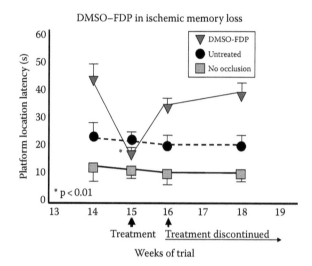

FIGURE 8.12 Old rats subjected to permanent bilateral carotid artery occlusion (2-VO) for 14 weeks. Rats with severest rate of visuospatial memory impairment selected by highest platform location latency (PLL) were given DMSO-FDP (triangles) for 7 days on weeks 14 and 15. Significant (54%) improvement is seen in PLL after DMSO-FDP as compared to untreated 2-VO in reference to non occlusion rats. Following discontinuation of DMSO-FDP on weeks 16–18, gained memory improvement is partially lost. Untreated rats with 2-VO (circles) show unchanging moderate rate of visuospatial memory impairment, while rats with no 2-VO (squares) show no memory deficits throughout the 18 weeks of trials. Bars are ± SEM. See text for details. (Adapted from de la Torre, J.C. et al., *Brain Res.*, 779, 285, 1998.)

3 weeks and rats were tested again in the water maze on weeks 16 and 18.[252] The results showed that after 14 weeks of 2-VO, rats given DMSO + FDP had a 54% improvement in their visuospatial memory task, which nearly reached controls who had no carotid artery occlusion (Figure 8.12). Moreover, when treatment of DMSO + FDP was discontinued, the marked memory improvement seen on weeks 15 and 16 regressed to almost pretreatment level[252] (Figure 8.12).

Postmortem microscopic and densitometric analysis of brain sections stained for neurons and axonal fibers showed only mild loss of hippocampal neurons in rats with 2-VO regardless of treatment, and an increase in glial density was observed only in untreated 2-VO animals.[252]

The results of this study imply that DMSO + FDP is an effective treatment for improving severe ischemic-induced visuospatial memory impairment in aging rats. The effect of DMSO + FDP on visuospatial memory improvement and its partial loss after discontinuation of therapy on week 16 is suggestive of a drug cause and effect. It also suggests that in order to maintain memory improvement during induced brain ischemia, this drug combination must be maintained. Studies have suggested that from 2 to 3 years before clinical dementia symptoms begin at a stage called mild cognitive impairment, there are already changes occurring in patients' slowdown of visuospatial skills involving speeded tasks that require coordination of eye and visual representation needed for normal short-term memory function.[256]

Since visuospatial memory deficit is one of the earliest signs of impending AD during the mild cognitive impairment stage,[257] DMSO + FDP may have interventional value in preventing irreversible memory impairment in patients at high risk of developing AD.

Confirmation of DMSO-FDP study was reported using the brain hypoperfusion model of bilateral carotid artery occlusion (2-VO) to test DMSO without FDP in rats.[256] DMSO at doses of 0.25 mL administered intraperitoneally for 5 consecutive days 11 weeks following 2-VO was seen to improve spatial memory learning as compared to nontreated 2-VO animals whose memory capacity was markedly worse.[258] In addition, microscopic examination of rat brains showed that 2-VO without postoperative treatment induced hippocampal and cortical neuronal loss in nontreated animals, but it was observed that in DMSO-treated rats, this loss of neurons was significantly reduced.[258]

An interesting study using Lurcher mice was recently reported by Markvartova and her colleagues.[259] Lurcher mutant mice represent one of the frequently used mouse models of olivocerebellar degeneration. It is caused by a mutation in the δ2 glutamate receptor subunit encoding gene.[260]

Lurcher mice suffer from almost complete degeneration of cerebellar Purkinje cells after birth and develop decreased density of granule, stellate, inferior olivary neurons, and basket cells but mild changes in deep cerebellar nuclei.[261]

With almost total disruption of cerebellocortical signal processing, loss of Purkinje cells makes walking progressively more difficult for heterozygous Lurchers and clearly abnormal by 30 days postnatally. Homozygous Lurchers die at birth.

Another characteristic of Lurcher mice is the link between cerebellar pathology and a physical activity that can influence the long-term potentiation (LTP) in the hippocampal region.

LTP is widely considered one of the major cellular mechanisms that underlie learning and memory.[262]

In this setting, DMSO was found to attenuate the age-related deterioration of some cognitive functions, including spatial learning abilities in 21-day-old treated Lurcher mice.[259]

Animals were tested in a water maze designed to measure their learning and short-term memory ability. Before each memory trial, mice received either saline (no treatment) or DMSO administered intraperitoneally 30 min before the experiment at a dose of 4 μL/g of the body weight. Lurcher mice showed spatial navigation and spatial memory deficits compared to normal wild mice. In DMSO-treated Lurcher mice, a substantial improvement was observed in the learning and memory ability as compared to Lurcher mice given saline.[259]

The possible mechanism for this improvement was not discussed by the authors. It is possible that DMSO may have imparted some usefulness to Lurcher mice by its ability to protect loss of function mutations like that which occurs in olivocerebellar neurodegeneration. The damaging mechanism could involve an attack of normal protein folding and lead to their degradation, a process that has been shown to be inhibited by DMSO with help from the neuron's own defense counterattack mechanisms.[176]

Additionally, DMSO is also capable of minimizing the formation of hydroxyl-induced sugar/base assault that is known to form near the vicinity of DNA damage during double-strand DNA breaks.[263,264] These mechanisms have not been studied in detail, and no final conclusions can be made with certainty. Finally, DMSO may be acting as a chemical chaperone, that is, as a low-molecular-weight compound whose function is to help correct conformational diseases characterized by protein misfolding, including neurodegenerative disorders.[265,266] This therapeutic approach would work by allowing the mutant proteins to escape the quality control systems of their genotype so that their function can be rescued by so-called folding agonists.

These small molecules or folding agonists like DMSO, glycerol, and deuterated water are osmotically active, a property thought to nonselectively stabilize the mutant proteins to facilitate their proper folding.[267]

DMSO IN EXPERIMENTAL SPINAL CORD INJURY

SCI refers to any physical damage that results in impaired function to movement or feeling below the level of trauma. Injury to the spinal cord is either complete, where sensory-motor function is lost below the injury site on both sides of the body, or incomplete, where some function remains and complete recovery is possible.

The level of injury is important because it determines what parts of the body will be affected. For example, a complete injury to the fourth cervical level of the cord (C-4) will result in quadriplegia with little function below the neck and will require a ventilator for breathing, whereas an injury to C-7 will retain the ability to breathe well and extend the arms but there may be trouble working the fingers with dexterity. Paraplegias can be complete or incomplete injuries that result from damage to the first thoracic level of the cord (T-1) and below and where the use of the arms is retained but not the legs or parts of the trunk.

The main causes of spinal cord injuries are from motor vehicle accidents, violence, falls, and sports, and CDC estimates about 11,000 spinal cord injuries occur annually who join a population of about 260,000 persons who already have an injury to the spinal cord.

There have been a number of advances made in the last 50 years in the study of injured spinal cord, including basic and clinical tools to examine the spinal cord objectively and as critical adjuncts of neurologic examinations.

Evoked potential is an important tool that can provide a quantitative measure of function in the respective sensory and motor tracts. Evoked potentials can also reveal the presence of unsuspected demyelinating disorders and their processes as well as monitor sensory-motor changes over time to assess neurologic progress and long-term prognosis.[268] Evoked potentials can be supplemented with newer techniques of evaluating spinal cord trauma such as transcranial magnetic motor-evoked potentials and dermatomal somatosensory-evoked potentials, which provide additional quantifiable information about the status of the cord.[269]

Aside from the use of neurophysiologic monitoring systems and methods that detail the morphology and contents of cord tissue, much of the experimentation conducted within the last 25 years has provided a better understanding and clinical therapeutic approach to the injured spinal cord than at any time before. Such work has exposed significant aspects in the biochemistry and vascular mechanics associated with trauma to the cord.[270] A growing and intriguing area of spinal injury research lies in probing the factors related to neuronal plasticity and regeneration of the cord tissue.

Despite hundreds of experimental treatments reported in the last half century on dozens of SCI animal models, no drug therapy thus far has shown effectiveness in reversing or arresting the pathologic damage, leading to paralysis when tested in randomized human clinical trials. The failure of testing DMSO to treat severe spinal cord trauma in humans, as will be seen, is a curious phenomenon given the positive evidence that will be presented here.

Instead, clinically, intravenous steroid infusions, which were available 40 years ago and have been compared experimentally to DMSO in rigorous animal experiments (see the following), are presently used in patients by many hospitals after a traumatic SCI in order to reduce the acute swelling that generally results after a severe contusion injury. This approach is baffling to many who treat spinal injuries. The reason is that strong evidence indicates that methylprednisolone, the steroid of choice, lacks evidence of effectiveness at the doses recommended following an SCI in humans and may in fact be harmful to the patient.[271] Interestingly, it was found that neurosurgeons in North America had been prescribing methylprednisolone for acute spinal cord trauma not out of conviction that the drug was effective but out of fear of being sued for malpractice.[272]

A recommendation by several Canadian studies was based on a wide search of citations that specifically addressed spinal cord injuries and the use of methylprednisolone in humans and that, after analyses, did not support the use of this drug as a standard treatment for spinal cord trauma.[271–273] The call into question surrounding methylprednisolone as a standard therapy for spinal cord injuries pointed to the dubious study design and inappropriate statistical analysis used in the original study.[274] As a result of this controversy, the FDA has decided not to approve methylprednisolone for SCI.

Surgery and decompression may be needed when the injury has caused the bones to be unstable or when there is pressure on the spinal cord or spinal nerves by broken or displaced vertebral bones, a blood clot, or a displaced disk. These bones can move and cause further injury if not stabilized.

The use of DMSO in SCI was introduced by de la Torre and his group[63,275–277] in 1973, as a result of experiments showing dramatic benefit of this drug in rhesus and squirrel monkeys following severe brain trauma and disabling brain ischemia.[63]

DMSO has been used in experimental SCI, but it has not been studied in a large human population.[278] The results of animal experiments indicate that if a severe spinal cord trauma is treated with intravenous DMSO within 2 h, paralysis may be prevented.[63,275,276,279] Most studies performed thus far are consistent with this action by DMSO.

When DMSO was examined for its action in experimental SCI and compared with other standard therapy such as steroids, HBO, mannitol, or urea, the findings indicated that DMSO fared far superior with respect to faster sensory-motor recovery, reduced neural damage to the cord, lower swelling of tissue after trauma, increased muscle tone return, and earlier return of somatosensory-evoked potentials.[63,280–282]

The membrane-protecting and membrane-repair qualities after DMSO administration have been described for axons and their myelin sheaths after spinal cord trauma.[280] An underlying mechanism in neuronal degeneration following trauma to the spinal cord is the inability of the tissue to repair itself in the face of membrane damage from edema, hemorrhage, and toxic molecules generated by the injury. DMSO was found to render significant improvement in resealing guinea pig axolemma following transection of the cord.[283]

This was physiologically demonstrated by monitoring membrane potential recovery and by the lack of extrusion of the marker horseradish peroxidase (HRP) given 60 min after injury.[283] It is difficult to explain how DMSO may achieve the protection or repair ability of tissue membranes, but a likely mechanism may rely on how DMSO can limit spinal cord tissue inflammation and tissue cavitation while increasing spinal cord blood flow.[281,282]

One of the original studies on the use of DMSO in SCI involved permanent paralysis induced in dogs from experimental trauma to the spinal cord.[63] Drug treatments consisting of mannitol(1 g/kg), dexamethasone 0.5 mg/kg, isotonic saline (5 mL), or DMSO (2.5 g/kg, 40% solution) were given intravenously 1 h after trauma and for the next 3 days after injury.[63] Dogs were evaluated daily for 90 days with respect to sensory-motor function and bladder control, and strict posttrauma veterinary care was provided for bladder, pain, and discomfort during recovery.

The results of this study are summarized in Figure 8.13. Essentially, a beneficial response to DMSO treatment was seen in 3 of 8 dogs 4 days after SCI,

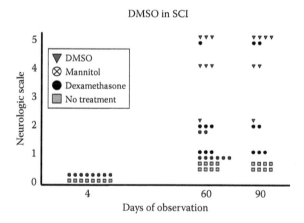

FIGURE 8.13 Neurological status using a 0–5 scale after SCI in dogs (N = 8/group) was recorded and graded daily for 90 days by an observer unaware of treatment. Each symbol represents a dog. Full recovery 90 days after SCI is seen in five dogs treated with DMSO (triangles), and two dogs treated with dexamethasone (dark circles) (p < 0.01). One other dog treated with DMSO also reached good recovery (grade 4), and one dog remained at grade 2, while another dog in this series died. Five dexamethasone-treated dogs did not reach a grade above 2 at the end of the 90-day observation period, one dog died in this series. Mannitol (circled cross) and no treatment (squares) showed no benefit and remained at grade 0–1 for 90 days after SCI. Scale: 0 = flaccid paraplegia, 1 = some muscle tone, 2 = reflex standing, 3 = spastic walking, 4 = walking, running with deficit, 5 = normal, full recovery.[63]

and a gradual improvement was noted in these animals so that 5 dogs in this series recovered completely and 1 dog showed walking and running ability with deficit.

By contrast, dexamethasone-treated dogs showed improvement late in the first and second month with 2 dogs showing full recovery at 90 days and 5 dogs remaining incapacitated (Figure 8.13). Dogs treated with mannitol or isotonic saline showed little to no improvement during the 90-day observation period. When somatosensory-evoked potentials were taken 4 h after trauma, only DMSO-treated dogs showed a partial response, which was later seen to be a positive correlation with the prognosis of eventual full recovery.[63]

This study by de la Torre and his team[63] indicated two key points. First, it established that DMSO, if given within an hour after an SCI, can protect cord tissue using this model from neurodegeneration, an effect that was evident 4 days following trauma; and second, DMSO appeared superior to steroid therapy, which is presently considered standard treatment at many U.S. centers that treat spinal trauma. These findings were never extended to explore how late in time the protective action by DMSO can be maintained after an SCI. The findings also imply that if DMSO were approved for the treatment of SCI and was seen to lose its neuroprotective activity after 1 h following trauma, it would be expedient for ambulance technicians to assess the patient at the place of injury and, if indicated, administer a bolus of DMSO intravenously.

Because DMSO has been found superior to methylprednisolone and similar steroid in cats, dogs, monkeys, and rodents sustaining a severe spinal cord trauma with respect to better motor performance, improved somatosensory-evoked potentials, and less cord tissue damage, it has been argued that substituting DMSO for methylprednisolone in the treatment of human SCI could not only prove to be neurologically superior to methylprednisolone but that DMSO would likely cause less postoperative medical complications that are seen with steroid therapy.[278,279] This argument is compelling for the following reasons.

DMSO has been shown in several studies to consistently improve somatosensory-evoked potentials following spinal cord trauma.[63,280–282] This action has not been reported following steroid treatment in spinal trauma. For example, it was found that DMSO given intraperitoneally 1 h after injury produced better motor control of rats' lower legs than methylprednisolone or naloxone 14 days after injury, a finding that was clinically supported by the improvement of the somatosensory-evoked potentials when all treatments were compared.[280]

Spinal trauma–induced dogs were given 0.63 g/kg of DMSO intravenously in a 40% solution 1 h after injury and once daily for the first two postoperative days.[281] In the same study, DMSO was compared to dexamethasone (2.2 mg/kg) at the same time schedules as DMSO. It was reported that DMSO-treated animals had a markedly significant higher group score than dexamethasone or no treatment in categories such as recovery of walking, running ability, cortical-evoked potentials, and histologically, with less cavitation, meningeal hyperplasia, and necrosis of the cord. The authors concluded that even though steroids are used clinically to treat SCI, DMSO may be a better treatment option as suggested by their findings and those of others.[281]

DMSO was examined by Gelderd and his colleagues[282] after complete spinal cord trauma, where recovery of function below the lesion site is unlikely. Rats underwent complete transection of the cord at T5 and after surgery; the animals were treated with either HBO, DMSO, or both. These investigators found that in animals receiving HBO for 54 consecutive days, 3 of 10 rats showed coordinated hind-limb movement and weight-bearing ability, whereas 6 of 10 animals given HBO + DMSO showed similar improvements and some sensory return. There were additionally less scarring and collagen formation with more undamaged nerve fibers in HBO + DMSO than other treatments.[282]

DMSO was observed to lower the neuropathic pain threshold in rats following chronic pressure on the sciatic nerve.[239] When 2% DMSO was injected intrathecally twice daily for 5 days, raised cytochrome oxidase-2 levels in the ipsilateral to injury dorsal horn resulted and significant analgesic effects were seen when compared to untreated animals.[239]

The neuroprotective effect of DMSO administration after spinal cord trauma is also observed by the ability of this drug to maintain better structural integrity than other potential neuroprotective agents, including steroids, when examined by electron microscopy. DMSO appears able to detoxify oxidative stress and peroxynitrite damage by lowering lipid peroxidase and nitrite–nitrate activity in the presence of cord ischemia.[284]

Oxidative stress is involved in a host of disorders, including brain trauma, stroke, AD, atherosclerosis, and cardiac disease,[285] while peroxynitrite is known to impair cell signal transduction, cause DNA strand breaks, and change protein structure.[286]

As reviewed in this chapter and in Chapters 1, 2, and 4, DMSO has been shown to play a positive role in brain trauma, ischemic stroke, experimental dementia, cardiac deficits, atherosclerosis, DNA strand breaks, and protein folding. It is safe to say that these properties are highly unusual for any drug that has received the intense scrutiny and experimental testing that DMSO has undergone in the past 50 years and which continues to this day. The specific bioactivities shown by DMSO in the face of central nervous system pathology together with its relative safety record should be a gold standard in pharmacotherapeutic research. These activities have been reviewed in this chapter and include its anti-inflammatory, anti-ischemic, antiedema, anticytotoxic, membrane protective, and ionic channel–stabilizing properties.

The results summarized in this chapter also suggest that DMSO in combination with an agent or agents that provide a positive biologic activity that is lacking in DMSO may result in unanticipated benefit to medical conditions refractory to other treatments. This is observed when DMSO and FDP are combined into one solution to treat brain ischemia in animals[70] or humans.[210] When brain ischemia or hypoperfusion occurs, there is a reduction of acute or chronic energy metabolism that generally affects Na^+, K^+-ATPase, a biochemical pump that is dependent on the generation of ATP derived from anaerobic glycolysis and oxidative phosphorylation.[287] Na^+, K^+-ATPase is critical for maintaining ionic concentration gradients, neuronal action potentials, neurotransmitter synthesis, and slow/fast axonal transport. In addition, low glucose turnover in brain resulting from continuing ischemia will downregulate anaerobic glycolysis by inhibiting the key regulatory enzyme controlling glycolysis, phosphofructokinase.[288]

DMSO has no known activity in improving anaerobic glycolysis or restoring the activity of the Embden–Meyerhof pathway to activate oxidative phosphorylation when administered during prolonged ischemic states. On the other hand, FDP can because it is a glycolytic intermediate in the Embden–Meyerhof cycle that can stimulate anaerobic glycolysis and increase the production of the energy fuel ATP. However, FDP is not known to increase cerebral blood flow or scavenge free radicals during ischemia as DMSO can. So, it is reasonable to pair off these two substances to augment the quality and potential of the treatment in conditions such as stroke, coronary artery disease, and vascular-dependent dementias. This appears to work in a rat model of dementia[70] as reported here, and it is only speculative, but intriguing, to theorize that this drug combination could be useful if administered as a neuroprotector in preventing or delaying the onset of AD or vascular dementia. It is only hypothetical that DMSO may be synergistic with many other standard drugs in increasing their effectiveness and reducing their dose and side effects.

DMSO AS A SOLVENT

One note of caution concerns the use of DMSO as a solvent for testing other potentially therapeutic drugs. Since DMSO is an excellent solvent, it is a common practice to dissolve an experimental drug in DMSO and then test the compound on a given disorder. However, even small doses of DMSO can be active or synergistic when mixed with other agents. A case in point is a study by Levine et al.[289] who used the matrix-metalloproteinase inhibitor GM6001 dissolved in DMSO to treat dogs that had sustained an SCI.[289] The dogs were evaluated for long-term motor-sensory recovery using a modified Frankel scale and were shown to have significant improvement when GM6001 + DMSO were used but not when GM6001 was used alone. These investigators attributed the improved neurological outcome to DMSO and not to GM6001.[289]

A similar situation arose when an epidermal growth factor inhibitor was dissolved in DMSO and tested in a rodent SCI model. When the inhibitor was compared to the DMSO vehicle, recovery of motor and bladder function was significantly greater in rats treated with DMSO relative to the inhibitor.[290]

In the realm where the politics of science exists, it seems an ethical dilemma that a simple, inexpensive, relatively safe, and highly effective agent like DMSO, which appears to be superior to standard therapy for brain trauma, brain ischemia, spinal injury, and many other traumatic conditions, has not been the object of more meticulous and comprehensive clinical studies that can determine the merits of this drug.

Since many good therapies throughout the times have languished in obscurity and apathy until a chance rediscovery has allowed them a second chance at helping control damage from disease and physical injury, it is hoped that DMSO is one of those drugs that will escape this indifference and be the object of some well-designed, randomized clinical trials that can determine its place in the history of medicine.

The simple molecular structure of DMSO also invites further research into the possibility of finding other chemically similar molecules that may be even more powerful and effective than DMSO in treating brain and spinal trauma.

REFERENCES

1. Selassie AW, Zaloshnja E, Langlois JA, Miller T, Jones P, Steiner C. Incidence of long-term disability following traumatic brain injury hospitalization, United States, 2003. *J Head Trauma Rehabil*. 2008;23:123–131.
2. Fischer H. U.S. Military Casualty Statistics: Operation New Dawn, Operation Iraqi Freedom, and Operation Enduring Freedom. CRS Report for Congress. Congressional Research Service, 2012;7-5700.
3. Coronado VG, McGuire LC, Sarmiento K et al. Trends in traumatic brain injury in the U.S. and the public health response: 1995–2009. *J Safety Res*. 2012;43:229–307.
4. Daneshvar DH, Riley DO, Nowinski CJ, McKee AC, Stern RA, Cantu RC. Long-term consequences: Effects on normal development profile after concussion. *Phys Med Rehabil Clin N Am*. November 2011;22(4):683–700, ix.
5. American Congress of Rehabilitation Brain Injury Interdisciplinary Special Interest Group. Information/education page. Preventing falls following brain injury. *Arch Phys Med Rehabil*. June 2013;94(6):1219–1220.e1.
6. Coronado VG, McGuire LC, Faul MF, Sugerman DE, Pearson WS. Traumatic brain injury epidemiology and public health issues. In *Brain Injury Medicine: Principles and Practice*, 2nd edn., eds. N.D. Zasler, D.I. Katz, R.D. Zafonte. Demos Medical Publishing, New York, 2012; pp. 84–100.
7. Zaloshnja E, Miller T, Langlois JA, Selassie AW. Prevalence of long-term disability from traumatic brain injury in the civilian population of the United States, 2005. *J Head Trauma Rehabil*. 2008;14(6):394–400.
8. Narayan RK, Michel ME, Ansell B et al. Clinical trials in head injury. *J Neurotrauma*. May 2002;19(5):503–557.
9. Overgaard J. Reflections on prognostic determinants in acute severe head injury. In *Head Injuries*, ed. R.L. McLaurin. Grunne & Stratton, New York, 1976; pp. 11–21.
10. Schwartz ML, Tator CH, Rowed DW, Reid SR, Meguro K, Andrews DF. The University of Toronto head injury treatment study: A prospective, randomized comparison of pentobarbital and mannitol. *Can J Neurol Sci*. 1984;11(4):434–440.
11. Roberts I, Sydenham E. Barbiturates for acute traumatic brain injury. *Cochrane Database Syst Rev*. December 12, 2012;12:CD000033.
12. Klatzo I. Pathophysiological aspects of brain edema. In *Steroids and Brain Edema*, eds. H.J. Reulen, K. Schumann. Springer-Verlag, New York, 1972; pp. 1–8.
13. James HE. Methodology for the control of intracranial pressure with hypertonic mannitol. *Acta Neurochir (Wien)*. 1980;51(3–4):161–172.
14. Georgiou AP, Manara AR. Role of therapeutic hypothermia in improving outcome after traumatic brain injury: A systematic review. *Br J Anaesth*. March 2013;110(3):357–367.
15. Javid M. Urea; new use of an old agent; reduction of intracranial and intraocular pressure. *Surg Clin North Am*. August 1958;38(4):907–928.
16. Härtl R, Bardt TF, Kiening KL, Sarrafzadeh AS, Schneider GH, Unterberg AW. Mannitol decreases ICP but does not improve brain-tissue pO2 in severely head-injured patients with intracranial hypertension. *Acta Neurochir Suppl*. 1997;70:40–42.
17. Bramlett HM, Dietrich WD. Pathophysiology of cerebral ischemia and brain trauma: Similarities and differences. *J Cereb Blood Flow Metab*. 2004;14(2):133–150.
18. Zammit C, Knight WA. Severe traumatic brain injury in adults. *Emerg Med Pract*. March 2013;15(3):1–28.
19. Kalimo H, Olsson Y, Paljari L, Soderfeldt B. Structural changes in brain tissue under hypoxic-ischemic conditions. *J Cereb Blood Flow Metab*. 1982;2(Suppl 1):S19–S22.
20. Cross JL, Meloni BP, Bakker AJ, Lee S, Knuckey NW. Modes of neuronal calcium entry and homeostasis following cerebral ischemia. *Stroke Res Treat*. 2010;2010:316862.
21. Lipton P. Ischemic cell death in brain neurons. *Physiol Rev*. 1999;79(4):1431–1568.

22. Lund CG, Aamodt AH, Russell D. Patient selection for intraarterial cerebral revascularization in acute ischemic stroke. *Acta Neurol Scand Suppl.* 2013;(196):65–68.

23. Zechariah A, ElAli A, Doeppner TR, Jin F, Hasan MR, Helfrich I, Mies G, Hermann DM. Vascular endothelial growth factor promotes pericyte coverage of brain capillaries, improves cerebral blood flow during subsequent focal cerebral ischemia, and preserves the metabolic penumbra. *Stroke.* June 2013;44(6):1690–1697.

24. White TE, Ford GD, Surles-Zeigler MC, Gates AS, Laplaca MC, Ford BD. Gene expression patterns following unilateral traumatic brain injury reveals a local pro-inflammatory and remote anti-inflammatory response. *BMC Genomics.* April 25, 2013;14:282.

25. Astrup J, Siesjo BK, Symon L. Thresholds in cerebral ischemia—The ischemic penumbra. *Stroke.* 1981;12(6):723–725.

26. Davis SM, Donnan GA. 4.5 hours: The new time window for tissue plasminogen activator in stroke. *Stroke.* 2009;40(6):2266–2267.

27. Mahapatra AK, Kumar R, Kamal R. *Textbook of Traumatic Brain Injury.* Jaypee Brothers Med Publishers Ltd, New Delhi, India, 2012; p. 107.

28. Heiss WD. Ischemic penumbra: Evidence from functional imaging in man. *J Cereb Blood Flow Metab.* 2000;20(9):1276–1293.

29. Lo EH. A new penumbra: Transitioning from injury into repair after stroke. *Nat Med.* 2008;14(5):497–500.

30. Pestalozza IF, Di Legge S, Calabresi M, Lenzi GL. Ischaemic penumbra: Highlights. *Clin Exp Hypertens.* October–November 2002;24(7–8):517–529.

31. Symon L, Lassen NA, Astrup J, Branston NM. Thresholds of ischaemia in brain cortex. *Adv Exp Med Biol.* 1977;4–7(94):775–782.

32. Phan TG, Wright PM, Markus R, Howells DW, Davis SM, Donnan GA. Salvaging the ischaemic penumbra: More than just reperfusion? *Clin Exp Pharmacol Physiol.* January–February 2002;29(1–2):1–10.

33. Fisher M, Takano K. The penumbra, therapeutic time window and acute ischaemic stroke. *Baillieres Clin Neurol.* August 1995;4(2):279–295.

34. Brumm KP, Perthen JE, Liu TT, Haist F, Ayalon L, Love T. An arterial spin labeling investigation of cerebral blood flow deficits in chronic stroke survivors. *Neuroimage.* July 1, 2010;51(3):995–1005.

35. de la Torre JC. Cerebral hemodynamics and vascular risk factors: Setting the stage for Alzheimer's disease. *J Alzheimers Dis.* 2012;32(3):553–567.

36. Park SJ, Nakagawa T, Kitamura H et al. IL-6 regulates in vivo dendritic cell differentiation through STAT3 activation. *J Immunol.* September 15, 2004;173(6):3844–3854.

37. Justicia C, Gabriel C, Planas AM. Activation of the JAK/STAT pathway following transient focal cerebral ischemia: Signaling through Jak1 and Stat3 in astrocytes. *Glia.* May 2000;30(3):253–270.

38. Suzuki S, Tanaka K, Nogawa S, Dembo T, Kosakai A, Fukuuchi Y. Phosphorylation of signal transducer and activator of transcription-3 (Stat3) after focal cerebral ischemia in rats. *Exp Neurol.* July 2001;170(1):63–71.

39. Loane DJ, Faden AI. Neuroprotection for traumatic brain injury: Translational challenges and emerging therapeutic strategies. *Trends Pharmacol Sci.* 2010;14(12):596–604.

40. Lei J, Gao GY, Jiang JY. Is management of acute traumatic brain injury effective? A literature review of published Cochrane Systematic Reviews. *Chin J Traumatol.* February 1, 2012;15(1):17–22.

41. Brown FD, Johns LM, Mullan S. Dimethyl sulfoxide in experimental brain injury, with comparison to mannitol. *J Neurosurg.* 1980;53:58–62.

42. Camp PE, James HE, Werner R. Acute dimethyl sulfoxide therapy in experimental brain edema: Part I. Effects on intracranial pressure, blood pressure, central venous pressure, and brain water and electrolyte content. *Neurosurgery.* 1981;9:28–33.

43. Tsuruda J, James HE, Camp PE, Werner R. Acute dimethyl sulfoxide therapy in experimental brain edema: Part 2. Effect of dose and concentration on intracranial pressure, blood pressure, and central venous pressure. *Neurosurgery*. March 1982;10(3):355–359.

43a. Del Bigio M, James HE, Camp PE, Werner R, Marshall L, Tung H. Acute dimethyl sulfoxide therapy in brain edema. Part 3: Effect of a 3-hour infusion. *Neurosurgery*. 1982;10:86–89.

44. Kulah A, Akar M, Baykut L. (Stuttg) Dimethyl sulfoxide in the management of patient with brain swelling and increased intracranial pressure after severe closed head injury. *Neurochirurgia*. 1990;33:177–180.

45. Karaca M, Bilgin UY, Akar M, de la Torre JC. Dimethyl sulphoxide lowers ICP after closed head trauma. *Eur J Clin Pharmacol*. 1991;40:113–114.

46. Waller FT, Tanabe C, Jacob SW, Paxton H. Dimethyl sulfoxide for control of intracranial pressure. *Neurosurgery*. 1979;5:583–586.

47. Hill PK, de la Torre JC, Thompson SM, Rosenfield-Wessels S, Beckett ML. Ultrastructural studies of rat fasciculi gracilis unmyelinated fibers after contusion and DMSO treatment. *Ann NY Acad Sci*. 1983;411:200–217.

48. de la Torre JC, Kawanaga HM, Hill PK, Crockard HA, Surgeon JW, Mullan S. Experimental therapy after middle cerebral artery occlusion in monkeys. *Surg Forum*. 1975;26:489–492.

49. Santos NC, Figueira-Coelho J, Martins-Silva J, Saldanha C. Multidisciplinary utilization of dimethyl sulfoxide: Pharmacological, cellular, and molecular aspects. *Biochem Pharmacol*. 2003;65(7):1035–1341.

50. Hulsmann SC, Greiner C, Kohling R, Wolfer J, Moskopp D, Riemann B, Lucke A, Wassmann H, Speckmann EJ. Dimethyl sulfoxide increases latency of anoxic terminal negativity in hippocampal slices of guinea pig in vitro. *Neurosci Lett*. 1999;261:1–4.

51. Ginsburg K, Narahash T. Time course and temperature dependence of allethrin modulation of sodium channels in rat dorsal root ganglion cells. *Brain Res*. 1999;847:38–49.

52. Repine JE, Pfenninger OW, Talmage DW, Berger EM, Pettijohn DE. Dimethyl sulfoxide prevents DNA nicking mediated by ionizing radiation or iron/hydrogen peroxide-generated hydroxyl radical. *Proc Natl Acad Sci USA*. 1981;78:1001–1003.

53. Regoli F, Winston GW. Quantification of total oxidant scavenging capacity of antioxidants for peroxynitrite, peroxyl radicals, and hydroxyl radicals. *Toxicol Appl Pharmacol*. 199;156:96–105.

54. Lu C, Mattson MP. Dimethyl sulfoxide suppresses NMDA- and AMPA-induced ion currents and calcium influx and protects against excitotoxic death in hippocampal neurons. *Exp Neurol*. 2001;170:180–185.

55. Zhao JB, Zhang Y, Li GZ, Su XF, Hang CH. Activation of JAK2/STAT pathway in cerebral cortex after experimental traumatic brain injury of rats. *Neurosci Lett*. 2011;498:147–152.

56. Mattson DH, Shimojo N, Cowan EP, Baskin JJ, Turner R, Shvetsky B, Coligan JE, Maloy W, Biddison WE. Differential effects of amino acid substitutions in the b-sheet floor and a-2 helix of HLA-A2 on recognition by alloreactive viral peptide-specific cytotoxic T lymphocytes. *J Immunol*. 1989;143:1101–1107.

57. Lim R, Mullan S. Enhancement of resistance of glial cells by dimethyl sulfoxide against sonic disruption. *Ann NY Acad Sci*. January 27, 1975;243:358–361.

58. Rosato MP, Iaffaldano N. Cryopreservation of rabbit semen: Comparing the effects of different cryoprotectants, cryoprotectant-free vitrification, and the use of albumin plus osmoprotectants on sperm survival and fertility after standard vapor freezing and vitrification. *Theriogenology*. 2013;79(3):508–516.

59. Reid TJ, LaRussa VF, Esteban G, Clear M, Davies L, Shea S, Gorogias M. Cooling and freezing damage platelet membrane integrity. *Cryobiology*. 1999;38:209–224.

60. Shimoda K, Nomura M, Kato M. Effect of antioxidants, anti-inflammatory drugs, and histamine antagonists on sparfloxacin-induced phototoxicity in mice. *Fundam Appl Toxicol*. 1996;31:133–140.

61. Weil ZM, Gaier KR, Karelina K. Injury timing alters metabolic, inflammatory and functional outcomes following repeated mild traumatic brain injury. *Neurobiol Dis*. 2014;70:108–116.

62. Tolias CM, Bullock MR. Critical appraisal of neuroprotection trials in head injury: What have we learned? *NeuroRx*. 2004;1:71–79.

63. de la Torre JC, Rowed DW, Kawanaga HM, Mullan S. Dimethyl sulfoxide in the treatment of experimental brain compression. *J Neurosurg*. 1973;38:345–354.

64. de la Torre JC, Kawanaga HM, Johnson CM, Goode DJ, Kajihara K, Mullan S. Dimethyl sulfoxide in central nervous system trauma. *Ann NY Acad Sci*. January 27, 1975;243:362–389.

65. Tung H, James HE, Laurin R, Marshall LF. Modification of the effect of dimethyl sulfoxide on intracranial pressure, brain water, and electrolyte content by indomethacin. *Acta Neurochir (Wien)*. 1983;68(1–2):101–110.

66. de la Torre JC, Surgeon JW, Ernest T, Wollmann R. Subacute toxicity of intravenous dimethyl sulfoxide in rhesus monkeys. *J Toxicol Environ Health*. January 1981;7(1):49–57.

67. Swadron SP, LeRoux P, Smith WS, Weingart SD. Emergency neurological life support: Traumatic brain injury. *Neurocrit Care*. September 2012;17(Suppl 1):S112–S121.

68. Eskandari R, Filtz MR, Davis GE, Hoesch RE. Effective treatment of refractory intracranial hypertension after traumatic brain injury with repeated boluses of 14.6% hypertonic saline. *J Neurosurg*. August 2013;119(2):338–346.

69. James HE, Camp PE, Harbaugh RD, Marshall LF, Werner R. Comparison of the effects of DMSO and pentobarbitone on experimental brain oedema. *Acta Neurochir (Wien)*. 1982;60(3–4):245–255.

70. de la Torre JC. Synergic activity of combined prostacyclin: Dimethyl sulfoxide in experimental brain ischemia. *Can J Physiol Pharmacol*. February 1991;69(2):191–198.

71. Rao CV. Differential effects of detergents and dimethylsulfoxide on membrane prostaglandin E1 and F2alpha receptors. *Life Sci*. 1977;20(12):2013–2022.

72. LaHann TR, Horita A. Effects of dimethyl sulfoxide (DMSO) on prostaglandin synthetase. *Proc West Pharmacol Soc*. 1975;18:81–82.

73. Cheng G, Kong RH, Zhang LM, Zhang JN. Mitochondria in traumatic brain injury and mitochondrial-targeted multipotential therapeutic strategies. *Br J Pharmacol*. October 2012;167(4):699–719.

74. Fiskum G. Mitochondrial participation in ischemic and traumatic neural cell death. *J Neurotrauma*. October 2000;17(10):843–855.

75. Wang CH, Wu SB, Wu YT, Wei YH. Oxidative stress response elicited by mitochondrial dysfunction: Implication in the pathophysiology of aging. *Exp Biol Med (Maywood)*. May 1, 2013;238(5):450–460.

76. Lagouge M, Larsson NG. The role of mitochondrial DNA mutations and free radicals in disease and ageing. *J Intern Med*. June 2013;273(6):529–543.

77. Soustiel JF, Zaaroor M. Mitochondrial targeting for development of novel drug strategies in brain injury. *Cent Nerv Syst Agents Med Chem*. June 2012;12(2):131–145.

78. Ikeda Y, Long DM. Comparative effects of direct and indirect hydroxyl radical scavengers on traumatic brain oedema. *Acta Neurochir Suppl (Wien)*. 1990;51:74–76.

79. McCord JM, Fridovich I. The utility of superoxide dismutase in studying free radical reactions. I. Radicals generated by the interaction of sulfite, dimethyl sulfoxide, and oxygen. *J Biol Chem*. November 25, 1969;244(22):6056–6063.

80. Chapman JD, Doern SD, Reuvers AP, Gillespie CJ, Chatterjee A, Blakely EA, Smith KC, Tobias CA. Radioprotection by DMSO of mammalian cells exposed to X-rays and to heavy charged-particle beams. *Radiat Environ Biophys*. February 23, 1979;16(1):29–41.

81. Raaz U, Toh R, Maegdefessel L, Adam M, Nakagami F, Emrich F, Spin JM, Tsao PS. Hemodynamic regulation of reactive oxygen species: Implications for vascular diseases. *Antioxid Redox Signal*. 2014;5:5214.

82. Sanmartín-Suárez C, Soto-Otero R, Sánchez-Sellero I, Méndez-Álvarez E. Antioxidant properties of dimethyl sulfoxide and its viability as a solvent in the evaluation of neuroprotective antioxidants. *J Pharmacol Toxicol Methods*. March–April 2011;63(2):209–215.

83. Bhatia RK, Pallister I, Dent C, Jones SA, Topley N. Enhanced neutrophil migratory activity following major blunt trauma. *Injury*. 2005;36:956–962.

84. Hall ED, Vaishnav RA, Mustafa AG. Antioxidant therapies for traumatic brain injury. *Neurotherapeutics*. 2010;7:51–61.

85. Lu J, Goh SJ, Tng PY, Deng YY, Ling EA. Systemic inflammatory response following acute traumatic brain injury. *Front Biosci*. 2009;14:3795–3813.

86. Lucas SM, Rothwell NJ, Gibson RM. The role of inflammation in CNS injury and disease. *Br J Pharmacol*. 2006;147(Suppl 1):S232–S240.

87. Bone RC. Toward a theory regarding the pathogenesis of the systemic inflammatory response syndrome: What we do and do not know about cytokine regulation. *Crit Care Med*. 1996;24:163–172.

88. Essani NA, Fisher MA, Jaesche H. Inhibition of NF-κB activation by dimethyl sulfoxide correlates with suppression of tnf-alfa formation, reduced icam-1 gene transcription, and protection against endotoxin-induced liver injury. *Shock*. 1997;7:90–96.

89. Simpson IA, Carruthers A, Vannucci SJ. Supply and demand in cerebral energy metabolism: The role of nutrient transporters. *J Cereb Blood Flow Metab*. November 2007;27(11):1766–1791.

90. Rapela CE, Green HD. Autoregulation of canine cerebral blood flow. *Circ Res*. 1964;15:I205–I211.

91. Boysen G. Cerebral hemodynamics in carotid surgery. *Acta Neurologica Scandinavica*. 1973;49(Suppl 52):3–86.

92. Pollock SS, Harrison MJ. Red cell deformability is not an independent risk factor for stroke. *J Neurol Neurosurg Psychiat*. 1982;45:369–371.

93. Tomiyama Y, Brian JE Jr, Todd MM. Plasma viscosity and cerebral blood flow. *Am J Physiol Heart Circ Physiol*. 2000;279:1949–1954.

94. Filosa JA, Blanco VM. Neurovascular coupling in the mammalian brain. *Exp Physiol*. 2007;92:641–646.

95. Leenders KL, Perani D, Lammertsma AA, Heather JD, Buckingham P, Healy MJ, Gibbs JM, Wise RJ, Hatazawa J, Herold S. Cerebral blood flow, blood volume and oxygen utilization. Normal values and effect of age. *Brain*. 1990;113(Pt 1):27–47.

96. Chen Y, Wolk DA, Reddin JS et al. Voxel-level comparison of arterial spin-labeled perfusion MRI and FDG-PET in Alzheimer disease. *Neurology*. 2011;77:1977–1985.

97. Zou Q, Wu CW, Stein EA, Zang Y, Yang Y. Static and dynamic characteristics of cerebral blood flow during the resting state. *Neuroimage*. 2009;48:515–524.

98. Davis PK, Musunuru H, Walsh M, Cassady R, Yount R, Losiniecki A, Moore EE, Wohlauer MV, Howard J, Ploplis VA. Platelet dysfunction is an early marker for traumatic brain injury-induced coagulopathy. *Neurocrit Care*. April 2013;18(2):201–208.

99. Sillesen M, Johansson PI, Rasmussen LS et al. Platelet activation and dysfunction in a large-animal model of traumatic brain injury and hemorrhage. *J Trauma Acute Care Surg*. May 2013;74(5):1252–1259.

100. Floccard B, Rugeri L, Faure A et al. Early coagulopathy in trauma patients: An on-scene and hospital admission study. *Injury*. January 2012;43(1):26–32.

101. Rosenblum WI, El-Sabban F. Dimethyl sulfoxide (DMSO) and glycerol, hydroxyl radical scavengers, impair platelet aggregation within and eliminate the accompanying vasodilation of, injured mouse pial arterioles. *Stroke*. January–February 1982;13(1):35–39.

102. Dujovny M, Rozario R, Kossovsky N, Diaz FG, Segal R. Antiplatelet effect of dimethyl sulfoxide, barbiturates, and methyl prednisolone. *Ann NY Acad Sci.* 1983;411:234–244.

103. Schiffer CA, Whitaker CL, Schmukler M, Aisner J, Hilbert SL. The effect of dimethyl sulfoxide on in vitro platelet function. *Thromb Haemost.* August 31, 1976;36(1):221–229.

104. Moncada S, Vane R. The role of prostacyclin in vascular tissue. *Fed. Proc.* 1979;38:66–71.

105. Needleman P, Moncada S, Vane R, Bunting S, Hamberg M, Samuelsson B. Identification of an enzyme in platelet microsomes which generates thromboxane A_2 from prostaglandin endoperoxides. *Nature.* 1976;261:558–560.

106. Lefer AM. Role of the prostaglandin-thromboxane system in vascular homeostasis during shock. *Circ. Shock.* 1979;6:297–303.

107. Bergstrom S, Farnebo L, Fuxe K. Effect of prostaglandin E_2 on central and peripheral catecholamine neurons. *Eur. J. Pharmacol.* 1973;21:362–368.

108. Gorman R, Bunting S, Miller O. Modulation of human platelet adenylate cyclase by prostacyclin (PGX). *Prostaglandins.* 1977;13:377.

109. Wieser PB, Zeiger MA, Fain JN. Effects of dimethylsulfoxide on cyclic AMP accumulation, lipolysis and glucose metabolism of fat cells. *Biochem Pharmacol.* April 15, 1977;26(8):775–778.

110. Di Pietrantonio F, Di Matteo E, Di Nicola M, Trubiani O, Di Primio R, Serra E, Spoto G. Cyclase and phosphodiesterase activity on pre-T lymphoid human cells, treated with dimethyl sulfoxide (DMSO). *Nucleosides Nucleotides Nucleic Acids.* October 2004;23(8–9):1241–1244.

111. Gresele P, Momi S, Falcinelli E. Anti-platelet therapy: Phosphodiesterase inhibitors. *Br J Clin Pharmacol.* 2011;72(4):634–646.

112. Hidaka H, Asano T. Human blood platelet 3′:5′-cyclic nucleotide phosphodiesterase. Isolation of low-Km and high-Km phosphodiesterase. *Biochim Biophys Acta.* 1976;429:485–497.

113. Rondina MT, Weyrich AS. Targeting phosphodiesterases in anti-platelet therapy. *Handb Exp Pharmacol.* 2012;210:225–238.

114. Panek AC, Poppe SC, Panek AD, Junqueira VB. Effect of dimethylsulfoxide on signal transduction in mutants of Saccharomyces cerevisiae. *Braz J Med Biol Res.* 1990;23(2):105–111.

115. Boss GR. cGMP-induced differentiation of the promyelocytic cell line HL-60. *Proc Natl Acad Sci USA.* September 1989;86(18):7174–7178.

116. Johnson M, Ramwell PW. Implications of protaglandins in hematology. In *Prostaglandins and Cyclic AMP*, eds. R.H. Kahn, W.E. Landis. Academic Press, New York, 1974; pp. 275–304.

117. Takeuchi T, Hata F, Yagasaki O. Role of cyclic AMP in prostaglandin-induced modulation of acetylcholine release from the myenteric plexus of guinea pig ileum. *Jpn J Pharmacol.* 1992;60(4):327–333.

118. de la Torre JC. Role of dimethyl sulfoxide in prostaglandin-thromboxane and platelet systems after cerebral ischemia. *Ann NY Acad Sci.* 1983;411:293–308.

119. Asmis L, Tanner FC, Sudano I, Luscher TF, Camici GG. DMSO inhibits human platelet activation through cyclooxygenase-1 inhibition. A novel agent for drug eluting stents? *Biochem Biophys Res Commun.* 2010;391:1629–1633.

120. Camici GG, Steffel J, Akhmedov A et al. Dimethyl sulfoxide inhibits tissue factor expression, thrombus formation, and vascular smooth muscle cell activation: A potential treatment strategy for drug-eluting stents. *Circulation.* 2006;114(14):1512–1521.

121. Steffel J, Akhmedov A, Greutert H, Luscher TF, Tanner FC. Histamine induces tissue factor expression: Implications for acute coronary syndromes. *Circulation.* 2005;112:341–349.

122. Andoh K, Pettersen KS, Filion-Myklebust C, Prydz H. Observations on the cell biology of tissue factor in endothelial cells. *Thromb Haemost.* 1990;63:298–302.

123. Konigsberg W, Kirchhofer D, Riederer MA, Nemerson Y. The TF: VIIa complex: Clinical significance, structure-function relationships and its role in signaling and metastasis. *Thromb Haemost.* 2002;86(3):757–771.

124. Thomas WS, Mori E, Copeland B, Yu JQ, Morrissey J, del Zoppo GJ. Tissue factor contributes to microvascular defects after focal cerebral ischemia. *Stroke.* 1993;24:847–853.

125. del Zoppo GJ, Schmid-Schonbein GW, Mori E, Copeland BR, Chang C-M. Polymorphonuclear leukocytes occlude capillaries following middle cerebral artery occlusion and reperfusion in baboons. *Stroke.* 1991;22:1276–1283.

126. Carter AJ. The importance of voltage-dependent sodium channels in cerebral ischaemia. *Amino Acids.* 1998;14:159–169.

127. Okuyama K, Kiuchi S, Okamoto M, Narita H, Kudo Y. A novel Na^+ and Ca^{2+} channel blocker, T-477, prevents brain edema following microsphere-induced permanent occlusion of cerebral arterioles in rats. *Jpn J Pharmacol.* 1999;81:170–175.

128. Squire IB, Lees KR, Pryse-Phillips W, Kertesz A, Bamford J. Efficacy and tolerability of lifarizine in acute ischemic stroke. A pilot study. *Ann NY Acad Sci.* 1995;765:317–318.

129. Ogura T, Shuba LM, McDonald TF. Action potentials, ionic currents and cell water in guinea pig ventricular preparations exposed to dimethyl sulfoxide. *J Pharmacol Exp Ther.* June 1995;273(3):1273–1286.

130. Repine JE, Eaton J, Anders MW, Hoidal JR, Fox RB. Generation of hydroxyl radicals by enzymes, chemicals, and human phagocytes in vitro. Detection with the anti-inflammatory agent, dimethyl sulfoxide. *J Clin Invest.* 1979;64:1642–1651.

131. Ashwood-Smith MJ. Current concepts concerning radioprotective and cryoprotective properties of dimethyl sulfoxide in cellular systems. *Ann NY Acad Sci.* January 27, 1975;243:246–256.

132. He S, Woods LC 3rd. Effects of dimethyl sulfoxide and glycine on cryopreservation induced damage of plasma membranes and mitochondria to striped bass (Morone saxatilis) sperm. *Cryobiology.* June 2004;48(3):254–262.

133. Gollan F. Effect of DMSO and THAM on ionizing radiation in mice. *Ann NY Acad Sci.* March 15, 1967;141(1):63–64.

134. Imada K, Leonard WJ. The Jak-STAT pathway. *Mol Immunol.* 2000;37:1–11.

135. Linnekin D, Mou S, Deberry CS et al. Stem cell factor, the JAK-STAT pathway and signal transduction. *Leuk Lymphoma.* 1997;27:439–444.

136. Ayad M, Eskioglu E, Mericle RA. Onyx: A unique neuroembolic agent. *Expert Rev Med Devices.* November 2006;3(6):705–715.

137. Levett JM, Johns LM, Grina NM, Mullan BF, Kramer JF, Mullan JF. Effects of dimethyl sulfoxide on systemic and cerebral hemodynamic variables in the ischemic canine myocardium. *Crit Care Med.* 1987;15:656–660.

138. Gerthoffer WT. Mechanisms of vascular smooth muscle cell migration. *Circ Res.* 2007;100(5):607–621.

139. Kassell NF, Sprowell JA, Boarini DJ, Olin JJ. Effect of dimethyl sulfoxide on the cerebral and systemic circulations of the dog. *Neurosurgery.* January 1983;12(1):24–28.

140. Albin MS, Bunegin L, Helsel P. Dimethyl sulfoxide and other therapies in experimental pressure-induced cerebral focal ischemia. *Ann NY Acad Sci.* 1983;411:261–268.

141. Lovell MR, Collins MW, Iverson GL, Johnston KM, Bradley JP. Grade 1 or 'ding' concussions in high school athletes. *Am J Sports Med.* January–February 2004;32(1):47–54.

142. De Beaumont L, Henry LC, Gosselin N. Long-term functional alterations in sports concussion. *Neurosurg Focus.* 2012;33(6):E8:1–7.

143. Koh JO, Cassidy JD, Watkinson EJ. Incidence of concussion in contact sports: A systematic review of the evidence. *Brain Inj.* October 2003;17(10):901–917.

144. Anderson T, Heitger M, Macleoud AD. Concussion and mild head injury. *Pract Neurol.* 2006;6:342–357.

145. Ma R, Miller CD, Hogan MV, Diduch BK, Carson EW, Miller MD. Sports-related concussion: Assessment and management. *J Bone Joint Surg Am.* 2012;94(17):1618–1627.

146. Jordan BD. The clinical spectrum of sport-related traumatic brain injury. *Nat Rev Neurol.* April 2013;9(4):222–230.

147. Nariai T, Inaji M, Tanaka Y, Hiura M, Hosoda C, Ishii K, Ohno K. PET molecular imaging to investigate higher brain dysfunction in patients with neurotrauma. *Acta Neurochir Suppl.* 2013;118:251–254.

148. Stern RA, Riley DO, Daneshvar DH, Nowinski CJ, Cantu RC, McKee AC. Long-term consequences of repetitive brain trauma: Chronic traumatic encephalopathy. *PM R.* October 2011;3(10 Suppl 2):S460–S467.

149. Giza CC, Hovda DA. The neurometabolic cascade of concussion. *J Athl Train* 2001;36:228–235.

150. Signoretti S, Lazzarino G, Tavazzi B, Vagnozzi R. The pathophysiology of concussion. *PM R.* October 2011;3(10 Suppl 2):S359–S368.

151. Fineman I, Hovda DA, Smith M, Yoshino A, Becker DP. Concussive brain injury is associated with a prolonged accumulation of calcium: A 45Ca autoradiographic study. *Brain Res.* 1993;624:94–102.

152. Golding EM, Steenberg ML, Contant CF Jr, Krishnappa I, Robertson CS, Bryan RM Jr. Cerebrovascular reactivity to CO(2) and hypotension after mild cortical impact injury. *Am J Physiol.* October 1999;277(4 Pt 2):H1457–H1466.

153. Grindel SH. Epidemiology and pathophysiology of minor traumatic brain injury. *Curr Sports Med Rep.* February 2003;2(1):18–23.

154. Kolias AG, Guilfoyle MR, Helmy A, Allanson J, Hutchinson PJ. Traumatic brain injury in adults. *Pract Neurol.* August 2013;13(4):228–35.

155. Shah KR, West M. The effect of concussion on cerebral uptake of 2-deoxy-D-glucose in rat. *Neurosci Lett.* 1983;40:287–291.

156. Echemendia RJ, Iverson GL, McCrea M, Macciocchi SN, Gioia GA, Putukian M, Comper P. Advances in neuropsychological assessment of sport-related concussion. *Br J Sports Med.* April 2013;47(5):294–298.

157. Katayama Y, Becker DP, Tamura T, Hovda DA. Massive increases in extracellular potassium and the indiscriminate release of glutamate following concussive brain injury. *J Neurosurg.* 1990;73:889–900.

158. Mark LP, Prost RW, Ulmer JL, Smith MM, Daniels DL, Strottmann JM, Brown WD, Hacein-Bey L. Pictorial review of glutamate excitotoxicity: Fundamental concepts for neuroimaging. *AJNR Am J Neuroradiol.* November–December 2001;22(10):1813–1824.

159. Mullan S, Jafar J, Hanlon K, Brown FD. Dimethyl sulfoxide on the management of postoperative hemiplegia. In *Cerebral Arterial Spasm,* ed. R.H. Wilkins. Williams & Wilkins, Baltimore, MD, 1980; pp. 646–653.

160. Jacob SW, de la Torre JC. Pharmacology of dimethyl sulfoxide in cardiac and CNS damage. *Pharmacol Rep.* March–April 2009;61(2):225–235.

161. de la Torre JC. Treatment of head injury in mice, using a fructose 1,6-diphosphate and dimethyl sulfoxide combination. *Neurosurgery.* August 1995;37(2):273–279.

162. Markov AK. Hemodynamics and metabolic effects of fructose 1,6-diphosphate in ischemia and shock: Experimental and clinical observations. *Ann Emerg Med.* 1986;15:1470–1477.

163. Farias LA, Smith EE, Markov AK. Prevention of ischemic-hypoxic brain injury and death in rabbits with fructose-1, 6-diphosphate. *Stroke.* 1990;21:606–613.

164. Markov AK, Lehan P, Figueroa A: Improvement of left ventricular function with fructose 1,6-diphosphate in clinical coronary artery disease. *Eur Heart J.* 1994;15(Suppl):277.

165. Hall ED, Yonkers P, McCall JM, Braughler JM. Effects of the 21-aminosteroid U74006F on experimental head injury in mice. *J Neurosurg.* 1988;68:456–461.

166. Sutton RL, Lescaudron L, Stein DG. Behavioral consequences of moderate and severe cortical contusion in the rat. *Soc Neurosci Abstr*. 1990;16:777.
167. Marcel GA. Red cell deformability: Physiological, clinical and pharmacological aspects. *J Med* 1979;10:409–416.
168. Heckler FR, Markov A, White TZ, Jone EW. Effect of fructose 1,6-diphosphate on canine hind limbs subjected to tourniquet ischemia. *J Hand Surg*. 1983;8:622–623.
169. Benesh R, Benesh RE. The effect of organic phosphates from the human erythrocyte on the allosteric properties of hemoglobin. *Biochem Biophys Res Commun*. 1967;26:162–167.
170. Benesh R, Benesh RE. Intracellular organic phosphates as regulators of oxygen release by hemoglobin. *Nature*. 1969;221:618–622.
171. Beharry S, Bragg PD. Interaction of beef-heart mitochondrial F1-ATPase with immobilized ATP in the presence of dimethyl sulfoxide. *J Bioenerg Biomembr*. 1992;24:507–514.
172. de la Torre JC, Hill PK. Ultrastructural studies on formation of brain edema and its treatment following experimental brain infarction in monkeys. In *Dynamics of Brain Edema*, eds. H. Pappius, W. Feindel. Springer-Verlag, Berlin, Germany, 1976; pp. 306–314.
173. Bickler PE, Buck LT. Effects of fructose-1,6-bisphosphate on glutamate release and ATP loss from rat brain slices during hypoxia. *J Neurochem*. October 1996;67(4):1463–1468.
174. Espanol MT, Litt L, Hasegawa K, Chang LH, Macdonald JM, Gregory G, James TL, Chan PH. Fructose-1,6-bisphosphate preserves adenosine triphosphate but not intracellular pH during hypoxia in respiring neonatal rat brain slices. *Anesthesiology*. February 1998;88(2):461–472.
175. Cárdenas A, Hurtado O, Leza JC, Lorenzo P, Bartrons R, Lizasoain I, Moro MA. Fructose-1,6-bisphosphate inhibits the expression of inducible nitric oxide synthase caused by oxygen-glucose deprivation through the inhibition of glutamate release in rat forebrain slices. *Naunyn Schmiedebergs Arch Pharmacol*. September 2000;362(3):208–212.
176. Welch G, Loscalzo J. Nitric oxide and the cardiovascular system. *J Cardiovasc Surg*. 1994;9:361–371.
177. Rosamond W, Flegal K, Furie K. Heart disease and stroke statistics—2008 update: A report from the American heart association statistics committee and stroke statistics subcommittee. *Circulation*. 2008;117(4):e25–e46.
178. Braeuninger S, Kleinschnitz C, Nieswandt B, Stoll G. Focal cerebral ischemia. *Methods Mol Biol*. 2012;788:29–42.
179. Campbell BC, Donnan GA, Davis SM. Vessel occlusion, penumbra, and reperfusion—Translating theory to practice. *Front Neurol*. 2014;5:194–195.
180. Ramos-Cabrer P, Campos F, Sobrino T, Castillo J. Targeting the ischemic penumbra. *Stroke*. January 2011;42(1 Suppl):S7–S11.
181. de la Torre JC, Fortin T, Park GA, Pappas BA, Richard MT. Brain blood flow restoration 'rescues' chronically damaged rat CA1 neurons. *Brain Res*. September 24, 1993;623(1):6–15.
182. Watson BD, Dietrich WD, Busto R, Wachtel MS, Ginsberg MD. Induction of reproducible brain infarction by photochemically initiated thrombosis. *Ann Neurol*. 1985;17:497–504.
183. O'Brien MD, Waltz AG. Transorbital approach for occluding the middle cerebral artery without craniectomy. *Stroke*. 1973;4:201–206.
184. Koizumi J, Yoshida Y, Nakazawa T, Ooneda G. Experimental studies of ischemic brain edema. I: A new experimental model of cerebral embolism in rats in which recirculation can be introduced in the ischemic area. *Jpn J Stroke*. 1986;8:1–8.
185. O'Collins VE, Macleod MR, Donnan GA, Howells DW. Evaluation of combination therapy in animal models of cerebral ischemia. *J Cereb Blood Flow Metab*. 2012;32(4):585–597.

186. Tamura A, Graham DI, McCulloch J, Teasdale GM. Focal cerebral ischaemia in the rat: 1. Description of technique and early neuropathological consequences following middle cerebral artery occlusion. *J Cereb Blood Flow Metab*. 1981;1:53–60.

187. Longa EZ, Weinstein PR, Carlson S, Cummins R. Reversible middle cerebral artery occlusion without craniectomy in rats. *Stroke*. 1989;20:84–91.

188. Kanemitsu H, Nakagomi T, Tamura A, Tsuchiya T, Kono G, Sano K. Differences in the extent of primary ischemic damage between middle cerebral artery coagulation and intraluminal occlusion models. *J Cereb Blood Flow Metab*. 2002;22:1196–1204.

189. Kong LQ, Xie JX, Han HB, Liu HD. Improvements in the intraluminal thread technique to induce focal cerebral ischaemia in rabbits. *J Neurosci Methods*. 2004;137:315–319.

190. Baskaya MK, Dogan A, Dempsey RJ. Application of endovascular suture occlusion of middle cerebral artery in gerbils to obtain consistent infarction. *Neurol Res*. 1999;21:574–578.

191. Freret T, Bouet V, Toutain J, Saulnier R, Pro-Sistiaga P, Bihel E, Mackenzie ET, Roussel S, Schumann-Bard P, Touzani O. Intraluminal thread model of focal stroke in the non-human primate. *J Cereb Blood Flow Metab*. 2008;28:786–796.

192. Durukan A, Tatlisumak T. Acute ischemic stroke: Overview of major experimental rodent models, pathophysiology, and therapy of focal cerebral ischemia. *Pharmacol Biochem Behav*. 2007;87(1):179–197.

193. Bacigaluppi M, Comi G, Hermann DM. Animal models of ischemic stroke. Part two: Modeling cerebral ischemia. *Open Neurol J*. 2010;4:34–38.

194. Carmichael ST. Rodent models of focal stroke: Size, mechanism, and purpose. *NeuroRx*. July 2005;2(3):396–409.

195. Liu F, McCullough LD. Middle cerebral artery occlusion model in rodents: Methods and potential pitfalls. *J Biomed Biotechnol*. 2011;2011:464701. doi: 10.1155/2011/464701.

196. Armstead WM, Ganguly K, Kiessling JW. Signaling, delivery and age as emerging issues in the benefit/risk ratio outcome of tPA for treatment of CNS ischemic disorders. *J Neurochem*. 2010;113(2):303–312.

197. Zivin JA, Fisher M, DeGirolami U, Hemenway CC, Stashak JA. Tissue plasminogen activator reduces neurological damage after cerebral embolism. *Science*. 1985;230:1289–1292.

198. NINDS tSTSG. The National Institute of Neurological Disorders and Stroke rt-PA Stroke Study Group Tissue plasminogen activator for acute ischemic stroke. *N Engl J Med*. 1995;333:1581–1587.

199. Audebert HJ, Saver JL, Starkman S, Lees KR, Endres M. Prehospital stroke care: New prospects for treatment and clinical research. *Neurology*. July 30, 2013;81(5):501–508.

200. Hainsworth AH, Markus HS. Do in vivo experimental models reflect human cerebral small vessel disease? A systematic review. *J Cereb Blood Flow Metab*. December 2008;28(12):1877–1891.

201. Fisher M, Feuerstein G, Howells DW, Hurn PD, Kent TA, Savitz SI, Lo EH, STAIR Group. Update of the stroke therapy academic industry roundtable preclinical recommendations. *Stroke*. June 2009;40(6):2244–2250.

202. Donnan GA, Davis SM, Parsons MW, Ma H, Dewey HM, Howells DW. How to make better use of thrombolytic therapy in acute ischemic stroke. *Nat Rev Neurol*. June 14, 2011;7(7):400–409.

203. Dirnagl U. Bench to bedside: The quest for quality in experimental stroke research. *J Cereb Blood Flow Metab*. 2006;26:1465–1478.

204. Philip M, Benatar M, Fisher M, Savitz SI. Methodological quality of animal studies of neuroprotective agents currently in phase II/III acute ischemic stroke trials. *Stroke*. 2009;40:577–581.

205. Rossi J. Nonhuman primate research: The wrong way to understand needs and necessity. *Am J Bioeth*. 2009;9:21–23.

206. Marshall JW, Cummings RM, Bowes LJ, Ridley RM, Green AR. Functional and histo-logical evidence for the protective effect of NXY-059 in a primate model of stroke when given 4 hours after occlusion. *Stroke*. 2003;34:2228–2233.
207. Marshall JW, Green AR, Ridley RM. Comparison of the neuroprotective effect of clo-methiazole, AR-R15896AR and NXY-059 in a primate model of stroke using histologi-cal and behavioural measures. *Brain Res*. 2003;972:119–126.
208. Shuaib A, Lees KR, Lyden P et al. NXY-059 for the treatment of acute ischemic stroke. *N Engl J Med*. 2007;357:562–571.
209. de la Torre JC, Surgeon JW. Dexamethasone and DMSO in experimental transorbital cerebral infarction. *Stroke*. November–December 1976;7(6):577–583.
210. Karaça M, Kiliç E, Yazici B, Demir S, de la Torre JC. Ischemic stroke in elderly patients treated with a free radical scavenger-glycolytic intermediate solution: A preliminary pilot trial. *Neurol Res*. January 2002;24(1):73–80.
211. Willmore LJ, Rubin JJ. The effect of tocopherol and dimethyl sulfoxide on focal edema and lipid peroxidation induced by isocortical injection of ferrous chloride. *Brain Res*. April 2, 1984;296(2):389–392.
212a. Demchenko IT, Boso AE, Bennett PB, Whorton AR, Piantadosi CA. Hyperbaric oxygen reduces cerebral blood flow by inactivating nitric oxide. *Nitric Oxide*. 2000;4(6):597–608.
212b. Tomiyama Y, Jansen K, Brian JE Jr, Todd MM. Hemodilution, cerebral O2 delivery, and cerebral blood flow: A study using hyperbaric oxygenation. *Am J Physiol*. April 1999;276(4 Pt 2):H1190–H1196.
212. Little JR, Spetzler RF, Roski RA, Selman WR, Zabramski J, Lesser RP. Ineffectiveness of DMSO in treating experimental brain ischemia. *Ann NY Acad Sci*. 1983;411:269–277.
213. Fukuda S, del Zoppo GJ. Models of focal cerebral ischemia in the nonhuman primate. *ILAR J*. 2003;44(2):96–104.
214. de Crespigny AJ, D'Arceuil HE, Maynard KI et al. Acute studies of a new primate model of reversible middle cerebral artery occlusion. *J Stroke Cerebrovasc Dis*. March–April 2005;14(2):80–87.
215. Cook DJ, Tymianski M. Nonhuman primate models of stroke for translational neuropro-tection research. *Neurotherapeutics*. April 2012;9(2):371–379.
216. Asplund K. Randomized clinical trials of hemodilution in acute ischemic stroke. *Acta Neurol Scand Suppl*. 1989;127:22–30.
217. Carson S, McDonagh M, Russman B, Helfand M. Hyperbaric oxygen therapy for stroke: A systematic review of the evidence. *Clin Rehabil*. December 2005;19(8):819–833.
218. McGraw CP. Treatment of cerebral infarction with dimethyl sulfoxide in the mongolian gerbil. *Ann NY Acad Sci*. 1983;411:278–285.
219. Shimizu S, Simon RP, Graham SH. Dimethylsulfoxide (DMSO) treatment reduces infarction volume after permanent focal cerebral ischemia in rats. *Neurosci Lett*. December 19, 1997;239(2–3):125–127.
220. Bardutzky J, Meng X, Bouley J, Duong TQ, Ratan R, Fisher M. Effects of intravenous dimethyl sulfoxide on ischemia evolution in a rat permanent occlusion model. *J Cereb Blood Flow Metab*. August 2005;25(8):968–977.
221. Gilboe DB, Kintner D, Fitzpatrick JH, Emoto SE, Esanu A, Braquet PG, Bazan NG. Recovery of postischemic brain metabolism and function following treatment with a free radical scavenger and platelet-activating factor antagonists. *J Neurochem*. 1991;56:311–319.
222. Nakahiro M, Arakawa O, Narahashi T, Ukai S, Kato Y, Nishinuma K, Nishimura T. Dimethyl sulfoxide (DMSO) blocks GABA-induced current in rat dorsal root ganglion neurons. *Neurosci Lett*. 1992;138:5–8.
223. Sawada M, Sato M. The effect of dimethyl sulfoxide on the neuronal excitability and cholinergic transmission in Aplysia ganglion cells. *Ann NY Acad Sci*. 1975;243:337–357.

224. Saver JL. Number needed to treat estimates incorporating effects over the entire range of clinical outcomes: Novel derivation method and application to thrombolytic therapy for acute stroke [published correction appears in *Arch Neurol*. 2004;61(10):1599]. *Arch Neurol*. 2004;61(7):1066–1070.

225. The National Institute of Neurological Disorders and Stroke rt-PA Stroke Study Group. Tissue plasminogen activator for acute ischemic stroke. *N Engl J Med*. 1995;333(24):1581–1587.

226. Laha RK, Dujovny M, Barrionuevo PJ, DeCastro SC, Hellstrom HR, Maroon JC. Protective effects of methylprednisolone and dimethyl sulfoxide in experimental middle cerebral artery embolectomy. *J Neurosurg*. 1978;49:508–516.

227. Demopoulos HB, Flamm E, Yoder M. Mechanisms of barbiturate protection of ischemic endothelium. A scanning electron microscopy study. *Am Assoc Neurrol Surg*. 1979;19:35–36.

228. Miller TR. Costs associated with gunshot wounds in Canada in 1991. *CMAJ*. November 1, 1995;153(9):1261–1268.

229. Sorani MD, Morabito D, Rosenthal G, Giacomini K, Manley G. Characterizing the dose–response relationship between mannitol and intracranial pressure in traumatic brain injury patients using a high-frequency physiological data collection system. *J Neurotrauma*. 2008;25(4):291–298.

230. Brown FD, Johns L, Mullan S. Dimethyl sulfoxide therapy following penetrating brain injury. *Ann NY Acad Sci*. 1983;411:245–252.

231. Marmarou A, Saad A, Aygok G, Rigsbee M. Contribution of raised ICP and hypotension to CPP reduction in severe brain injury: Correlation to outcome. *Acta Neurochir Suppl*. 2005;95:277–280.

232. Struchen MA, Hannay HJ, Contant CF, Robertson CS. The relation between acute physiological variables and outcome on the glasgow outcome scale and disability rating scale following severe traumatic brain injury. *J Neurotrauma*. 2001;18:115–125.

233. Manley G, Knudson MM, Morabito D, Damron S, Erickson V, Pitts L. Hypotension, hypoxia, and head injury: Frequency, duration, and consequences. *Arch Surg*. 2001;136:1118–1123.

234. Pigula FA, Wald SL, Shackford SR, Vane DW. The effect of hypotension and hypoxia on children with severe head injuries. *J Pediatr Surg*. 1993;28:310–14.

235. Haddad SH, Arabi YM. Critical care management of severe traumatic brain injury in adults. *Scand J Trauma Resusc Emerg Med*. 2012;20:12–29.

236. U.S. Census Bureau, Population Division. December 2012. Table 2. Projections of the Population by Selected Age Groups and Sex for the United States: 2015 to 2060 (NP2012-T2).

237. Viswanathan M, Kim SK, Berdichevsky A, Guarente A. A role for SIR-2.1 regulation of ER stress response genes in determining C. elegans life span. *Dev Cell*. 2005;9:605–615.

238. Kampkotter A, Nkwonkam CG, Zurawski RF, Timpel C, Chovolou Y. Investigations of protective effects of the flavonoids quercetin and rutin on stress resistance in the model organism Caenorhabditis elegans. *Toxicology*. 2007;234:113–123.

239. Wang G, Huang C, Wang Y, Guo Q, Jiang H, Wen J. Changes in expression of cyclooxygenase-2 in the spinal dorsal horn after intrathecal p38MAPK inhibitor SB203580 on neuropathic pain in rats. *Zhong Nan Da Xue Xue Bao Yi Xue Ban*. July 2013;38(7):686–690.

240. de la Torre JC. Alzheimer's disease: How does it start? *J Alzheimers Dis*. December 2002;4(6):497–512.

241. de la Torre JC, Mussivand T. Can disturbed brain microcirculation cause Alzheimer's disease? *Neurol Res*. June 1993;15(3):146–153.

242. de la Torre JC. Is Alzheimer's disease a neurodegenerative or a vascular disorder? Data, dogma, and dialectics. *Lancet Neurol*. 2004;3:184–190.

243. de la Torre JC. Critical threshold cerebral hypoperfusion causes Alzheimer's disease. *Acta Neuropathol*. 1999;98:1–8.

244. Stürchler-Pierrat C, Abramowski D, Duke M et al. Two amyloid precursor protein transgenic mouse models with Alzheimer disease-like pathology. *Proc Natl Acad Sci USA.* 1997;94:13287–13292.

245. McGowan E, Eriksen J, Hutton M. A decade of modeling Alzheimer's disease in transgenic mice. *Trends Genet.* 2006;22:281–289.

246. Davies P, Maloney AJ. Selective loss of central cholinergic neurons in Alzheimer's disease. *Lancet.* 1976;2:1403.

247. Ebert U, Kirch W. Scopolamine models of dementia: Electroencephalogram findings and cognitive performance. *Eur J Clin Invest.* 1998;28:944–949.

248. Sergeant N, Buée L. TAU models. In *Animal Models of Dementia*, eds. P.P. De Deyn, D. Van Dam. Springer Science + Business Media, New York, 2010; pp. 449–468.

249. Castañé A, Theobald DE, Robbins TW. Selective lesions of the dorsomedial striatum impair serial spatial reversal learning in rats. *Behav Brain Res.* 2010;210:74–83.

250. Levitan D, Greenwald I. Facilitation of lin-12-mediated signalling by sel-12, a Caenorhabditis elegans S182 Alzheimer's disease gene. *Nature.* 1995;377:351–354.

251. de la Torre JC, Fortin T, Park G, Butler K, Kozlowski P, Pappas B, de Socarraz H, Saunders J. Chronic cerebrovascular insufficiency induces dementia-like deficits in aged rats. *Brain Res.* 1992;582:186–195.

252. de la Torre JC, Nelson N, Sutherland RJ, Pappas BA. Reversal of ischemic-induced chronic memory dysfunction in aging rats with a free radical scavenger-glycolytic intermediate combination. *Brain Res.* 1998;779:285–288.

253. de la Torre JC, Nelson N, Sutherland RJ. Combined fructose 1,6-diphosphate-dimethyl sulfoxide reverses chronic memory deficits in brain ischemic rats. In *Neurochemistry*, eds. A. Teelken, J. Korf, Plenum Press, New York, 1997; pp. 173–176.

254. Hofman A, Ott A, Breteler MM, Bots ML, Slooter AJ, van Harskamp F, van Duijn CN, Van Broeckhoven C, Grobbee DE. Atherosclerosis, apolipoprotein E, and prevalence of dementia and Alzheimer's disease in the Rotterdam Study. *Lancet.* January 18, 1997;349(9046):151–154.

255. van Oijen M, de Jong FJ, Witteman JC, Hofman A, Koudstaal PJ, Breteler MM. Atherosclerosis and risk for dementia. *Ann Neurol.* May 2007;61(5):403–410.

256. Johnson DK, Morris JC, Galvin JE. Verbal and visuospatial deficits in dementia with Lewy bodies. *Neurology.* October 25, 2005;65(8):1232–1238.

257. Iachini I, Iavarone A, Senese VP, Ruotolo F, Ruggiero G. Visuospatial memory in healthy elderly, AD and MCI: A review. *Curr Aging Sci.* March 2009;2(1):43–59.

258. Farkas E, Institóris A, Domoki F, Mihály A, Luiten PG, Bari F. Diazoxide and dimethyl sulphoxide prevent cerebral hypoperfusion-related learning dysfunction and brain damage after carotid artery occlusion. *Brain Res.* May 22, 2004;1008(2):252–260.

259. Markvartova V, Cendelin J, Vozeh F. Effect of dimethyl sulfoxide in cerebellar mutant Lurcher mice. *Neurosci Lett.* May 24, 2013;543:142–145.

260. Porras-Garcia E, Cendelin J, Dominguez-del-Toro E, Vozeh F, Delgado-Garcia JM. Purkinje cell loss affects differentially the execution acquisition and prepulse inhibition of skeletal an facial motor responses in Lurcher mice. *Eur. J. Neurosci.* 2005;21:979–988.

261. Caddy KW, Biscoe TJ. Structural and quantitative studies on the normal C3H and Lurcher mutant mouse. *Philos Trans R Soc Lond B Biol Sci.* 1979;287(1020):167–201.

262. Barcal J, Cendelín J, Korelusová I, Tůma J, Vozeh F. Glutamate receptor block in Lurcher mutant mice during ontogeny and its effect on hippocampal long-term potentiation. *Prague Med Rep.* 2010;111(2):127–134.

263. deLara CM, Jenner TJ, Townsend KM, Marsden SJ, O'Neill P. The effect of dimethyl sulfoxide on the induction of DNA double-strand breaks in V79-4 mammalian cells by alpha particles. *Radiat Res.* October 1995;144(1):43–49.

264. Cigarrán S, Barrios L, Caballín MR, Barquinero JF. Effect of DMSO on radiation-induced chromosome aberrations analysed by FISH. *Cytogenet Genome Res.* 2004;104(1–4):168–172.

265. Gregersen N, Bross P, Vang S, Christensen JH. Protein misfolding and human disease. *Annu Rev Genomics Hum Genet.* 2006;7:103–124.

266. Arakawa T, Ejima D, Kita Y, Tsumoto K. Small molecule pharmacological chaperones: From thermodynamic stabilization to pharmaceutical drugs. *Biochim Biophys Acta.* 2006;1764(11):1677–1687.

267. Papp E, Csemely P. Chemical chaperones: Mechanisms of action and potential use. In *Molecular Chaperones in Health and Disease*. Handbook of Experimental Pharmacology, Vol. 172, Gadstel, M (Ed.), Springer-Verlag, Berlin, Heidelberg, 2006; pp. 405–416.

268. Curt A, Ellaway PH. Clinical neurophysiology in the prognosis and monitoring of traumatic spinal cord injury. *Handb Clin Neurol.* 2012;109:63–75.

269. Shields CB, Ping Zhang Y, Shields LB, Burke DA, Glassman SD. Objective assessment of cervical spinal cord injury levels by transcranial magnetic motor-evoked potentials. *Surg Neurol.* November 2006;66(5):475–483.

270. de la Torre JC. Spinal cord injury. Review of basic and applied research. *Spine.* 1981;6(4):315–335.

271. Hurlbert RJ. Methylprednisolone for acute spinal cord injury: An inappropriate standard of care. *J Neurosurg.* 2000;93:1–7.

272. Hurlbert RJ, Hamilton MG. Methylprednisolone for acute spinal cord injury: 5-year practice reversal. *Can J Neurol Sci.* March 2008;35(1):41–45.

273. Hurlbert RJ, Moulton R. Why do you prescribe methylprednisolone for acute spinal cord injury? A Canadian perspective and a position statement. *Can J Neurol Sci.* August 2002;29(3):236–239.

274. Pandya KA, Weant KA, Cook AM. High-dose methylprednisolone in acute spinal cord injuries: Proceed with caution. *Orthopedics.* May 2010;33(5):327–331.

275. Kajihara K, Kawanaga H, de la Torre JC, Mullan S. Dimethyl sulfoxide in the treatment of experimental acute spinal cord injury. *Surg Neurol.* January 1973;1(1):16–22.

276. Mullan S, de la Torre JC, Kajihara K, Kawanaga HM. Use of dimethyl sulfoxide in experimental head and spinal cord injuries. *J Neurol Neurosurg Psychiat.* 1973;36:153–154.

277. de la Torre JC, Kajihara K, Rowed DW, Kawanaga HM, Mullan JF. Modification of experimental head and spinal cord injury. *Trans Amer Neurol Assoc.* 1973;97:230–233.

278. Goldsmith HS. Application of the omentum to brain and spinal cord. In *The Omentum, Application to Brain and Spinal Cord*, ed. H.S. Goldsmith. Forefront Publishing, Wilton, CT, 2000; pp. 25–43.

279. Goldsmith HS. Can the standard treatment of acute spinal cord injury be improved? Perhaps the time has come. *Neurol Res.* 2007;29;16–20.

280. Zileli M, Ovul I, Dalbasti T. Effects of methyl prednisolone, dimethyl sulphoxide and naloxone in experimental spinal cord injuries in rats. *Neurol Res.* 1988;10:232–235.

281. Rucker NC, Lumb WV, Scott RJ. Combined pharmacologic and surgical treatments for acute spinal cord trauma. *Ann NY Acad Sci.* 1983;411:191–199.

282. Gelderd JB, Fife WP, Bowers DE, Deschner SH, Welch DW. Spinal cord transection in rats: The therapeutic effects of dimethyl sulfoxide and hyperbaric oxygen. *Ann NY Acad Sci.* 1983;411:218–233.

283. Shi R, Qiao X, Emerson N, Malcom A. Dimethylsulfoxide enhances CNS neuronal plasma membrane resealing after injury in low temperature or low calcium. *J Neurocytol.* September–October 2001;30(9–10):829–839.

284. Turan NN, Akar F, Budak B, Seren M, Parlar AI, Sürücü S, Ulus AT. How DMSO, a widely used solvent, affects spinal cord injury. *Ann Vasc Surg.* January 2008; 2(1):98–105.

285. Blokhina O, Virolainen E, Fagerstedt KV. Antioxidants, oxidative damage and oxygen deprivation stress: A review. *Ann Bot.* January 2003;91 Spec No:179-94.
286. Szabó C, Ohshima H. DNA damage induced by peroxynitrite: Subsequent biological effects. *Nitric Oxide.* October 1997;1(5):373–385.
287. Erecinska M, Silver IA. ATP and brain function. *J Cereb Blood Flow Metab.* 1989;9:2–19.
288. Clarke DD, Sokoloff L. Circulation and energy metabolism of the brain. In *Basic Neurochemistry*, eds. G.J. Siegel, B.W. Agranoff, R.W. Albers, P.B. Molinoff. Raven Press, New York, 1994; pp. 645–680.
289. Levine JM, Cohen ND, Heller M et al. Efficacy of a metalloproteinase inhibitor in spinal cord injured dogs. *PLoS One.* 2014;9(5):e96408.
290. Sharp K, Yee KM, Steward O. A re-assessment of the effects of treatment with an epidermal growth factor receptor (EGFR) inhibitor on recovery of bladder and locomotor function following thoracic spinal cord injury in rats. *Exp Neurol.* 2012;233:649–659.

9 DMSO in Clinical Neuroprotection

OVERVIEW OF CLINICAL TRAUMATIC BRAIN INJURY

Head injuries in humans can involve trauma to the scalp, skull, or brain. They can be described as mild, moderate, or severe (Figure 9.1). Head injuries can be penetrating where the skull is broken, or closed, where the brain may be injured but the skull remains intact.

Penetrating injuries are generally caused by auto accidents, blows to the head, and gunshot wounds. Closed head injuries are caused mainly by vehicular accidents, acts of violence, falls, and sports-related trauma.[1] The severity of closed head injuries range from mild concussions to disabling brain trauma leading to total incapacitation or death. Traumatic brain injuries typically involve young adult men.

Brain trauma can damage tissue at the site of impact or at the opposite poles of impact due to sudden acceleration–deceleration movement of the brain, which acts like a mass of jelly moving forward and backward inside the skull. This type of brain injury is called a *coup–contercoup* injury because damage occurs at the site of impact and on the direct, opposite side as the brain moves forward and backward. Symptoms after a traumatic brain injury vary according to the severity of the injury as well as what part of the brain is affected and whether the injury is localized to one part of the brain or is diffuse. Symptoms can range from a simple headache with no loss of consciousness (LOC) to fatigue, dizziness, nausea, and vomiting, to severe cognitive impairment, mood changes resulting in confusion, restlessness, agitation,[2] to motor-sensory deficits involving speech, vision, muscle coordination, and coma.[3]

Focal and diffuse cerebral injuries are descriptive ways to classify the extent of structural or functional damage to the brain. As the term implies, focal injury occurs in a specific brain location (Figure 9.1), while diffuse injury can involve more widespread regions of the brain (Figure 9.1, arrows). An example of a focal injury is an epidural or subdural hematoma where a blood clot forms on the surface of the brain (epidural) or below the epidural membrane, typically as a result of a blow to the head.

Diffuse axonal injuries can follow a mild or moderate head injury, and recovery is often possible when the lesions are few and diffusivity is contained.[4,5] Diffuse axonal injuries are generally caused by acceleration–deceleration of the head, which may or may not contact any external object in particular since the injury can occur from violent brain shaking. This shaking within the skull can burst blood vessels and severely injure brain tissue. One of the outcomes of diffuse injuries involves the shearing of axonal tracts and tissue white matter that result as brain tissues of different densities slide over each other much like tectonic plates under the earth slipping past each other and colliding to create earthquakes. Diffuse axonal injuries rarely

Severity of traumatic brain injury

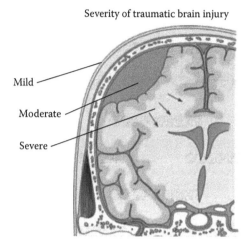

Mild

Moderate

Severe

FIGURE 9.1 **(See color insert.)** A mild head injury refers to the severity of the initial blunt trauma, which is generally a concussion affecting the scalp or the skull and involves brief (<30 min) changes in mental status or consciousness as opposed to a severe injury, which results in extended LOC, coma, and posttraumatic amnesia (PTA) lasting days, months, or years. This injury has a high risk of mortality. LOC after a moderate head injury does not generally exceed 6 h, and PTA may extend up to 24 h. A moderate-to-severe head injury can result in an epidural hematoma, subdural hematoma, or intraparenchymal hemorrhage.

induce death, but they can be devastating as the cause of permanent vegetative state and they are considered the most significant determinant of morbidity in patients with brain trauma.[6] It is estimated that only 10% of people affected by severe diffuse axonal injury regain consciousness and only a small percentage of these will attain near-normal recovery of neurological function.[7] Significant improvement after 1 year is rare.[8]

Diffuse axonal injuries are maybe difficult to detect by imaging techniques unless ventricular hemorrhage, abnormal foci, or multiple petechial hemorrhages are detected (Figure 9.1). These injuries frequently involve the frontal and inferior temporal lobes and the brainstem but spare major damage to basic life functions that control cardiac and respiratory systems. Often, the diagnosis of diffuse axonal injury is made when the patient's symptoms are disproportionate to the imaging findings.[9]

The treatment of focal or diffuse brain trauma is aimed at immediate measures to reduce swelling inside the brain and prevent secondary damage and a deadly outcome. Brain swelling creates dangerous increases in intracranial pressure (ICP) because the cranial vault is a rigid container and the tissue, with nowhere to go except to squeeze through the foramen magnum, can be compressed to quickly diminish cerebral perfusion and establish ischemia and brain infarction.[10] The herniated brain tissue *coning* through the foramen magnum can also put fatal pressure on regions of the brainstem, which are vital to life.[11]

Brain trauma can affect cerebral autoregulation. Cerebral autoregulation is a homeostatic mechanism that maintains a constant level of cerebral blood flow (CBF) in the presence of changing cerebral perfusion pressures (CPPs). Autoregulation fails when the mean arterial pressure (MAP) falls below 60 mmHg or rises above 150 mmHg. When cerebral autoregulation fails, the chances for survival are considered poor. A gold standard for measuring cerebral autoregulation is not available at the present time, and a search of the literature shows considerable divergence in the methods used. This fact is not surprising given that cerebral autoregulation is more a concept than a physically measurable entity. However, neuromonitoring of cerebral autoregulation is essential in guiding therapy and assessing the metabolic requirement of the cerebral tissues at a time of physiological crisis.[12]

Because the primary damage cannot be controlled due to the fact that a direct mechanical injury has already occurred, prevention of secondary damage is the primary goal of neurocritical care and treatment. Secondary damage develops within hours or sometimes days and is dependent on whether treatment can reverse or minimize impaired cerebrovascular autoregulation, cerebral metabolic dysfunction, tissue edema, inflammation, and inadequate brain oxygenation.[13–15] Figure 9.2 summarizes the initial clinical approach to the patient sustaining a severe head injury as recommended by a joint section on neurotrauma and critical care from the American Association of Neurological Surgeons.[16] The chronological order of this approach can vary according to the circumstances of the head injury. These guidelines can also be used for moderate and even mild head injuries when deemed appropriate.

Extended tissue destruction spreading from the primary site of head trauma is the first pathophysiological stage characterizing secondary brain damage. The extended tissue damage will kill more brain cells as it spreads through the brain tissue mainly from impaired cerebral perfusion, growing edema, and as cytotoxic free radicals and excitatory neurotransmitters are released.

Basic clinical approach to the
head-injured patient

➤ Provide oxygenation/ventilation (when needed)

➤ Reduce ICP

➤ Restore CBF

➤ Maintain normal blood pressure

➤ Identify associated injuries

➤ Prevent secondary brain damage

FIGURE 9.2 Emergency approach to the head-injured patient includes oxygenation to prevent hypoxemia, reduction of high ICP, prevention of low CBF that is associated with poor outcome, and prevention of systolic blood pressure drop below 90 mmHg, ruling out secondary injuries and providing treatment to prevent secondary damage. (From American Association of Neurological Surgeons, *J. Neurotrauma*, 24, S-1, 2007.)

This pathologic process will progress if untreated and further involve cellular DNA fragmentation and inhibition of DNA repair.[17] The final outcome from this tissular cataclysm is the obliteration of vascular and cellular structures leading to necrotic or programmed cell death.[17] The clinical outcome of uncontrolled secondary damage following brain trauma will likely result in severe neurologic impairment or death.[11]

The most commonly used method for grading the severity of brain injury is the Glasgow coma score (GCS), a neurological scale that attempts to provide an objective and reliable way of quantifying a person's conscious level following traumatic brain injury and a possible prediction of outcome. The GCS is used primarily because it is simple and has a relatively high degree of interobserver reliability, but it should be used only after any needed resuscitative measures are applied.

Developed by Teasdale and Jennett,[18] the GCS scale classifies patients with mild, moderate, or severe head injuries using three aspects of neurologic function: eye opening, verbal response, and motor response. The presence or absence of these functions is scored using a scale ranging from 3 to 15, which correspond to a mild injury (13–15), moderate injury (9–12), and severe injury (3–8).[18] The GCS allows serial evaluations of patients' progress and can also provide an assessment of the value of a given treatment. The difference between a GCS of 3–4 and 7–8 can mean the difference between permanently poor outcome (3–4) to potentially good recovery and moderate disability. Patients arriving at the hospital with a GCS of 3 and bilateral, fixed dilated pupils have little chance of survival as opposed to patients with the same GCS score and reactive pupils.

It has been argued that the motor response of the GCS yields similar prediction rates of outcome as the summed GCS score due to the fact the motor component is not only linearly related to survival but also preserves almost all the predictive power of the GCS.[19,20]

Using the motor component of the GCS also allows intubated patients to be evaluated without the verbal response of the GCS. It appears then that the best motor response is probably the most significant predictor of outcome. Besides the GCS motor response, other classifications of injury severity focus on the duration of posttraumatic amnesia (PTA) and duration of LOC. The PTA correlates with the overall severity of the injury and can be assessed with the Galveston orientation and amnesia test (GOAT) or the orientation log (O-Log).[21]

The guidelines recommended for the clinical management of patients (Figure 9.2) and the GCS are particularly useful in patients with severe brain trauma who clearly benefit from prompt and effective emergency care because most of the pathology that determines long-term outcome will be present in the first hours after a head injury.[22] Therefore, a concise, clear, and practical approach to the evaluation and treatment of patients with severe brain trauma is necessary to avoid delays in management and to minimize secondary brain damage.

In the general U.S. population, roughly 1.7 million brain injuries are reported annually, leading to more than 52,000 deaths.[23] The majority of these injuries are minor bumps on the head that require no treatment, but even mild head injuries can lead to lifelong disability or even death. Beyond the effects of acute injury, troubling new findings indicate that even minor brain injury can predispose to neurodegeneration and dementia in later life.[24]

American involvement in recent wars indicates that as many as 320,000 U.S. troops in Iraq and Afghanistan have suffered head injuries mostly from improvised explosive devices (IEDs).[25] Many of these blast injuries, while not immediately life threatening, can have a long-term impact that fosters personality changes, impulse-control problems, and dementia in later life.[25] It may also explain the high incidence of suicide among veterans returning home from combat.[26]

HEAD INJURY IN CHILDREN

Head injury occurs frequently in children and accounts for approximately 500,000 emergency department visits annually in the United States. Nearly 75% of these injuries are ultimately diagnosed as mild in severity, and the vast majority of cases will have no long-term neurological sequelae. However, even mild head injuries will sustain intracranial damage and long-term consequences.[27] Although it is a challenge to triage children with a mild head injury with respect to possible intracerebral lesions on CT scanning, several indices such as admission GCS score, focal neurological deficits, and fractures detected by skull radiography have been found reliable predictors of positive findings when CT is performed.[28]

Diagnosing severe traumatic brain injury in children is most important because long-term neurodisability and death can quickly follow and early neurosurgical intervention and/or pharmacotherapy may improve the outcome.[29] Children who survive severe traumatic brain injury are at high risk for problems in behavior, adaptive functioning, and educational performance.[30]

COST AND CONSEQUENCES OF TRAUMATIC BRAIN INJURY

According to the World Health Organization, traumatic brain injury is a problem that affects 10 million people annually worldwide.[31] Among these, 5.3 million Americans live with long-term disabilities related to traumatic brain injury.[32]

In the United States, traumatic brain injury–associated direct and indirect costs have been estimated to be more than 76.5 billion dollars a year.[33] This cost does not include 11 years of warfare in Iraq and Afghanistan, where, according to a Congressional Research Service report into military casualty statistics, approximately 300,000 troops have experienced some form of brain trauma.[25]

Aside from cognitive and physical disability problems encountered by most people with traumatic brain injury, this injury is believed to be a risk factor in the development of Alzheimer's disease, sometimes occurring decades after the trauma.[34] It is now clear that a history of brain trauma predisposes the individual 2–4 times to an increased risk of Alzheimer's disease when compared to individuals with no history of brain trauma.[35]

The long-term neuropathologic process evolving after single traumatic brain event of sufficient severity to produce LOC is poorly understood, but it has been speculated that chronic cerebral hypoperfusion may play a major role.[36] This conclusion is supported by arterial spin-labeling magnetic resonance imaging (MRI) of people with a prior moderate-to-severe traumatic brain injury who showed cerebral hypoperfusion in regions of the brain such as the posterior cingulate cortex, later associated with neuropathological changes, characteristic of Alzheimer's dementia.[37] Reduced CBF during aging is known to be a marker for cognitive impairment prior to the onset of Alzheimer's dementia.[38,39]

CHRONIC TRAUMATIC ENCEPHALOPATHY

Boxers who have had repeated blows to the head for years were shown to have cerebral hemodynamic impairment affecting cerebral perfusion, and these changes were associated with a higher incidence of neurocognitive dysfunction corresponding to a lack of sustained attention, difficulty concentrating on tasks at hand, disorientation, and not remembering newly learned material when compared to a similar group without a history of brain trauma.[40] Originally, this type of injury was described as *punch drunk* by Martland in 1927 and later given the more descriptive term *dementia pugilistica* by Millspaugh.[41]

The realization was slow in coming that the same repetitive type of head knocking in boxers could also occur in players of other contact sports such as football, soccer, lacrosse, rugby, and hockey. This brain injury is now known as chronic traumatic encephalopathy (CTE). At the beginning, CTE may be asymptomatic or start slowly with symptoms initially involving problems in memory and concentration. Months and years after the initial symptoms of concussion have faded, new symptoms occur as gradual degeneration of brain function occurs with episodes of disorientation, confusion, dizziness, severe neurocognitive disturbances, and dementia.

Although early diagnosis of CTE is often difficult due to an absence of bleeding or major structural abnormalities on a CT scan, grossly, CTE in the final stages is characterized by the enlargement of the ventricles and atrophy of the hippocampus, amygdala, and frontal, temporal, parietal, and entorhinal cortices.[42] Histologically, there is significant neuronal loss and increase of reactive glial cells usually accompanied by neurofibrillary tangle formation in the hippocampus and entorhinal cortex, two regions associated with learning and memory and the primary targets for Alzheimer's disease pathologic lesions.[42,43]

These findings suggest that structural brain lesions, both focal and diffuse, are the primary contributors to the abnormal cerebral perfusion changes seen in chronic traumatic brain injury and CTE and that CBF quantitation may serve as an early marker to assess the degree of damage to neurons and their synapses.

Besides neuroimaging measurements of CBF, neurocognitive tests designed to assess mental status are now used to complement any structural or physiological brain abnormality in people at risk of CTE. These tests can evaluate decision-making ability, reaction time, attention, and memory of an athlete after sustaining a concussion or following repetitive knocks to the head from the participation of sports-related activities. The goal of these measures is to identify and prevent further damage to the brain in athletes from the particular sport activity before it becomes irreversible if head pounding is continued or ignored.

TREATING TRAUMATIC BRAIN INJURY

The initial focus of managing an acute traumatic brain injury is to stabilize the injured person in order to minimize complications from secondary damage. This approach generally follows the clinical guidelines outlined in Figure 9.1.

Mild traumatic brain injuries usually require no treatment other than rest and over-the-counter pain relievers to treat a headache. However, it is estimated that up to 3% of mild head injuries progress to more serious consequence, so careful evaluation of the injury and its potential to worsen is mandatory. Most mild head injuries may not require hospitalization and can be closely monitored at home by a responsible caretaker given written instructions for any persistent worsening of initial symptoms or new symptoms.

For moderate-to-severe traumatic brain injuries, the general rule is to stabilize the patient to minimize secondary damage such as the development of plateau waves. Plateau waves during traumatic brain injury are abrupt increases of ICP above 40 mmHg that can be due to lower CBF, cardiac output, and brain tissue oxygenation. These plateau waves can be controlled when CPP and oxygenation can be restored in head-injured patients.[44]

This can be done initially with intravenous (IV) fluids to maintain the patient in a state of euvolemia or mild hypervolemia.[45] Fluid resuscitation has replaced the old rule of fluid restriction since hypovolemia has been shown to reduce cardiac output. When cardiac output is decreased following brain trauma, there is often a reduction in cerebral perfusion that may cause an increase in cerebral edema and ICP.[36] Thus, fluid restriction is contraindicated in patients with raised ICP and brain edema. The brain trauma patient may also require sedation to prevent agitation as well as pain medication as required.

The preferred treatment for increased ICP is currently the use of osmotherapy, such as mannitol and hypertonic saline.[46] This treatment however is problematic especially when it is extended for long periods of time since it has been shown to promote electrolyte abnormalities, especially hypernatremia.[47] Other adverse effects after osmotherapy are associated with cardiac failure, bleeding diathesis, and phlebitis. No benefits of small-volume resuscitation with hypertonic saline have been shown in patients with traumatic brain injury.[48]

Consequently, in adults, there is insufficient evidence to support the use of hypertonic saline or mannitol for managing ICP because of the risks of severe hypernatremia, and caution has been advised with the combined use of mannitol and hypertonic saline.[49] Careful control of fluid balance, electrolyte status, and serum osmolarity (<320 mmol/L) is mandatory when hypertonic agents are used.

Many prehospital care systems use advanced-level prehospital care providers in the belief that prevention or early correction of these insults will improve patient outcomes.

The main goal of prehospital management is to prevent hypoxia and hypotension, because these systemic insults are prone to induce secondary brain damage.[50] When patient status was recorded by trained paramedics at the scene of injury before hospital admission, oxygen saturation below 90% was found in half of all cases reviewed and hypotension was present in about 25% of head injury patients.[51,52]

Trauma renders the brain more vulnerable to these secondary insults,[52] and hypoxia and hypotension are strongly associated with poor outcome.[53]

Consequently, the introduction of a prehospital system capable of normalizing oxygenation and blood pressure has been associated with improved outcome.[54,55] To our knowledge however, only one randomized control led trial (RCT) has compared physician and nonphysician teams in prehospital brain trauma management.[55] This single RCT supported the value of trained paramedical personnel able to perform bag-valve-mask ventilation, apply external compression for hemorrhage, intubate orally without adjuvant neuromuscular blockade, and establish IV lines to administer crystalloid or other solutions.[56]

Are there pharmacological treatments now clinically available that will stop and reverse the pathology that follows a moderate or severe traumatic brain injury? In a word, no.[57]

This grim conclusion is supported by the sobering fact that the best and worst hospitals in the United States report similar morbidity and mortality rates of patients admitted with moderate-to-severe traumatic brain injuries.[58] This finding stresses the predictably poor outcome in this group of patients and the lack of an effective treatment or intervention to control the catastrophic sequelae associated with this injury.

Lu et al.[59] conducted a comprehensive literature review of RCTs in adults sustaining a traumatic brain injury over the past 30 years. RCTs are the gold standard of clinical trials due to their rigorous way of determining whether a cause–effect relation exists between treatment and outcome.

Lu et al.[59] dramatically discovered that of 55 acute-phase trials of interventions for traumatic brain injury, 40 showed either no effect or adverse effects when used on brain trauma patients. These acute-phase trials were defined as interventions that occurred within 24 h following injury to the brain.[59] The few trials with limited effectiveness in treating brain trauma patients were nonpharmacological and included early nutritional therapy, cognitive rehabilitation, and physical rehabilitation. Treatments such as decompressive craniotomy, hyperventilation, therapeutic hypothermia, and osmotic pharmacotherapy led to either ineffective or mixed results.[59]

For example, decompressive craniotomy, hyperosmotic therapy including mannitol or hypertonic saline, and hypothermia have shown inconsistent effects, with some studies reporting positive results[60,61] and others reporting negative or no results.[62–64]

Lu's review[59] of 32 RCTs examining drugs targeting neuronal excitotoxicity such as magnesium, insulin, or corticosteroids and drugs blocking lipid peroxidation either failed to show a positive treatment effect or led to adverse effects. Of the 32 clinical RCTs, only 3 drugs showed positive treatment effects for acute traumatic brain injury, including 1 involving methylphenidate, a dopamine agonist and psychostimulant.[65]

However, a recent Cochrane Central Register analysis of 20 traumatic brain injury RCTs using dopamine agonists including methylphenidate showed no trend toward efficacy and safety when measures of cognitive and neuropsychological function were applied.[66] One major concern of these RCTs was the high risk of bias involved in each investigation, which consisted mainly of selection reporting bias and the lack of information to determine the efficacy of randomization.[66] This search analysis of the relevant literature concluded that dopamine agonists such as methylphenidate, amantadine, and bromocriptine cannot be recommended as part of the acute treatment regimen to preserve cognitive function following traumatic brain injury.[66]

The other two drugs found to have efficacy in the review of Lu et al.[59] of brain trauma treatments were the calcium channel antagonist nimodipine and progesterone, a reproductive hormone involved in pregnancy and the menstrual cycle. The activities of progesterone and DMSO in traumatic brain injury (TBI) are discussed in detail in the "DMSO Compared to Progesterone for Traumatic Brain Injury" section. This section includes two most recent clinical trials reported in December 2014, showing no benefit to patients treated for TBI using progesterone.

As for the calcium channel blockers, Harders et al.[67] reported on a prospective, randomized, double-blind, placebo-controlled study of nimodipine on 123 patients to treat subarachnoid hemorrhage secondary to traumatic brain injury and found that 6 months after this treatment, significantly less unfavorable results in terms of severe disability, vegetative state, or death resulted when compared to a placebo-treated group.

Again, however, a systematic review of RCTs using calcium channel blockers, including nimodipine for acute traumatic head injury, showed considerable uncertainty over its effects. The use of nimodipine after blunt head injury revealed an increase in adverse reactions suffered by the group receiving this treatment, a finding that implies the drug is harmful for some patients.[68]

With regard to progesterone, two RCTs, reported in 2007–2008, showed positive data for this drug. In the first trial, 159 randomized patients were randomized to receive progesterone (N = 82) or a placebo (N = 77) within 8 h of injury. All patients were evaluated after 3 and 6 months of follow-up. Improved neurological outcome and lower mortality rate were reported for the progesterone-treated group at both time points.[69]

The second trial using progesterone was a randomized, double-blind, placebo-controlled trial consisting of 100 adult brain trauma patients who were treated within 11 h of injury with a postresuscitation GCS of 4–12. After 30 days postinjury, only moderately injured patients (GCS 9–12) showed lower mortality rate than placebo and *possible* signs of benefit when compared to the placebo group. The mortality rate reduction was marginally significant when the treatment group was compared with placebo.[70] Stein et al.[71] theorized that the possible mechanisms of progesterone neuroprotection may lie in its ability to lower brain edema and downregulate the inflammatory cascade associated with traumatic-induced brain swelling.

More recently, a Cochrane Injuries Group's Specialised Register analyzed 3 small RCTs involving 315 people treated with progesterone or placebo for traumatic brain injury and concluded that progesterone may improve the neurologic outcome of patients with traumatic brain injury but cautioned that evidence for effectiveness remains insufficient.[72] This Cochrane analysis recommended that multicenter RCTs are needed to examine whether any benefit for head injury is provided by progesterone therapy.[72]

DMSO IN TRAUMATIC BRAIN INJURY

Although much has been learned in the past 25 years about the cellular and molecular pathologies associated with traumatic brain injury, the hunt for effective treatments has been a historical disappointment. This failure to find an effective treatment for traumatic brain injury may have more to do with choice of the experimental model than with the choice of what end points may be relevant to the human injury.

For example, there are two major flaws in correlating the effectiveness of a therapy for brain trauma using small animals as the injury model. The first flaw is that the animal models used to test potential treatments for head injury, such as rats, cats, rabbits, sheep, pigs, and mice,[73] all have cerebrovascular supplies that are inherently dissimilar to humans or nonhuman primates and this dissimilarity creates a variable tolerance when trauma is applied to their brains.

Human and nonhuman primate brain is highly organized and lacks the collateralization of cerebral vessels found in small mammals, so its vulnerability to brain trauma or ischemia is much greater than that of small mammals. The second major flaw may be an inherent ineffectiveness of the treatment being tested. These two experimental flaws may help explain why dramatic therapeutic responses often reported as breakthroughs in the head injury field using small animals subjected to head injury later are shown to have almost zero correlation to the human counterpart.

Consequently, extrapolations relating the success of these therapies in small animal models to humans are at best uncertain and impossible to predict.

To complicate matters, traumatic brain injury in humans can result in focal pathology due to the formation of hematomas and vessel infarction, or diffuse injury involving spreading edema, activation of inflammatory cytokines, release of excitatory neurotransmitters, and diffuse axonal injury.[74] Some but not all of the pathological processes are easy to create in a small animal model but difficult to clinically assess posttrauma after a treatment is delivered due to the variability and tolerance associated with each animal species.

From this perspective, DMSO had a distinct advantage when it was first tested for traumatic brain injury in the early 1970s because nonhuman primates were used. For reasons discussed in Chapter 8, using nonhuman primates, whose brains very closely resemble human brain, means that the injury will closely reflect the human injury and if the treatment works in the nonhuman primate, it is also likely to work in people.[75–78] The only caveat in the use of a nonhuman primate is that human traumatic brain injury is more complex than the controlled laboratory injury monkeys experience.

The results of these brain injury studies showed DMSO had a dramatic positive effect on ICP, CPP, cerebral metabolic rate of oxygen, CBF, and survival.[75–78]

CPP is defined as the net pressure gradient that drives CBF to perfuse the brain. It is generally measured as the MAP minus ICP (MAP – ICP). CPP in adults is normally between 70 and 100 mmHg, while the normal ICP ranges from 5 to 15 mmHg.[79,80]

It is important that CPP be maintained at narrow limits during trauma to the brain, because if it is too low, it can contribute to both brain ischemia and irreversible brain damage and if it is too high, it can raise ICP and increase the risk of complications or death.[79] One standard technique of lowering ICP after brain trauma in people is to induce controlled hyperventilation in order to maintain $PaCO_2$ at the low end of normal, that is, 35–40 mmHg.[79]

The basic studies on the use of DMSO for traumatic brain injury previously reported in monkeys[75–78] led to several pilot clinical trials using IV DMSO on head-injured patients. One early study by Waller et al.[81] tested DMSO on 11 adult patients with high ICP and a GCS score of 4–6 following brain trauma or subarachnoid hemorrhage. All 11 patients had failed to respond to standard therapy and were considered to be on the verge of dying. DMSO was given as a rapid bolus intravenously in a 40% or 10% solution at 1 g/kg. This dose was repeated every 6 h as needed to maintain the ICP below 20 mmHg.[81]

In all patients, DMSO effectively increased diuresis and reduced the ICP where standard therapy had failed. DMSO treatment showed that three patients who were expected to die survived with good recovery; the other eight patients died despite their initial normalization of the intracranial hypertension, which could not be controlled after several days.[81]

The ability of DMSO to maintain CPP above 60 mmHg as shown by Brown et al.[77,78] after severe brain trauma may explain the benefit seen in Waller's et al.[81] results on patients with lethal ICPs. This conclusion has support from a preliminary study showing the importance of maintaining CPP from sinking too low in the presence of high ICP.[77]

The findings from the study by Waller et al.[81] indicated that aggressive management to keep CPP above 60 mmHg can lead to good neurological outcomes despite the presence of extremely high ICP. Such a process may work by lowering cerebrovascular resistance, which will increase CBF to supply low perfusion brain regions where undamaged but dysfunctional penumbral brain cells at risk of dying can be rescued.

A study by Marshall et al.[82] was reported several years after the Waller et al.[81] clinical trial using DMSO for intractable, severe head injury. Six patients refractory to barbiturate and other standard treatments were given 10% DMSO intravenously at a dose of 1 g/kg.[82] Because of difficulties with fluid overload reflected by electrolyte disturbances and increased pulmonary wedge pressure, DMSO was titrated against ICP and readministered as a 20% solution. DMSO was observed to rapidly lower ICP below 20 mmHg in all but one patient within an average of 6 h, three patients responding after 3–4 h and the rest after 12–18 h. It was not clear from this report how many of the six patients died when ICP could no longer be controlled, but it must be assumed that poor survival resulted.[82] Both at 10% and 20% DMSO volumes, electrolyte disturbances involving hypernatremia developed in all patients treated.

These investigators reported that DMSO was a *rapid and effective agent* in lowering uncontrolled ICP where standard therapy had failed but were also discouraged from handling DMSO because they needed to use special plastic IV bags and IV tubing that were not made of polyvinyl chloride.[82] The reason for not using polyvinyl chloride tubing and IV bags is that DMSO is known to leach this plastic when it comes in contact with it.[76] However, this problem can be readily overcome by using glass dispensing IV bottles and teflon IV tubing or teflon catheters.[83] The electrolyte problems encountered in the Marshall et al.[82] study were likely due to the high IV volumes given at 10% and 20% solutions at a concentration of 1 g/kg.

A brief communication by Gumerlock and Neuwelt[83] shortly following the clinical study by Marshall et al.[82] involved five closed head injury patients treated with IV DMSO for increased ICP. DMSO 20% solution was given at 100 mg/kg as an initial bolus followed by repeated hourly infusions of 100 mg/kg tapering off as ICP was lowered. A rapid response was observed after the DMSO bolus and ICP was reduced to normal levels using this regimen.[83] The aftermath of these treatments with DMSO consisted of two outcomes. The first outcome involved two children given IV DMSO. The first was a 1(1/2)-year-old girl admitted with a GCS score of 7 (severe) and an ICP of 30 mmHg. This patient received DMSO for 84 h after being unresponsive to mannitol treatment and recovered slowly over a 3-week period when she was discharged from the hospital. Her electrolytes during treatment remained normal, including serum sodium values.[83]

The second patient was a 7-year old child admitted with a GCS of 5 (severe) and ICP of 25 mmHg.[83] Using the same aforementioned regimen, DMSO was given for 36 h and ICP was controlled during that time. The patient recovered after 8 weeks and was discharged from the hospital. Serum electrolytes remained normal during treatment, including sodium values. The remaining three patients were adults ranging in age from 17 to 52 years old who were admitted with GCS scores of 3–5 (severe).[83] Two of these three patients had ICPs above 50 mmHg. Although all three patients initially responded to the DMSO regimen used before, the ICP escaped control and all three died. The sodium values in these DMSO nonresponders ranged from 159 to 172 mmol/L.[83]

Although the number of patients in this trial was small, it is of interest to note that age and ICP values may have played a role in the response and final outcome to DMSO and to the elevated serum sodium concentrations. The very high ICP seen in the three adult patients in the Gumerlock and Neuwelt[83] study and the time it took to administer DMSO may have determined the irreversibility of the injury and the inability by DMSO to affect the eventual outcome recorded.

OPTIMAL DMSO DOSE FOR TRAUMATIC BRAIN INJURY

The findings published by Marshall et al.[82] and by Gumerlock and Neuwelt using DMSO in human head injury and the experimental findings by de la Torre et al.[75,76] and Brown et al.[77,78] using rhesus monkeys and by others using smaller mammals subjected to a variety of brain insults[84–86] indicated that the *optimal* dose of IV DMSO to be given after traumatic brain injury should be 28% solution at a staring concentration of 1–1.5 g/kg.

The aforementioned volume and concentration of DMSO were theorized in order to avoid fluid overload, electrolyte disturbances, and significant hemolysis while ensuring a safer and more effective response to increased ICP. The DMSO fluid delivery time would also optimally range between a fast bolus and a fast drip as originally suggested by de la Torre et al.[76]

The recommended dosing schedule for DMSO was adopted in a preliminary study by Turkish investigators. IV DMSO was administered as a first-line treatment to 10 patients with severe closed head injury by Karaça et al.[87] using a 28% solution at a concentration of 1.12 g/kg delivered as a fast drip. On admission, the mean ICP recorded was 73 mmHg and ranged from 40 to 127 mmHg.

The DMSO dose was reduced by half if and when the ICP reached <20 mmHg. DMSO was continued for 10 days or until stabilization of the ICP or full recovery was observed. These investigators reported a dramatic lowering of ICP in all but three patients after a 6-day treatment. Neurological improvement and survival were observed in the remaining seven patients followed up for 3 months.[87] Two patients in this series died, and one patient had poor neurologic outcome.

The modified DMSO dose and its application to severe head injury patients shortly after admission differed substantially from the previous trials that used DMSO as a second-line treatment. These findings using IV DMSO as a primary treatment for severe closed head injury could be described as *dramatic* due to the reduction in permanent disability, 75% survival rate, and 25% mortality that historically has a 60%–70% mortality using standard therapy.

Following the encouraging findings reported by Karaça et al.[87] using DMSO for severe head trauma, a second study was quickly undertaken by Kulah et al.[88] working from the same Turkish hospital and neurosurgical unit. This group gave IV DMSO as a bolus (50 mL of a 28% solution mixed in 5% dextrose) to 10 patients with ICPs exceeding 25 mmHg and GCS between 3 and 6. A marked decrease in the ICP and an increase in CPPs occurred within 10 min in most of the patients after the start of DMSO administration. No rebound effect was noted as it commonly occurs with mannitol treatment.[88]

The increase in CPP was observed to parallel the decline in ICP in all cases.[88] No serious adverse events were seen after DMSO treatments.[88] Blood pressure, cardiac output and central venous pressure, and electrolytes were unaffected by DMSO infusions. When CBF was measured in the head injury patients, an average increase of 20% was observed following DMSO administration.[88] Brisk diuresis was noted in most patients shortly after the initiation of each DMSO administration.[88] The mortality rate using DMSO was 30%, and six patients survived with good-to-excellent neurological outcome after a 3-month follow-up.[88]

Classically, a GCS score of 3–6 on presentation at the intensive care unit in patients with severe blunt trauma to the brain has been recognized as an ominous prognostic factor. The reported mortality rate in these patients ranges from 45% to 100%.[89] A poor neurological outcome is observed in 90% of patients with a GCS of 3–6 when followed up for a year.[90]

Thus, a survival rate of 70% with good functional recovery in 60% of the patients with severe blunt trauma given IV DMSO as reported by teams led by Karaça et al.[87] and by Kulah et al.[88] must be considered a highly significant improvement over standard therapy in the management of traumatic brain injury. It is clear from these findings that the mixed results reported by Waller et al.,[81] Marshall et al.,[82] and Gumerlock and Neuwelt[83] using DMSO for severe head injury differed from those obtained by Karaça et al.[87] and by Kulah et al.[88] who also used DMSO for similar severe head injuries but who reported much better outcomes and responses to treatment. These important differences need to be more closely examined.

COMPARING DMSO DOSES, DURATION OF TREATMENT, AND USE AS A FIRST- OR SECOND-LINE TREATMENT FOR TRAUMATIC BRAIN INJURY

Analyzing the findings of the studies on severe head injury reported by Waller et al.,[81] Marshall et al.,[82] Gumerlock and Neuwelt,[83] Karaça et al.,[87] and Kulah et al.[88] with regard to DMSO therapy as discussed earlier, several fundamental factors may partly explain the mixed results.

First, it is an axiomatic in neurosurgery that the sooner a patient with moderate-to-severe traumatic brain injury is *successfully* treated to lower high ICP, the better the chances of a good response and a favorable outcome. This rule relies on the pathophysiological events that secondary damage may quickly introduce after brain trauma.

Secondary damage appears in the form of the *four H syndrome*: hypotension, hypoperfusion, hypoxia, and hypercarbia. These systemic changes alter the local cerebral milieu at the cellular–molecular levels that involve the release of inflammatory cytokines, excitotoxic amino acids' induction of ionic fluxes, oxidative stress, and energy metabolic failure.

Consequently, effective therapy that can target the local and systemic secondary sequelae that rise rapidly after brain trauma may be able to protect to some degree against devastating damage and brain cell death and thus help in the recovery of the brain trauma patient.

In the Waller et al.,[81] Marshall et al.,[82] and Gumerlock and Neuwelt[83] preliminary studies, much precious time was spent administering standard therapy, which failed to control ICP before DMSO was administered. In the Karaça et al.[87] and Kulah et al.[88] studies, DMSO was given *as a first-line treatment* to severe head injury patients upon arrival at the hospital. This was a critical difference from previous studies in the management of severe head injury using DMSO.

This lapse of time difference, which ranged from 8 to 24 h, reasonably explains the mixed outcome results reported in the Waller et al.,[81] Marshall et al.,[82] and Gumerlock and Neuwelt[83] trials and the more favorable outcomes reported by Karaça et al.[87] and by Kulah et al.[88] when DMSO is given as a primary agent within 6 h after trauma. The other important difference in comparing these trials is the DMSO concentration and the dosing time interval.

The 10%–20% solution of DMSO administered in the Marshall et al.[82] trial may not have been effective due to fluid overload and a suboptimal DMSO concentration, whereas in the Karaça and Kulah studies, the 28% DMSO solution given avoided fluid overload and prevented significant hemolysis and electrolyte disturbances.

The IV administration of DMSO that was continued until absolute control of the ICP was attained may also have been a key factor in achieving the better survival rate and improved neurological outcome reported in the Karaça–Kulah studies.[87,88]

Both investigators reported that continuous or intermittent DMSO administration extended for 2–7 days of treatment *if needed* generally prevented the return of increased ICP to the original levels seen at admission.[87,88]

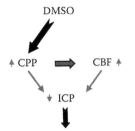

FIGURE 9.3 The main therapeutic effect by DMSO aims at lowering ICP (↓) following severe traumatic brain injury (TBI). This action increases CPP (↑) and CBF (↑). Increasing CPP–CBF may prevent or diminish secondary damage involving inadequate cerebral oxygenation, cerebrovascular dysautoregulation, neurometabolic impairment, inflammatory responses, and excitotoxic cell death. (From Werner, C. and Engelhadt, K., *Br. J. Anaesth.*, 99, 4, 2007.)

Data from these two preliminary studies indicate that early, aggressive management of CPP to increase CBF by DMSO can lead to improved neurological outcome despite extremely high ICPs[91] (Figure 9.3).

There are two key hemodynamic changes that may explain DMSO's efficacy in controlling ICP following traumatic brain injury: (1) the ability to significantly increase CBF and (2) the ability to improve CPP. Both CBF and CPP values are lowered following brain insults.[36] Aggressive CPP therapy should be performed and maintained even though apparently lethal ICP levels may be present. It thus seems reasonable to correlate the increase in CPP and CBF by DMSO with the reduction of ICP.

Aggressive CPP therapy should be performed and maintained even though apparently lethal ICP levels may be present. It seems reasonable to correlate the increase in CPP and CBF by DMSO with the reduction of ICP. The reason is that traumatic brain injury is widely known to cause dynamic changes in CBF and secondary brain insults have been reported to decrease CPP.[36]

In patients with moderate-or-severe head injury, CPP has also been shown to correlate with patients' functional outcome. This action ostensibly may lead to better tissue oxygenation, better control of brain autoregulation, reduced glutamate excitotoxicity, and stabilization of impaired neurometabolism and inflammatory responses.[17] These pathological processes are schematized in Figure 9.4, with respect to DMSO, and a general scheme of how DMSO may affect outcome following head injury is summarized in Figure 9.3.

The most consequential detail to recognize about DMSO therapy in traumatic brain injury is that this apparently useful drug does not act as a miraculous *lazaroid*, that is, an imaginary agent that can wake up dead neurons. This is an unrealistic expectation which investigators often assume is the inability of a drug to reverse the symptoms and pathophysiological consequences of brain trauma.

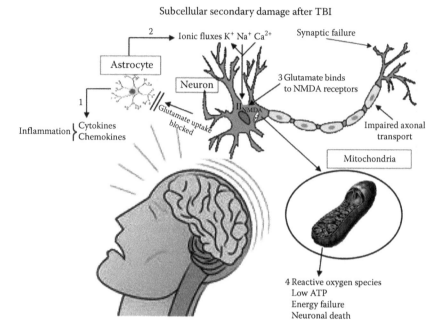

Subcellular secondary damage after TBI

FIGURE 9.4 **(See color insert.)** Diagram shows major subcellular events associated with secondary damage following primary traumatic brain injury (TBI). Astrocytes undergo reactive transformation in response to blunt head injury and release cytotoxic cytokines and chemokines. Reactive astrocytes are unable to uptake glutamate. Glutamate, in turn, accumulates in the extracellular space at the site of injury where it binds and opens NMDA channel receptors in neurons. This action by glutamate creates an influx of calcium (Ca^{2+}) and sodium (Na^+) ions and an efflux of potassium (K^+) ions, which result in the depolarization of neurons. This ionic dyshomeostasis damages intracytoplasmic mitochondria that promote the formation of reactive oxygen species, reduced ATP synthesis, energy failure, and cell death. Studies have shown that DMSO is able to antagonize the formation of cytokines and chemokines in injured tissue (1), block Na^+ and Ca^{2+} influx into cells after neuronal depolarization (2), prevent glutamate induction of excitotoxic neuronal death by suppressing NMDA ion currents (3), and scavenge the formation of reactive oxygen species and the damage caused by these oxidizing radicals. See text for details.

If function is regained after a severe traumatic brain injury with or without therapy, it implies that surviving brain cells peripheral to the injury site have recuperated sufficiently either through an external intervention or through unknown internal factors. It also means that in the case of intervention, extension of brain cell damage beyond the penumbral region bordering the injury core has been prevented.

Very little information is presently available to explain the recovery of penumbral neurons, that is, those nerve cells bordering the central core of injury that remain undamaged but nonfunctional. Needless to say, brain cells rarely recover by themselves after neural tissue damage and they are, by definition, at risk of sudden or protracted death by apoptosis or necrosis.

Astrocytes undergo reactive transformation in response to physical brain injury by a process known as reactive gliosis, which may impede neural recovery. Astrocytes, like other brain cells, are vulnerable to reactive oxygen species generated by traumatic brain injury and ischemia.[92]

Intact brain studies[93] suggest that the rise of extracellular glutamate in brain that occurs during energy failure secondary to brain trauma may be due to failure of astrocytes to uptake extracellular glutamate, which then accumulates excessively at the site of injury and binds to N-methyl-D-aspartate (NMDA) receptors in neurons where an influx of calcium and sodium ions and an efflux of potassium ions result in the depolarization of neurons.[94] This ionic imbalance damages neuronal mitochondria that leads to cytotoxic reactive oxygen species, reactive nitrogen species, mitochondrial ATP loss, energy failure, and neuronal death (Figure 9.4).

Findings suggest that DMSO administration in traumatic brain injury appears to target several of these pathophysiological processes to either block or reverse their activity (Figure 9.4).

For example, in normal brain metabolism, glutamate, the most abundant excitatory neurotransmitter, is taken up by astrocytes, which convert it to glutamine and deliver it to neurons where it is converted back to glutamate as a source of energy. All glutamate receptors are cation channels that allow the influx of sodium (Na^+) and calcium (Ca^{2+}) ions and the efflux of potassium (K^+) ions in response to glutamate binding to NMDA membrane receptors.[95] In traumatic brain injury, astrocytes are unable to take up glutamate, which can accumulate outside brain cells.

This excitotoxic process causes K^+ ions to exit and Na^+ and Ca^{2+} ions to enter nerve cells via NMDA receptor channels located on the neuronal membrane, a pathologic process leading to neuronal depolarization, damage, and eventual neuronal cell death (Figure 9.4). Ionic dyshomeostasis will uncouple mitochondrial ATP synthesis, induce oxidative stress, reactive oxygen species, and neuronal cell death.[96,97]

The mechanism of neuronal cell death includes damage to mitochondria from excessively high intracellular Ca^{2+}, while the production of reactive oxygen species will lead to lowered ATP energy fuel and eventual energy failure (Figure 9.4). Energy failure will disrupt axonal transport mechanisms and impede neurotransmission. The excitotoxic ischemic cascade will also create inflammatory cytokine and chemokine secretions from damaged astrocytes.[97] The delicate balance of glutamate metabolism in brain implies that too much or too little glutamate can be harmful. If glutamate levels are too low, cognitive and behavioral problems can develop, but if glutamate levels are too high, the overstimulation of nerve cells can lead to cell death.

Regulation of the subcellular events associated with traumatic brain injury is therefore vital for controlling the pathophysiological cascade and neurological dysfunction that will result following brain trauma.

There are reports that DMSO is able to block Na^+ and Ca^{2+} ions entry into cells following tissue injury[98,99] and block Na^+ channel activation after neuronal depolarization.[99,100] Drugs like DMSO that prevent abnormal sodium influx into injured cells provide effective protection against Na^+ and Ca^{2+} overload. However, it remains unclear whether DMSO is capable of exerting this effect in normalizing ionic dyshomeostasis in traumatized brain tissue. The mechanisms exerted by DMSO on Na^+ and Ca^{2+} channels need to be further investigated in mammalian models of traumatic brain injury since the results of such studies could produce extremely useful and relatively safe agents in treating neurotrauma.

Another critically important action shown by DMSO is the prevention of glutamate induction of excitotoxic neuronal death by the suppression of NMDA ion currents[101] (Figure 9.4). DMSO was observed to suppress NMDA-induced cation currents and excessive Ca^{2+} influx into plated hippocampal neurons within minutes after its introduction, sparing brain cells from excitotoxic death.[101] This action by DMSO on cultured neurons did not appear to involve changes in gene expression or protein translation. It suggested instead a direct effect by DMSO on neuronal membrane receptors mediating glutamate receptor channels and excitatory transmission.[101]

The decrease in NMDA ionic currents by DMSO could also be affected by the antioxidant action of DMSO, which is capable of modulating glutamate receptor activity.

An important phase of traumatic brain injury is the inflammatory reaction seen with blunt head injuries. Although the peak inflammatory response to brain trauma can evolve after several days, cytokines and chemokines (Figure 9.4) are released from astrocytes, microglia, and polymorphonuclear cells within hours after injury, a process that leads to opening of the blood–brain barrier and to complement-mediated activation of cell death.[102,103]

Prevention or control of cytokine release and accumulation in the extracellular space after brain injury, especially cytotoxic concentrations of several important pro-inflammatory interleukins, such as the IL-1 family, IL-6, and TNF-α, may be life saving in head injury if specific therapy can be applied to diminish their pathogenic effect.[104]

In this vein, the actions of DMSO on inflammatory cytokines have been reported in intestinal cells exposed to a variety of inflammatory cytokines *in vitro*.[105]

DMSO at 0.5% was shown to significantly decrease mRNA levels of inflammatory proteins, including IL-1α, IL-1β, and IL-6.[105] The IL-1 family constitutes a complex network of cytokines involved in the initiation of inflammatory responses resulting from tissue injury, including brain ischemia and stroke.[106] These interleukins have become a therapeutic target that when neutralized can result in a sustained reduction of ischemic trauma severity.[107]

IL-6 plays a major role in many undesired effects of the immune system, and its expression has been found to increase quickly at the sites of axonal injury.[108,109]

In addition to suppressing cytokine production, DMSO was reported to decrease secretions of a macrophage chemoattractant protein-1 (MCP-1) in a dose-dependent manner in rodent intestinal cells.[105] MCP-1 is a major chemokine involved in macrophage migration and has been shown to play important roles in tumor growth and metastasis. Chemokines are small proteins that direct the trafficking of immune cells to sites of inflammation and activate the production and secretion of inflammatory cytokines such as the IL-1 family.[110] MCP-1 is involved in inflammation and has been isolated from monocyte and macrophages during an inflammatory reaction.[110]

It remains to be seen whether the actions of DMSO in suppressing inflammatory cytokines and chemokines in intestinal cells also apply to the formation and accumulation of these small proteins in brain following neurotrauma.

Traumatic brain injury may result in a dynamic, pathological process that can evolve quickly if not aggressively treated.

Thus, it seems clear from the collective findings relevant to traumatic brain injury that once secondary damage appears and remains uncontrolled, extensive neuronal cell populations will die, and the prognosis will be poor despite the best efforts of any successful therapy.

For these reasons, it seems logical and proper to implement new procedures to treat moderate-to-severe head injuries as an emergency, much like a heart attack or stroke that requires immediate attention to improve the odds of diminishing returns. It has been suggested that prehospital treatment may be started by qualified ambulance paramedics trained to intubate head-injured patients whose airway may be compromised at the scene of an accident.[55]

However, unless the patient is unconscious, intubation may require neuromuscular blocking agents or anesthetic drugs, which may pose a hurdle for even trained ambulance personnel. This dilemma is a problem that requires further studies to resolve because if rapid and appropriate prehospital treatment can be applied to the head-injured patient, as the Karaça et al.[87] and Kulah et al.[88] studies imply, major success in preventing poor neurologic outlook and death following brain trauma could replace the grim statistics presently associated with this brain insult.

Thus, prehospital neurotrauma care could address the 4H syndrome, hypotension, hypoxia, hypoperfusion, and hypercarbia seen in most head-injured patients and provide a preplanned therapeutic approach at the scene of accident to counteract the negative conditions that arise after blunt trauma to the brain.

DMSO IN INTRACRANIAL ANEURYSM HEMORRHAGE

Intracranial aneurysms are bulging lesions where blood collects like an expanding balloon due to a weak area in the wall of an artery that supplies blood to the brain. Although aneurysms can occur in any weakened artery, brain aneurysms commonly occur at branch points of major cerebral arteries.

When an aneurysm ruptures, a subarachnoid hemorrhage results as blood accumulates into the subarachnoid space where it compresses the brain. Depending on the amount of the bleed, the resulting intracranial hemorrhage can cause sudden, severe headache, nausea, vomiting, stiff neck, LOC, and signs of a stroke, including hemiplegia.[111] Rupture of intracranial aneurysms is associated with high morbidity and mortality if not immediately treated.[112] Nonpharmacological treatment of a bleeding aneurysm includes clipping the neck of the aneurysm or inserting a coil inside the aneurysms to create a blood clot that stops the bleeding.

Pharmacological treatments are given to manage the symptoms only and include vasopressors and calcium channel blockers that may alleviate the vascular resistance caused by the spastic, narrowed blood vessels and provide improved blood flow through these vessels. Once an intracranial aneurysm has bled, there is a very high chance that it will rebleed unless treatment is applied.

Vasospasm, reduced blood flow to the brain, and hemiplegia remain the leading causes of morbidity in patients who survive the initial subarachnoid hemorrhage (Figure 9.5).

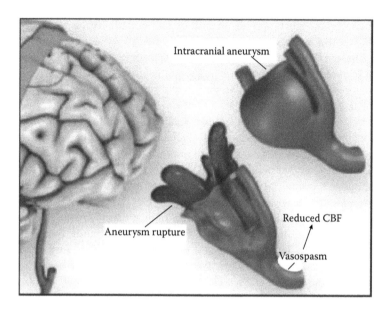

FIGURE 9.5 (See color insert.) Rupture of intracranial aneurysm, depending on type, size, and location, can cause subarachnoid hemorrhage leading to serious and life-threatening complications including hemiplegia, aphasia, regional vasospasm, and brain swelling. Treatment with IV DMSO of ruptured aneurysms in people is reported to reverse progressive hemiplegia, vasospasm and resulting low CBF, and improve hemispheric motor deficits. (From Mullan, S. et al., Dimethyl sulfoxide in the management of postoperative hemiplegia, in: Wikins, R.H., ed., *Arterial Spasm*, William & Wilkins, Baltimore, MD, 1980, pp. 646–653.)

Mullan et al.[113] reported on nine cases of patients who suffered partial or total hemiplegia following surgical repair of intracranial aneurysms. The rationale for using DMSO to manage postoperative hemiplegia secondary to aneurysmal bleeding was prompted by the ability of this drug to improve the neurological status of rhesus monkeys subjected to traumatic brain injury, high missile injury, and experimental stroke.[75–78]

These reports indicated that DMSO treatment of rhesus monkeys subjected to various CNS insults was able to increase CPP, CBF, and metabolic rate of oxygen while reducing cerebral edema, elevated ICP, and significant mortality in all treated animals when compared to standard treatments.[75–78]

A summary of how patients typically responded to DMSO treatment after subarachnoid hemorrhage secondary to aneurysm rupture[113] provides insight into how, and when, the proper use of DMSO can help achieve a successful outcome.

Case 1: A 61-year-old man suffered a subarachnoid hemorrhage and underwent clipping of two aneurysms located on the right middle cerebral artery (MCA) and right internal carotid artery without incident. After surgery, the blood pressure climbed spontaneously and mannitol was given every 4 h. Despite this regimen, left-sided paralysis developed and continued without improvement. A diagnosis of severe arterial spasm was made, and a decision to use DMSO was made in order to avoid permanent paralysis. DMSO at 1 g/kg bolus was given every 8 h on day 1. The affected right hemispheric blood flow increased 20% and the contralateral left hemisphere by 11% some 30 min after the initial DMSO infusion. After 24 h of DMSO, the clinical condition was satisfactory and DMSO was continued and stopped after 5 days when the patient's condition was much improved.[113]

On day 6, the patient developed low-grade fever, progressive weakness on the left side, and confusion. DMSO was restarted at 3 g/24 h and reduced to 1 g/24 h for the next 3 days when left side strength and alertness returned. On day 9, DMSO was reduced to 0.5 g/kg/8 h and stopped on day 12. The patient recovered progressively and was discharged on day 18.[113]

Case 2: A 67-year-old woman was diagnosed with a left MCA aneurysm and elected to have it clipped. After surgery, she was alert and had full strength in all four limbs. Shortly thereafter, she became aphasic and had right-sided hemiplegia. Mannitol was given for the next 8 h, but no improvement was seen. CBF was 49 mL/100 g/min on both hemispheres. DMSO was begun at a rate of 0.5 g/kg and increased to 1 g/kg/8 h. After 45 min, she was alert and strength returned to the right side. Two hours after the initial DMSO dose, her CBF was 60 on the right and 66 on the left, a 37% increased blood flow in the affected hemisphere. DMSO was continued for another 3 days at a dose of 1 g/kg daily. Her motor strength returned to normal levels within 12 h, and CBF remained at 60 mL/100 g/min until discharge.[113]

In the same series of cases reported by Mullan et al.,[113] two patients treated were relatively young adults.

Case 3: A 25-year-old woman was hospitalized with severe headache and high systolic pressure. Neuroimaging showed a left MCA aneurysm with spasm in the internal carotid artery. After clipping the aneurysm without incident and postsurgical recovery, the patient developed significant right-sided weakness, right-leg paralysis, and dysphasia 12 days postoperative. Mannitol was given, but after 8 h, all signs worsened. DMSO 1 g/kg was given over the first hour. Ninety minutes following DMSO, the patient could lift her right leg off the bed and by the following day her dysphasia had cleared. DMSO was continued and reduced to 2 g/kg/day until discharge 3 days later when she was symptom free.[113]

Case 4: A 28-year-old woman was admitted with excruciating headache and weakness on the right side. After 5 days, she developed hemiparesis and aphasia. A CT scan showed blood in her sylvian fissure and after craniotomy an MCA aneurysm was discovered and clipped without incident. Steroids and mannitol were given for severe internal carotid artery spasm, but no improvement resulted. DMSO was given 1 g/kg/4 h and reduced to 1 g/kg/24 h for the next 3 days. A total recovery was seen at discharge.[113]

The remaining five cases of aneurysmal hemorrhage were treated with DMSO using a similar regimen as the previous patients, and all except one patient had excellent recoveries. The patient who died had a history of hypertension, cardiac deficits, seizures, and heavy smoking and presented with giant aneurysm at the internal carotid artery bifurcation, which penetrated into the hypothalamus and right hemisphere. Despite initial success with DMSO in raising CBF 27% on the affected hemisphere, the patient developed high systolic blood pressure and seizures, which did not respond to treatment, and the patient expired after 25 days.[113]

The authors of this study concluded that little can be said for certain after this nine-patient minitrial, but the speed of action shown by DMSO administration to reverse specific motor deficits suggested that this drug was strikingly effective in preventing almost certain progressive hemiplegia and restoring hemispheric function at the early stages of treatment.[113] The authors also pointed out that DMSO appeared safe at the doses used and the most likely reason for its beneficial effect may be have been its ability to increase CBF.[113]

The minitrial using DMSO to treat intracerebral hemorrhage and ensuing arterial spasm by Mullan et al.[113] indicates that a DMSO bolus or very fast drip at doses of 1 g/kg/8 h in a recommended 28% solution appears as a safe and effective regimen. As deficits resolve, the DMSO doses can be reduced as reflected by the patient's status. Also, hematuria from red cell osmotic hemolysis was seen in all patients only after the initial loading dose of DMSO. Hematuria was seen to stabilize after subsequent doses of DMSO.[113] The loading DMSO dose has no other consequences other than lowering the hematocrit about 25%, a temporary reaction that paradoxically lowers blood viscosity and vascular resistance while increasing CBF.

DMSO COMPARED TO PROGESTERONE FOR TRAUMATIC BRAIN INJURY

At the time of this writing, one of the most promising therapies proposed to combat traumatic brain injury is the sex hormone progesterone. In 1996, Stein's group[114] reported that subcutaneous injection of progesterone (4 mg/kg) given 1 h after bilateral frontal lobe contusion in male and pre-estrus female rats resulted in significantly less brain edema in the female rats when compared to male rats 24 h after brain contusion. The conclusion from this study suggested that higher progesterone levels in pre-estrus females were protective after induced brain trauma.[115] Other evidence in rodents indicated that progesterone was beneficial in additional forms of neurotrauma, such as penetrating brain injury,[116] improved cognitive performance,[117] and spinal cord injury.[118]

Reports of improved neurological outcomes and decreased cortical infarcts after induction of stroke in rats indicated a beneficial effect after progesterone whether it was administered preinjury or postinjury.[119,120] Curiously, one study reported using progesterone dissolved in 10% DMSO for middle cerebral artery occlusion (MCAO) in rats and showed significant improvement in reducing infarct volume when progesterone–DMSO combination was compared to progesterone–saline solution.[121]

The same authors[122] used 10% DMSO as a vehicle for 5′-deoxy-5-iodotubercidin (5′d-5IT), an adenosine kinase inhibitor in a rat MCAO model, and found similar results, that is, a smaller infarct volume and reduced neurological deficits with DMSO-5′d-5IT when compared with no treatment. Improved neurologic scores were seen in both DMSO-5′d-5IT rat groups, with MCAO and in DMSO given alone.[122]

In 2007, Wright et al.[123] reported a phase II randomized controlled study comparing progesterone to placebo in 100 patients (77:23) with traumatic brain injury. The primary measure of benefit was 30-day mortality in patients whose GCS ranged from 4 (severe) to 12 (mild). Thirty days postinjury, patients with severe traumatic

TABLE 9.1
Clinical Comparison of Treatments Using DMSO or Progesterone after Severe Traumatic Brain Injury (TBI) in Humans

	Lowers ICP	Increases CPP	Improves Clinical Outcome	Reduces Mortality
DMSO (N = 42)	Yes	Yes	Yes	Yes
Progesterone (N = 80)	No	No	Yes	Yes

Note: Progesterone is reported to improve clinical outcome and reduce mortality but has no effect on ICP or CPP. DMSO is reported to lower both ICP and CPP, improve outcome, and reduce mortality. See text for details.

brain injury given progesterone had a lower mortality rate than the placebo group, but due to the small number of patients enrolled in the study, these differences did not reach significant levels.[123] Two smaller clinical studies on progesterone followed.

Despite the many reports discussing the benefits of progesterone treatment in a variety of rodent CNS injuries,[124] this hormone has been used in only three small human studies for traumatic brain injury.[72]

The three studies included 315 patients, of which 180 received progesterone (Table 9.1). Two studies involving 215 patients were done in China and one in the United States with 100 patients.[123,125,126] All three studies reported the results of progesterone on mortality as compared to placebo following traumatic brain injury.[72] The general protocol for progesterone administration was 0.71 mg/kg IV infusion for the first hour and 0.5 mg/kg for the next 11 h.[123] Progesterone 0.5 mg/kg was continued every 12 h for 3 days. The patients were divided according to their GCS, which ranged from 4–8, severe trauma, to 9–12, moderate trauma.[123]

Results from these studies reported no differences between progesterone and placebo for ICP or CPP. An improved mortality rate for progesterone after a 30-day[123] or 6-month[125,126] follow-up was observed. Paradoxically, disability in surviving U.S. patients was deemed to be better after placebo compared to progesterone, a detail that was attributed to slightly higher survival rate in the progesterone group.[123]

However, disability was reported to be less for progesterone-treated patients compared to controls after 6 months follow-up in the China trials using a modified functional-independent measure score.[125,126] The reason for this disparity may be a slightly higher dose of progesterone used in the Chinese trials.

None of the three studies reported the effects of progesterone on CBF, cerebrovascular resistance, or cerebral metabolic rate of oxygen (Table 9.1).

A Cochrane data analysis[72] of the three progesterone trials concluded that the evidence appeared insufficient to recommend the drug for isolated traumatic brain injury and that further multicenter RCTs were needed.

It is not known how progesterone is able to lower elevated ICP and reduce brain edema although its neurosteroidal, anti-inflammatory actions may play an important role.[114,127]

A comparison of the clinical effects for using progesterone versus DMSO treatments in severe traumatic brain injury is summarized in Table 9.1. In this summary, it is seen that although progesterone is reported to reduce mortality and neurologic outcome, it does so without increasing CPP and, by inference, CBF. This is a curious twist because increasing CPP to optimize CBF delivery has become a standard method to lower the secondary injury pathologic cascade after traumatic brain injury.[128,129] The secondary damage cascade fundamentally involves the development of ischemia, hypoxia, and neuronal energy dysfunction, which have become primary determinants of secondary brain injury outcome.

Consequently, considerable evidence now indicates that maintaining adequate CBF will improve the delivery of oxygen and energy substrates that are lowered during the period of ischemia and hypoxia that follows a moderate-to-severe traumatic brain injury.[130–133] This clinical goal has become a standard of practice for neurocritical care when using therapeutic interventions after brain trauma. It follows that managing CPP and CBF may have a beneficial effect in lowering ICP after brain injury and could also affect the molecular changes that promote Ca^{2+} and Na^+ cell influx, neuronal membrane instability, glutamate excitotoxicity, and mitochondrial dysfunction that are known to progress after brain tissue damage[129,134,135] (see Figures 8.3 and 9.4).

In contrast to progesterone therapy, DMSO improves CPP, CBF, and cerebrovascular resistance while increasing oxygen availability to ischemic tissue.[75–78,101,113]

These actions by DMSO may limit the damage caused by severe head injury, intracranial hemorrhage, and vasospasm secondary to aneurysm rupture in human patients.[76,87,88,113] Moreover, the cell and molecular pathology resulting from brain trauma is also targeted by DMSO and may help to explain the significant increase in survival and reduced morbidity following severe traumatic brain injury.[87,88]

DMSO and progesterone are strong anti-inflammatory agents,[136–139] but only DMSO is reported to block glutamate excitotoxicity following brain tissue trauma and stabilize Na^+ and Ca^{2+} influx into nerve cells that can destabilize neuronal membranes and cause cell death.[101]

Since clinical evidence indicates that both DMSO and progesterone treatment are reported to successfully lower mortality and improve neurologic outcome following severe traumatic brain injury, it would seem reasonable to further compare these two drugs in multicenter RCTs to assess the value of each for a life-threatening condition that has no efficacious intervention at the present time. Such an RCT would establish the safety and efficacy of each compound and could provide the first effective treatment available for brain and spinal cord insults.

Moreover, we cannot overstate that an RCT of DMSO and progesterone could result in a major medical breakthrough in preventing mortality and the psychosocial burden suffered by traumatized patients and their caretakers. A drug that could limit the damage caused by brain trauma would also be cost-effective, in view of the billions of dollars expended globally on traumatic brain injury, especially for survivors with special needs and constant care.

However, on December 25, 2014, as this book went to press, two phase III multicenter, double-blind clinical trials using progesterone were carried out on moderate to severe TBI patients.[101a,101b] These trials demonstrated no significant clinical benefit or functional improvement following progesterone administration when compared to placebo in the patients treated for TBI.[101a,101b]

DMSO + FRUCTOSE-1,6-DISPHOSPHATE (FDP)
FOR ISCHEMIC STROKE

Stroke is the third leading cause of death and severe disability in the United States. In the mid-1990s, great hope was created when tissue plasminogen activator (tPA) was approved by the United States Food and Drug Administration (FDA) for use as an IV agent in ischemic stroke.[140] This drug is a clot buster that works by catalyzing the conversion of plasminogen to the clot-breaking plasmin. tPA is the only drug currently authorized by the FDA for the treatment of acute ischemic stroke.

Despite almost a decade of use after FDA approval, tPA is greatly underused and it is reported that in the United States about 1%–7% of potentially eligible patients actually receive this therapy.[141,142] The major reasons for underusing tPA is the tight 3 h window of treatment required to ensure effectiveness in patients with ischemic stroke and, also, the fear of inducing cerebral hemorrhage. When one computes the delay time often involved in reaching a hospital following an acute ischemic stroke and the time required to rule out a hemorrhagic stroke with a CT scan, this 3 h window of treatment may be difficult to attain.

It is estimated that ischemic stroke strikes about 75% of people over the age of 55.[143] Elderly stroke patients are also more susceptible to neurologic disability and adverse events related to drug treatment following a cerebrovascular insult. Partly for these reasons, elderly patients are rarely included in large randomized clinical stroke trials involving investigational drug therapy.[144]

The pathobiology associated with ischemic stroke is not unlike that seen in traumatic brain injury since it involves numerous processes, including increased intracellular Ca^{2+} influx into cells, glutamate excitotoxicity, loss of cell ion homeostasis, free radical–mediated cell damage, cytokine-induced cytotoxicity, complement activation, reactive gliosis, energy failure, and infiltration of leukocytes through an impaired blood–brain barrier.[145]

As with traumatic brain injury, early management of ischemic stroke stresses restoration of CBF to the ischemic core and surrounding region as a principal goal that can prevent further tissue damage and possibly reverse some or all lost hemispheric function.[146,147]

The region bordering the stroke core is called the ischemic penumbra, an area encompassing as much as half of the total lesion volume during the initial stages of stroke.[147] The penumbral region therefore represents a neuronal population of dysfunctional but structurally undamaged brain cells, which may be salvaged via poststroke therapy.[148] Unfortunately, no such therapy has yet been approved for rescuing penumbral neurons although studies in rhesus monkeys have shown the feasibility of neuronal rescue after experimental stroke using DMSO.[149–151]

The rationale for using combined DMSO–FDP on ischemic stroke patients is based on clinical and experimental data indicating the beneficial properties of DMSO or FDP monotherapy following brain and cardiac ischemia.[152–154] Because neither DMSO nor FDP alone is able to target the multifactorial metabolic and physiological abnormalities seen experimentally in animals or clinically in humans, it was considered that their combined use might provide improved neuroprotection following ischemic stroke. Neither drug had been used for human ischemic stroke before.

In previous studies, DMSO has shown a number of biological activities that could be useful in embolic stroke therapy. These include acting as a powerful free radical scavenger,[155–158] blocking glutamate excitotoxicity,[101] suppressing tissue factor activity,[98] preventing vascular smooth muscle migration during cardiac ischemia,[98] inhibiting the formation of thrombin,[98] blocking sodium channel activation,[99,100] deaggregating platelets in brain microvessels,[158–160] and stabilizing cell membranes.[161–163]

Additionally, DMSO has the ability to increase CBF when given for a variety of vascular insults.[75–78,113,164] DMSO has also been reported to improve neurologic and functional outcome after stroke in animals, including nonhuman primates.[155,157,159] Unlike FDP, however, DMSO does not act as a substrate for aerobic glycolysis, nor does it reduce ischemic-induced cellular acidosis, or directly increases the synthesis of the energy-carrying molecule ATP, during tissue ischemia.

These are biological activities that FDP is able to carry out as an energy substrate glycolytic intermediate. During anaerobic glycolysis, FDP is capable of boosting the yield of net ATP twice as much as glucose on a per mole basis.[165] FDP has been shown to reduce the harmful effects of cellular acidosis promoted by lactic acid production following cerebral ischemia[166] and to diminish the effects of brain ischemic hypoxia.[167] These activities shown by FDP are highly desirable in combating ischemic stroke.

Because previous experimental findings using a combination of DMSO and FDP showed marked beneficial results when this treatment was applied for blunt head injury and for induced cerebral hypoperfusion in rodents,[87,88,168,169] this dual drug combination was selected as a treatment for human stroke by Karaça et al.[170]

A total of 16 patients suffering from acute and subacute cerebral infarction was selected to receive DMSO–FDP or standard therapy.[170] The patients ranged in age from 45 to 92 with a median age of 65.[170] Eleven patients were given IV DMSO–FDP (DMSO 560 mg/kg in a 28% solution, FDP 200 mg/kg) mixed in 5% dextrose twice daily for an average of 12 days. Five patients were given standard treatment consisting of fluids, oxygen, and osmotic agents when indicated.[170]

Safety and tolerability to DMSO–FDP were evaluated by recording any clinical adverse effects to drug therapy, by MRI and by laboratory data. DMSO–FDP drug efficacy was assessed by a reduction in infarct size using MRI, as well as by any mass effect, midline shift or extent of edema, and by magnetic resonance angiographic evidence of improved perfusion in affected vessels.[170]

A neurologic and medical evaluation was performed on every patient immediately before, during, and after treatments and continued after 7, 30, and 90 days following DMSO–FDP and standard treatments.[170] All but 1 of the 11 DMSO–FDP-treated patients was followed up for 3 months after discharge from the hospital. Neurologic assessment included patients' sensory-motor function and level of consciousness. A modified Rankin scale was used to predict outcome, and a five-point rating scale adapted from Tazaki et al.[171] was used to assess neurologic recovery.

All patients were rated as to their daily living condition as follows: (1) markedly improved, (2) improved, (3) slightly improved, (4) no change, and (5) deteriorated or died. There were no deaths that could be attributed to any therapy in this series of patients.

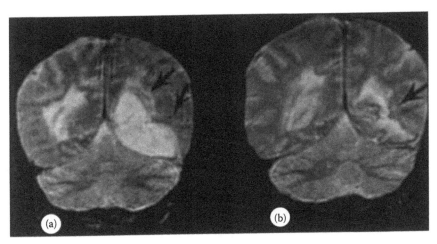

FIGURE 9.6   (a) T2-weighted coronal view MRI of posterior cerebral artery (PCA) territory infarct before treatment. White matter changes are seen with lateral mass effect on lateral ventricle, left thalamus involvement, and widespread edema involving the brainstem (arrows). (b) After 11 days of daily DMSO–FDP IV administration, dramatic reduction of edema and lower signal intensity are seen with an apparent improvement of gray matter and thalamic involvement (arrow). (From Karaça, M. et al., *Neurol. Res.*, 24(1), 73, 2002.)

Figure 9.6 shows the MRI of a 63-year-old patient who had sustained a large infarct of the left posterior cerebral artery territory 8 h before treatment. After 11 days of twice daily DMSO–FDP infusions, MRI showed a dramatic reduction of edema and apparent reduction of thalamic involvement and lower signal intensity in gray matter.[170]

A magnetic resonance angiogram is shown in Figure 9.7 of an 80-year-old patient who was diagnosed with a right MCA infarct that resulted in a mild mass effect and large hematoma affecting the basal ganglia territory. This patient was treated with DMSO–FDP twice daily after 48 h poststroke and showed improved perfusion in the MCA territory and reduced basal ganglia involvement.

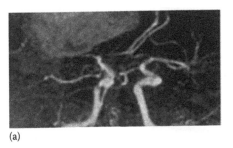
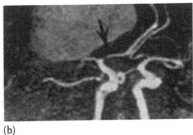

(a)   (b)

FIGURE 9.7   Sagittal view of magnetic resonance angiograph of internal carotid artery territory. A right MCA infarction with mild mass effect, large hematoma, and midline shift involving basal ganglia before treatment is seen (a) in an 80-year-old patient 48 h before DMSO–FDP treatment. (b) Eight days of twice daily DMSO–FDP infusions revealed no change in hematoma size, but there appeared an increased perfusion of the MCA ischemic territory, increased blood flow of MCA (arrow), and lessened basal ganglia involvement. (From Karaça, M. et al., *Neurol. Res.*, 24(1), 73, 2002.)

Neurologic evaluation of the 11 patients treated with DMSO–FDP showed that 7 of 11 patients had moderate or negligible disability (neurologic score, 1–2) after a 3-month follow-up. In the standard-treated patients at 3 months poststroke, one of five treated showed moderate disability (neurologic score, 3) and the others showed poor outcomes (neurologic score, 3–4).

Neurologic reduction of deficits in the DMSO–FDP-treated patients included significant improvement of sensation, motor strength, higher cortical function (speech, alertness, and comprehension), and ambulation in follow-ups conducted at 30 and 90 days after discharge from the hospital.[170] Significant motor improvement was recorded in 54% of the patients 1 month after receiving DMSO–FDP. This was especially evident in the three subjects treated within 12 h after their stroke symptoms appeared.[170]

In comparison, only 20% of patients showed minor motor improvement at 1 month poststroke when the standard treatment was given. The fact that 5 of 11 patients treated with DMSO–FDP improved their neurologic status even when treatment was administered 12 h after their stroke symptoms first appeared suggests that DMSO–FDP therapy may provide neuroprotective action to dysfunctional but surviving penumbral neurons by increasing CBF. This conclusion is supported by experimental findings that show CBF restoration after 2 weeks of chronic brain ischemia in rodents can rescue dysfunctional neurons from progressive damage and, at the same time, reverse reactive astrocytosis and preserve cognitive function from deteriorating, presumably by preventing glutamate excitotoxicity and the pathobiological changes that follow.[164]

In addition, since the average age of the seven patients achieving a markedly improved or moderately improved status was 63 years, it is reasonable to assume that the dual drug treatment is safe and effective in elderly subjects.

In summary, DMSO–FDP administration to mostly elderly subjects was well tolerated when given twice daily for an average of 12 days, following ischemic stroke. Moreover, DMSO–FDP might have provided neuroprotection to some of the patients in this small population group during a 3-month follow-up. Although this preliminary study of DMSO–FDP treatment for acute and subacute ischemic stroke examined a relatively small population, there was evidence that the dual treatment was more effective as a neuroprotector than standard treatment during the recovery period following an ischemic stroke.

While no hard conclusions can be drawn from these observational findings, the preliminary impression on the use of DMSO–FDP for ischemic stroke is that the dual drug combination may have improved recovery of neurologic deficits in the majority of treated patients. This effect by DMSO in elderly subjects following stroke has not been reported for any other treatment.

The remarkable finding of this study is that elderly patients who are known to be most vulnerable to ischemic stroke damage and more refractory to therapy than younger patients responded well to DMSO–FDP IV treatment with no major adverse effects and revealed an ostensible benefit to neurological outcome considerably past the 3 h window of opportunity offered by tPA therapy. The collective neuroimaging evidence gathered from the 11 patients treated for ischemic stroke with DMSO–FDP points to an apparent improvement of cerebral perfusion in the affected brain regions.

The DMSO–FDP findings also signal the need for a larger population RCT to test the value of this drug combination under a more controlled setting.

REFERENCES

1. Santiago LA, Oh BC, Dash PK, Holcomb JB, Wade CE. A clinical comparison of penetrating and blunt traumatic brain injuries. *Brain Inj.* 2012;26(2):107–125.
2. Reeves RR, Laizer JT. Traumatic brain injury and suicide. *J Psychosoc Nurs Ment Health Serv.* March 2012;50(3):32–38.
3. Marion DW. *Traumatic Brain Injury.* Thieme, New York, 1999; pp. 1–302.
4. Mittl RL, Grossman RI, Hiehle JF, Hurst RW, Kauder DR, Gennarelli TA, Alburger GW. Prevalence of MR evidence of diffuse axonal injury in patients with mild head injury and normal head CT findings. *AJNR Am J Neuroradiol.* September 1994;15(8):1583–1589.
5. Inglese M, Makani S, Johnson G, Cohen BA, Silver JA, Gonen O, Grossman RI. Diffuse axonal injury in mild traumatic brain injury: A diffusion tensor imaging study. *J Neurosurg.* August 2005;103(2):298–303.
6. Paterakis K, Karantanas AH, Komnos A, Volikas Z. Outcome of patients with diffuse axonal injury: The significance and prognostic value of MRI in the acute phase. *J Trauma.* December 2000;49(6):1071–1075.
7. Thomas M, Dufour L. Challenges of diffuse axonal injury diagnosis. *Rehabil Nurs.* September–October 2009;34(5):179–180.
8. Alberico AM, Ward JD, Choi SC, Marmarou A, Young HF. Outcome after severe head injury: Relationship to mass lesions, diffuse injury, and ICP course in pediatric and adult patients. *J Neurosurg.* 1987;67:648–656.
9. Smith DH, Meaney DF, Shull WH. Diffuse axonal injury in head trauma. *J Head Trauma Rehabil.* 2003;18(4):307–316.
10. Haddad SH, Arabi YM. Critical care management of severe traumatic brain injury in adults. *Scand J Trauma Resusc Emerg Med.* February 3, 2012;20:12.
11. Gruen P. Surgical management of head trauma. *Neuroimag Clin North Am.* 2002;12(2):339–343.
12. Budohoski KP, Czosnyka M, de Riva N, Smielewski P, Pickard JD, Menon DK, Kirkpatrick PJ, Lavinio A. The relationship between cerebral blood flow autoregulation and cerebrovascular pressure reactivity after traumatic brain injury. *Neurosurgery.* September 2012;71(3):652–660.
13. Jeager M, Schuhmann MU, Soehle M, Meixensberger J. Continuous assessment of cerebrovascular autoregulation after traumatic brain injury using brain tissue oxygen pressure reactivity. *Crit Care Med.* 2006;34:1783–1788.
14. Marrmarou A, Fatouros P, Barzo P. Contribution of edema and cerebral blood volume to traumatic brain swelling in head-injured patients. *J Neurosurg.* 2000;93:183–193.
15. Nortje J, Menon DK. Traumatic brain injury: Physiology, mechanisms, and outcome. *Curr Opin Neurol.* 2004;17:711–718.
16. American Association of Neurological Surgeons. Guidelines for the management of severe traumatic brain injury. *J Neurotrauma.* 2007;24:S-1–S-106.
17. Werner C, Engelhadt K. Pathophysiology of traumatic brain injury. *Br J Anaesth.* 2007;99:4–9.
18. Teasdale G, Jennett B. Assessment of coma and impaired consciousness. A practical scale. *Lancet.* 1974;2:81–84.
19. McNett M. A review of the predictive ability of Glasgow Coma Scale scores in head-injured patients. *J Neurosci Nurs.* April 2007;39(2):68–75.
20. Healy C. Improving the Glasgow Coma Scale score: Motor score alone is a better predictor. *Trauma.* April 2003;54(4):671–678.
21. Frey KL, Rojas DC, Anderson CA, Arciniegas DB. Comparison of the O-Log and GOAT as measures of posttraumatic amnesia. *Brain Inj.* 2007;21(5):513–520.
22. Zinc B. Traumatic brain injury outcome: Concept in emergency care. *Ann Emerg Med.* 2001;37(3):318–332.

23. Faul M, Xu L, Wald M, Coronado VG. *Traumatic Brain Injury in the United States: Emergency Department Visits, Hospitalizations, and Deaths.* Centers for Disease Control and Prevention, National Center for Injury Prevention and Control, Atlanta, GA, 2010.

24. Rona RJ. Long-term consequences of mild traumatic brain injury. *Br J Psychiat.* 2012;201(3):172–174.

25. Miller G. Neuropathology. Blast injuries linked to neurodegeneration in veterans. *Science.* 2012;336:790–791.

26. Brenner LA, Ignacio RV, Blow FC. Suicide and traumatic brain injury among individuals seeking Veterans Health Administration services. *J Head Trauma Rehabil.* July–August 2011;26(4):257–264.

27. Quayle KS. Minor head injury in the pediatric patient. *Pediatr Clin North Am.* December 1999;46(6):1189–1199.

28. Thiruppathy SP, Muthukumar N. Mild head injury: Revisited. *Acta Neurochir (Wien).* October 2004;146(10):1075–1082.

29. Fuller GW, Woodford M, Lawrence T, Coats T, Lecky F. The accuracy of alternative triage rules for identification of significant traumatic brain injury: A diagnostic cohort study. *Emerg Med J.* 2013;30(10):876.

30. Fletcher JM, Levin HS, Lachar D, Kusnerik L, Harward H, Mendelsohn D, Lilly MA. Behavioral outcomes after pediatric closed head injury: Relationships with age, severity, and lesion size. *J Child Neurol.* 1996;11:283–290.

31. Hyder AA, Wunderlich CA, Puvanachandra P, Gururaj G, Kobusingye OC. The impact of traumatic brain injuries: A global perspective. *NeuroRehabilitation.* 2007;22(5):341–353.

32. Binder S, Corrigan JD, Langlois JA. The public health approach to traumatic brain injury: An overview of CDC's research and programs. *J Head Trauma Rehabil.* May–June 2005;20(3):189–195.

33. Coronado VG, McGuire LC, Faul M, Sugerman DE, Pearson WS. Traumatic brain injury epidemiology and public health issues. In *Brain Injury Medicine: Principles and Practice*, Zasler ND (Ed.), Demos Medical, 2nd edition, New York. 2012; p. 84.

34. Roberts GW, Gentleman SM, Lynch A, Murray L, Landon M, Graham DI. Beta amyloid protein deposition in the brain after severe head injury: Implications for the pathogenesis of Alzheimer's disease. *J Neurol Neurosurg Psychiat.* 1994;57:419–425.

35. Johnson VE, Stewart W, Smith DH. Widespread tau and amyloid-beta pathology many years after a single traumatic brain injury in humans. *Brain Pathol.* 2012;22:142–149.

36. Jeremitsky E, Omert L, Dunham CM, Protetch J, Rodriguez A. Harbingers of poor outcome the day after severe brain injury: Hypothermia, hypoxia, and hypoperfusion. *J Trauma.* February 2003;54(2):312–319.

37. Kim J, Whyte J, Patel S, Europa E, Slattery J, Coslett HB, Detre JA. A perfusion fMRI study of the neural correlates of sustained-attention and working-memory deficits in chronic traumatic brain injury. *Neurorehabil Neural Repair.* September 2012;26(7):870–880.

38. Liu H, Zhang J, Yang Y, Zhang L, Zeng X. Decreased cerebral perfusion and oxidative stress result in acute and delayed cognitive impairment. *Curr Neurovasc Res.* August 2012;9(3):152–158.

39. de la Torre JC. Cerebral hemodynamics and vascular risk factors: Setting the stage for Alzheimer's disease. *J Alzheimers Dis.* 2012;32(3):553–567.

40. Bailey DM, Jones DW, Sinnott A, Brugniaux JV, New KJ, Hodson D, Marley CJ, Smirl JD, Ogoh S, Ainslie PN. Impaired cerebral haemodynamic function associated with chronic traumatic brain injury in professional boxers. *Clin Sci (Lond).* 2013;124:177–189.

41. Millspaugh JA. Dementia pugilistica. *US Naval Med Bull.* 1937;35:297–303.

42. McKee AC, Cantu RC, Nowinski CJ, Hedley-Whyte ET, Gavett BE, Budson AE, Santini VE, Lee HS, Kubilus CA, Stern RA. Chronic traumatic encephalopathy in athletes: Progressive tauopathy after repetitive head injury. *J Neuropathol Exp Neurol*. July 2009;68(7):709–735.

43. Frisoni GB, Bianchetti A, Geroldi C, Trabucchi M, Beltramello A, Weiss C. Measures of medial temporal lobe atrophy in Alzheimer's disease. *J Neurol Neurosurg Psychiat*. November 1994;57(11):1438–1439.

44. Dias C, Maia I, Cerejo A, Varsos G, Smielewski P, Paiva JA, Czosnyka M. Pressures, flow, and brain oxygenation during plateau waves of intracranial pressure. *Neurocrit Care*. 2014;21(1):124–132.

45. Giza CC, Hovda DA. The neurometabolic cascade of concussion. *J Athl Train*. 2001;36:228–235.

46. Namjoshi DR, Good C, Cheng WH, Panenka W, Richards D, Cripton PA, Wellington CL. Towards clinical management of traumatic brain injury: A review of models and mechanisms from a biomechanical perspective. *Dis Model Mech*. 2013;6(6):1325–1338.

47. Grände PO, Romner B. Osmotherapy in brain edema: A questionable therapy. *J Neurosurg Anesthesiol*. October 2012;24(4):407–412.

48. Cooper DJ, Myles PS, McDermott FT, Murray LJ, Laidlaw J, Cooper G, Tremayne AB, Bernard SS, Ponsford J, HTS Study Investigators. Prehospital hypertonic saline resuscitation of patients with hypotension and severe traumatic brain injury: A randomized controlled trial. *JAMA*. 2004;291(11):1350–1357.

49. Adamides AA, Winter CD, Lewis PM, Cooper DJ, Kossmann T, Rosenfeld JV. Current controversies in the management of patients with severe traumatic brain injury. *ANZ J Surg*. March 2006;76(3):163–174.

50. McHugh GS, Engel DC, Butcher I. Prognostic value of secondary insults in traumatic brain injury: Results from the IMPACT study. *J Neurotrauma*. 2007;24:287–293.

51. Silverstone P. Pulse oxymetry at the roadside: A study of pulse oxymetry in immediate care. *BMJ*. 1989;298:711–713.

52. Stocchetti N, Furlan A, Volta F. Hypoxemia and arterial hypotension at the accident scene in head injury. *J Trauma*. 1996;40:764–767.

53. De Witt DS, Jenkins LW, Prough D. Enhanced vulnerability to secondary ischemic insults after experimental traumatic brain injury. *New Horiz*. 1995;3:376–383.

54. Rudehill A, Bellander BM, Weitzberg E, Bredbacka S, Backheden M, Gordon E. Outcome of traumatic brain injuries in 1,508 patients: Impact of prehospital care. *J Neurotrauma*. July 2002;19(7):855–868.

55. Garner AA, Crooks J, Lee A, Bishop R. Efficacy of prehospital critical care teams for severe blunt head injury in the Australian setting. *Injury*. 2001;32(6):455–460.

56. Garner AA, Fearnside M, Gebski V. The study protocol for the Head Injury Retrieval Trial (HIRT): A single centre randomised controlled trial of physician prehospital management of severe blunt head injury compared with management by paramedics. *Scand J Trauma Resusc Emerg Med*. September 14, 2013;21(1):69.

57. Ziebell JM, Morganti-Kossmann MC. Involvement of pro- and anti-inflammatory cytokines and chemokines in the pathophysiology of traumatic brain injury. *Neurotherapeutics*. January 2010;7(1):22–30.

58. Jennett B. Outcome of severe damage to the central nervous system. Scale, scope and philosophy of the clinical problem. *Ciba Found Symp*. 1975;34:3–21.

59. Lu J, Gary KW, Neimeier JP, Ward J, Lapane KL. Randomized controlled trials in adult traumatic brain injury. *Brain Inj*. 2012;26:1523–1548.

60. Cruz J, Minoja G, Okuchi K. Major clinical and physiological benefits of early high doses of mannitol for intraparenchymal temporal lobe hemorrhages with abnormal pupillary widening: A randomized trial. *Neurosurgery*. 2002;51:628–637.

61. Qiu W, Zhang Y, Sheng H, Zhang J, Wang W, Liu W, Chen K, Zhou J, Xu Z. Effects of therapeutic mild hypothermia on patients with severe traumatic brain injury after craniotomy. *J Crit Care*. 2007;22:229–235.

62. Cooper DJ, Rosenfeld JV, Murray L et al. Decompressive craniectomy in diffuse traumatic brain injury. *N Engl J Med*. 2011;364:1493–1502.

63. Clifton GL, Miller ER, Choi SC, Levin HS, McCauley S, Smith KR, Muizelaar J, Marion DW, Luerssen TG. Hypothermia on admission in patients with severe brain injury. *J. Neurotrauma*. 2002;19:293–301.

64. Gründe PO, Romner B. Osmotherapy in brain edema: A questionable therapy. *J Neurosurg Anesthesiol*. 2012;24(4):407–412.

65. Moein H, Khalili HA, Keramatian K. Effect of methylphenidate on ICU and hospital length of stay in patients with severe and moderate traumatic brain injury. *Clin Neurol Neurosurg*. 2006;108:539–542.

66. Frenette AJ, Kanji S, Rees L, Williamson DR, Perreault MM, Turgeon AF, Bernard F, Fergusson DA. Efficacy and safety of dopamine agonists in traumatic brain injury: A systematic review of randomized controlled trials. *J Neurotrauma*. 2012;29(1):1–18.

67. Harders A, Kakarieka A. Braakman R. Traumatic subarachnoid hemorrhage and its treatment with nimodipine. German tSAH Study Group. *J Neurosurg*. 1996;85:82–89.

68. Langham J, Goldfrad C, Teasdale G, Shaw D, Rowan K. Calcium channel blockers for acute traumatic brain injury. *Cochrane Database Syst Rev*. 2003;(4):CD000565.

69. Xiao G, Wei J, Yan W, Wang W, Lu Z. Improved outcomes from the administration of progesterone for patients with acute severe traumatic brain injury: A randomized controlled trial. *Crit Care*. 2008;12:R61.

70. Stein DG, Wright DW. Progesterone in the clinical treatment of acute traumatic brain injury. *Expert Opin Investig Drugs*. 2010;19(7):847–857.

71. Stein DG, Wright DW, Kellermann AL. Does progesterone have neuroprotective properties? *Ann Emerg Med*. February 2008;51(2):164–172.

72. Ma J, Huang S, Qin S, You C. Progesterone for acute traumatic brain injury. *Cochrane Database Syst Rev*. October 17, 2012;10:CD008409.

73. Xiong Y, Mahmood A, Chopp M. Animal models of traumatic brain injury. *Nat Rev Neurosci*. February 2013;14(2):128–142.

74. Andriessen TM, Jacobs B, Vos PE. Clinical characteristics and pathophysiological mechanisms of focal and diffuse traumatic brain injury. *J Cell Mol Med*. October 2010;14(10):2381–2392.

75. de la Torre JC, Rowed DW, Kawanaga HM, Mullan S. Dimethyl sulfoxide in the treatment of experimental brain compression. *J Neurosurg*. March 1973;38(3):345–354.

76. de la Torre JC, Kawanaga HM, Johnson CM, Goode DJ, Kajihara K, Mullan S. Dimethyl sulfoxide in central nervous system trauma. *Ann NY Acad Sci*. January 27, 1975;243:362–389.

77. Brown FD, Johns LM, Mullan S. Dimethyl sulfoxide in experimental brain injury, with comparison to mannitol. *J Neurosurg*. July 1980;53(1):58–62.

78. Brown FD, Johns L, Mullan S. Dimethyl sulfoxide therapy following penetrating brain injury. *Ann NY Acad Sci*. 1983;411:245–252.

79. Carlson KK. (ed.) *AACN Advanced Critical Care Nursing*. Elsevier Saunders, St Louis, MO, 2009; p. 554.

80. Campbell WW. *DeJong's the Neurologic Examination*, 6th edn. Lippincott Williams & Wilkins, Philadelphia, PA, 2005.

81. Waller FT, Tanabe CT, Paxton HD. Treatment of elevated intracranial pressure with dimethyl sulfoxide. *Ann NY Acad Sci*. 1983;411:286–292.

82. Marshall LF, Camp PE, Bowers SA. Dimethyl sulfoxide for the treatment of intracranial hypertension: A preliminary trial. *Neurosurgery*. June 1984;14(6):659–663.

83. Gumerlock MK, Neuwelt EA. Treating intracranial pressure with dimethyl sulfoxide. *J Neurosurg*. 1984;14:662–663.

84. Camp PE, James HE, Werner R. Acute dimethyl sulfoxide therapy in experimental brain edema: Part I. Effects on intracranial pressure, blood pressure, central venous pressure, and brain water and electrolyte content. *Neurosurgery.* 1981;9:28–33.

85. Del Bigio M, James HE, Camp PE, Werner R, Marshall L, Tung H. Acute dimethyl sulfoxide therapy in brain edema. Part 3: Effect of a 3-hour infusion. *Neurosurgery.* 1982;10:86–89.

86. Tsuruda J, James HE, Camp PE, Werner R. Acute dimethyl sulfoxide therapy in experimental brain edema: Part 2. Effect of dose and concentration on intracranial pressure, blood pressure, and central venous pressure. *Neurosurgery.* 1982;10:355–359.

87. Karaça M, Bilgin UY, Akar M, de la Torre JC. Dimethyl sulphoxide lowers ICP after closed head trauma. *Eur J Clin Pharmacol.* 1991;40:113–114.

88. Kulah A, Akar M, Baykut L. Dimethyl sulfoxide in the management of patient with brain swelling and increased intracranial pressure after severe closed head injury. *Neurochirurgia (Stuttg).* 1990;33:177–180.

89. Chamoun R, Robertson CS, Gopinath SP. Outcome in patients with blunt head trauma and a Glasgow Coma Scale score of 3 at presentation. *J Neurosurg.* October 2009;111(4):683–687. doi: 10.3171/2009.2.JNS08817.

90. Leitgeb J, Mauritz W, Brazinova A, Majdan M, Janciak I, Wilbacher I, Rusnak M. Glasgow Coma Scale score at intensive care unit discharge predicts the 1-year outcome of patients with severe traumatic brain injury. *Eur J Trauma Emerg Surg.* June 2013;39(3):285–292.

91. Rosenfeld JV, Maas AI, Bragge P, Morganti-Kossman M, Manley G, Gruen R. Early management of severe traumatic brain injury. *Lancet.* 2012;380:1088–1098.

92. Chen Y, Chan PH, Swanson RA. Astrocytes overexpressing Cu, Zn superoxide dismutase have increased resistance to oxidative injury. *Glia.* 2001;33:343–347.

93. Seki Y, Feustel PJ, Keller RW, Tranmer BI, Kimelberg HK. Inhibition of ischemia-induced glutamate release in rat striatum by dihydrokinate and an anion channel blocker. *Stroke.* 1999;30:433–440.

94. Rossi DJ, Oshima T, Attwell D. Glutamate release in severe brain ischemia is mainly by reversed uptake. *Nature.* 2000;403:316–321.

95. Weber JT. Altered calcium signaling following traumatic brain injury. *Front Pharmacol.* 2012;3:60.

96. Yi JH, Hazell AS. Excitotoxic mechanisms and the role of astrocytic glutamate transporters in traumatic brain injury. *Neurochem Int.* 2006;48:394–403.

97. Bullock R, Zauner A, Woodward J, Myseros J, Choi S, Ward JD, Marmarou A, Young HF. Factors affecting excitatory amino acid release following severe human head injury. *J Neurosurg.* 1998;89:507–518.

98. Camici GG, Steffel J, Akhmedov A, Schafer N, Baldinger J, Schulz U, Shojaati K. Dimethyl sulfoxide inhibits tissue factor expression, thrombus formation, and vascular smooth muscle cell activation: A potential treatment strategy for drug-eluting stents. *Circulation.* 2006;114:1512–1521.

99. Hulsmann S, Greiner C, Kohling R. Dimethyl sulfoxide increases latency of anoxic terminal negativity in hippocampal slices of guinea pig in vitro. *Neurosci Lett.* 1999;261:1–4.

100. Ginsburg K, Narahashi T. Time course and temperature dependence of allethrin modulation of sodium channels in rat dorsal root ganglion cells. *Brain Res.* 1999;847:38–49.

101. Lu C, Mattson MP. Dimethyl sulfoxide suppresses NMDA- and AMPA-induced ion currents and calcium influx and protects against excitotoxic death in hippocampal neurons. *Exp Neurol.* 2001;170:180–185.

101a. Wright DW, Yeatts SD, Silbergleit R, Palesch YY, Hertzberg VS, Frankel M, Goldstein FC, Caveney AF, Howlett-Smith H, Bengelink EM, Manley GT, Merck LH, Janis LS, Barsan WG; the NETT Investigators. Very early administration of progesterone for acute traumatic brain injury. *N Engl J Med.* 2014;371:2457–2466.

101b. Skolnick BE, Maas AI, Narayan RK, van der Hoop RG, MacAllister T, Ward JD, Nelson NR, Stocchetti N; the SYNAPSE Trial Investigators. A clinical trial of progesterone for severe traumatic brain injury. *N Engl J Med.* 2014;371:2467–2476.

102. Mayer CL, Huber BR, Peskind E. Traumatic brain injury, neuroinflammation, and posttraumatic headaches. *Headache.* October 2013;53(9):1523–1530.

103. Hsieh CL, Kim CC, Ryba BE, Niemi EC, Bando JK, Locksley RM, Liu J, Nakamura MC, Seaman WE. Traumatic brain injury induces macrophage subsets in the brain. *Eur J Immunol.* August 2013;43(8):2010–2022.

104. Lin Y, Wen L. Inflammatory response following diffuse axonal injury. *Int J Med Sci.* 2013;10(5):515–521.

105. Hollebeeck S, Raas T, Piront N, Schneider YJ, Toussaint O, Larondelle Y, During A. Dimethyl sulfoxide (DMSO) attenuates the inflammatory response in the in vitro intestinal Caco-2 cell model. *Toxicol Lett.* 2011;206:268–275.

106. Dinarello CA. A clinical perspective of IL-1β as the gatekeeper of inflammation. *Eur J Immunol.* May 2011;41(5):1203–1217.

107. Dinarello CA. Interleukin-1 in the pathogenesis and treatment of inflammatory diseases. *Blood.* April 7, 2011;117(14):3720–3732.

108. Scheller J, Chalaris A, Schmidt-Arras D. The pro- and anti-inflammatory properties of the cytokine interleukin-6. *Biochim Biophys Acta.* 2011;1813:878–888.

109. Jawa RS, Anillo S, Huntoon K. Interleukin-6 in surgery, trauma, and critical care part II: Clinical implications. *J Intensive Care Med.* 2011;26:73–87.

110. Mantovani A, Sica A, Sozzani S, Allavena P, Vecchi A, Locati M. The chemokine system in diverse forms of macrophage activation and polarization. *Trends Immunol.* 2004;25:677–686.

111. Mak CH, Lu YY, Wong GK. Review and recommendations on management of refractory raised intracranial pressure in aneurysmal subarachnoid hemorrhage. *Vasc Health Risk Manage.* 2013;9:353–359.

112. Pierot L, Wakhloo AK. Endovascular treatment of intracranial aneurysms: Current status. *Stroke.* July 2013;44(7):2046–2054.

113. Mullan S, Jafar J, Hanlon K, Brown F. Dimethyl sulfoxide in the management of postoperative hemiplegia. In *Arterial Spasm*, ed. R.H. Wikins. William & Wilkins, Baltimore, MD, 1980; pp. 646–653.

114. Roof RL, Duvdevani R, Heyburn JW, Stein DG. Progesterone rapidly decreases brain edema: Treatment delayed up to 24 hours is still effective. *Exp Neurol.* April 1996;138(2):246–251.

115. Stein DG. Progesterone exerts neuroprotective effects after brain injury. *Brain Res Rev.* March 2008;57(2):386–397.

116. García-Estrada J, Luquín S, Fernández AM, Garcia-Segura LM. Dehydroepiandrosterone, pregnenolone and sex steroids down-regulate reactive astroglia in the male rat brain after a penetrating brain injury. *Int J Dev Neurosci.* April 1999;17(2):145–151.

117. O'Connor CA, Cernak I, Johnson F, Vink R. Effects of progesterone on neurologic and morphologic outcome following diffuse traumatic brain injury in rats. *Exp Neurol.* May 2007;205(1):145–153.

118. Thomas AJ, Nockels RP, Pan HQ, Shaffrey CI, Chopp M. Progesterone is neuroprotective after acute experimental spinal cord trauma in rats. *Spine.* October 15, 1999;24(20):2134–2138.

119. Chen J, Chopp M, Li Y. Neuroprotective effects of progesterone after transient middle cerebral artery occlusion in rat. *J Neurol Sci.* December 1, 1999;171(1):24–30.

120. Kumon Y, Kim SC, Tompkins P, Stevens A, Sakaki S, Loftus CM. Neuroprotective effect of postischemic administration of progesterone in spontaneously hypertensive rats with focal cerebral ischemia. *J Neurosurg.* May 2000;92(5):848–852.

121. Jiang N, Chopp M, Stein D, Feit H. Progesterone is neuroprotective after transient middle cerebral artery occlusion in male rats. *Brain Res.* 1996;735:101–107.
122. Jiang N, Kowaluk EA, Lee CH, Mazdiyasni H, Chopp M. Adenosine kinase inhibition protects brain against transient focal ischemia in rats. *Eur J Pharmacol.* February 12, 1997;320(2–3):131–137.
123. Wright DW, Kellermann AL, Hertzberg VS, Clark PL, Frankel M, Goldstein FC. ProTECT: A randomized clinical trial of progesterone for acute traumatic brain injury. *Ann Emerg Med.* 2007;49:391–402.
124. Deutsch ER, Espinoza TR, Atif F, Woodall E, Kaylor J, Wright DW. Progesterone's role in neuroprotection, a review of the evidence. *Brain Res.* September 12, 2013;1530:82–105.
125. Xiao GM, Wei J, Wu ZH. Clinical study on the therapeutic effects and mechanism of progesterone in the treatment for acute severe head injury. *Zhonghua Wai Ke Za Zhi.* 2007;45(2):106–108.
126. Wei J, Xiao GM. The neuroprotective effects of progesterone on traumatic brain injury: current status and future prospects. *Acta Pharmacol Sin.* 2013;34(12):1485–1490.
127. Shear DA, Galani R, Hoffman SW, Stein DG. Progesterone protects against necrotic damage and behavioral abnormalities caused by traumatic brain injury. *Exp Neurol.* November 2002;178(1):59–67.
128. Mangat HS. Severe traumatic brain injury. *Continuum (Minneap Minn).* June 2012;18(3):532–546.
129. McConeghy KW, Hatton J, Hughes L, Cook AM. A review of neuroprotection pharmacology and therapies in patients with acute traumatic brain injury. *CNS Drugs.* July 1, 2012;26(7):613–636.
130. Jaeger M, Dengl M, Meixensberger J, Schuhmann MU. Effects of cerebrovascular pressure reactivity-guided optimization of cerebral perfusion pressure on brain tissue oxygenation after traumatic brain injury. *Crit Care Med.* 2010;3:1343–1347.
131. Menon DK, Coles JP, Gupta AK, Fryer TD, Smielewski P, Chatfield DA, Aigbirhio F, Skepper JN, Minhas PS, Hutchinson PJ. Diffusion limited oxygen delivery following head injury. *Crit Care Med.* 2004;3:1384–1390.
132. Oddo M, Levine JM, Mackenzie L, Frangos S, Feihl F, Kasner SE, Katsnelson M, Pukenas B, Macmurtrie E, Maloney-Wilensky E. Brain hypoxia is associated with short-term outcome after severe traumatic brain injury independently of intracranial hypertension and low cerebral perfusion pressure. *Neurosurgery.* 2011;3:1037–1045.
133. Glenn TC, Kelly DF, Boscardin WJ, McArthur DL, Vespa P, Oertel M, Hovda DA, Bergsneider M, Hillered L, Martin NA. Energy dysfunction as a predictor of outcome after moderate or severe head injury: Indices of oxygen, glucose, and lactate metabolism. *J Cereb Blood Flow Metab.* 2003;3:1239–1250.
134. Unterberg AW, Stover J, Kress B, Kiening KL. Edema and brain trauma. *Neuroscience.* 2004;129(4):1021–1029.
135. Bor-Seng-Shu E, Kita WS, Figueiredo EG, Paiva WS, Fonoff ET, Teixeira MJ, Panerai RB. Cerebral hemodynamics: Concepts of clinical importance. *Arq Neuropsiquiatr.* May 2012;70(5):352–356.
136. Jacob SW, de la Torre JC. Pharmacology of dimethyl sulfoxide in cardiac and CNS damage. *Pharmacol Rep.* March–April 2009;61(2):225–235.
137. Jia Z, Zhu H, Li Y, Misra HP. Potent inhibition of peroxynitrite-induced DNA strand breakage and hydroxyl radical formation by dimethyl sulfoxide at very low concentrations. *Exp Biol Med (Maywood).* May 2010;235(5):614–622.
138. Sanmartín-Suárez C, Soto-Otero R, Sánchez-Sellero I, Méndez-Álvarez E. Antioxidant properties of dimethyl sulfoxide and its viability as a solvent in the evaluation of neuroprotective antioxidants. *J Pharmacol Toxicol Methods.* 2011;63(2):209–215.

139. Stein DG. Is progesterone a worthy candidate as a novel therapy for traumatic brain injury? *Dialog Clin Neurosci.* 2011;13(3):352–359.
140. NINDS Group. Tissue plasminogen activator for acute ischemic stroke. *N Engl J Med.* 1995;333:1581–1587.
141. Cocho D, Belvis R, Marti-Fabregas J et al. Reasons for exclusion from thrombolytic therapy following acute ischemic stroke. *Neurology.* 2005;64:719–720.
142. Katzan IL, Furlan AJ, Lloyd LE et al. Use of tissue-type plasminogen activator for acute ischemic stroke: The Cleveland area experience. *JAMA.* 2000;283:1151–1158.
143. Macciocchi SN, Diamond PT, Alves WM, Mertz T. Ischemic stroke: Relation of age, lesion location, and initial neurologic deficit to functional outcome. *Arch Phys Med Rehabil.* 1998;79:1255–1257.
144. Beyth RJ, Shorr RI. Epidemiology of adverse drug reactions in the elderly by drug class. *Drugs Aging.* 1999;14:231–239.
145. Dirnagl U, Iadecola C, Moskowitz MA. Pathobiology of ischaemic stroke: An integrated view. *Trends Neurosci.* 1999;22:391–397.
146. Mehta SL, Manhas N, Rahubir R. Molecular targets in cerebral ischemia for developing novel therapeutics. *Brain Res Rev.* 2007;54:34–66.
147. Fisher M, Bastan B. Identifying and utilizing the ischemic penumbra. *Neurology.* September 25, 2012;79(13 Suppl 1):S79–S85.
148. Heiss WD. The ischemic penumbra: How does tissue injury evolve? *Ann NY Acad Sci.* September 2012;1268:26–34.
149. de la Torre JC, Surgeon JW. Dexamethasone and DMSO in experimental transorbital cerebral infarction. *Stroke.* November–December, 1976;7(6):577–583.
150. de la Torre JC, Kawanaga HM, Hill PK, Crockard HA, Surgeon JW, Mullan S. Experimental therapy after middle cerebral artery occlusion in monkeys. *Surg Forum.* 1975;26:489–492.
151. de la Torre JC, Kawanaga HM, Hill PK, Crockard HA, Surgeon JW, Mullan S. Experimental therapy after middle cerebral artery occlusion in monkeys. *Surg Forum.* 1975;26:489–492.
152. Kaakinen T, Heikkinen J, Dahlbacka S, Alaoja H. Fructose-1,6-bisphosphate supports cerebral energy metabolism in pigs after ischemic brain injury caused by experimental particle embolization. *Heart Surg Forum.* 2006;9(6):E828–E835.
153. Farias LA, Willis M, Gregory GA. Effects of fructose-1,6-diphosphate, glucose, and saline on cardiac resuscitation. *Anesthesiology.* December 1986;65(6):595–601.
154. Markov AK, Brumley MA, Figueroa A, Skelton TN, Lehan PH. Hemodynamic effects of fructose 1,6-diphosphate in patients with normal and impaired left ventricular function. *Am Heart J.* May 1997;133(5):541–549.
155. Gilboe DD, Kintner D, Fitzpatrick JH, Emoto SE, Bazan NG. Treatment of post-ischemic brain with a free radical scavenger and a platelet activating factor antagonist: Recovery of metabolism and function. In *Pharmacology of Cerebral Ischemia*, eds. H. Oberpichler, J. Kriegistein. Wissenschaftliche-Verlag, Stuttgart, Germany, 1990; pp. 399–407.
156. Phillis JW, Estevez AY. Protective effects of the free radical scavengers, dimethyl sulfoxide and ethanol, in cerebral ischemia in gerbils. *Neurosci Lett.* 1998;244:109–111.
157. Shimizu S, Simon RP, Graham SH. Dimethyl sulfoxide treatment reduces infarction volume after permanent focal cerebral ischemia in rats. *Neurosci Lett.* 1997;239:125–127.
158. Bruck R, Aeed H, Shirin H. The hydroxyl radical scavengers dimethyl sulfoxide and dimethyl urea protect rats against thioacetamide-induced fulminant hepatic failure. *J Hepatol.* 1999;31:27–38.
159. Dujovny M, Rozario R, Kossovsky N, Diaz FG, Segal R. Antiplatelet effect of dimethyl sulfoxide, barbiturates, and methyl prednisolone. *Ann NY Acad Sci.* 1983;411:234–244.

160. Asmis L, Tanner FC, Sudano I, Lüscher TF, Camici GG. DMSO inhibits human platelet activation through cyclooxygenase-1 inhibition. A novel agent for drug eluting stents? *Biochem Biophys Res Commun*. January 22, 2010;391(4):1629–1633.
161. Lim R, Mullan S. Enhancement of resistance of glial cells by dimethyl sulfoxide against sonic disruption. *Ann NY Acad Sci*. January 27, 1975;243:358–361.
162. Lyman GH, Papahadjopoulos D, Preisler HD. Phospholipid membrane stabilization by dimethylsulfoxide and other inducers of Friend leukemic cell differentiation. *Biochim Biophys Acta*. October 19, 1976;448(3):460–473.
163. Pamphilon D, Mijovic A. Storage of hemopoietic stem cells. *Asian J Transfus Sci*. July 2007;1(2):71–76.
164. de la Torre JC, Fortin T, Park GA, Pappas BA. Brain blood flow restoration 'rescues' chronically damaged rat CA1 neurons. *Brain Res*. September 24, 1993;623(1):6–15.
165. Markov AK. Hemodynamics and metabolic effects of fructose 1,6-diphosphate in ischemia and shock: Experimental and clinical observations. *Ann Emerg Med*. 1986;15:1470–1477.
166. Ramos RC, de Meis L. Glucose 6-phosphate and fructose 1,6-diphosphate used as ATP-generating systems by cerebellum Ca^{2+}-transport ATPase. *J Neurochem*. 1999;72:81–86.
167. Espanol MT, Litt L, Hasegawa K. Fructose 1,6-diphosphate preserves ATP but not intracellular pH during hypoxia in respiring neonatal rat brain slices. *Anesthesiology*. 1998;88:461–472.
168. de la Torre JC, Nelson NB, Sutherland RJ, Pappas BA. Reversal of ischemic-induced chronic memory dysfunction in aging rats with a free radical scavenger–glycolytic intermediate combination. *Brain Res*. 1998;779:285–288.
169. de la Torre JC. Treatment of head injury in mice using a fructose 1,6-diphosphate-dimethyl sulfoxide combination. *Neurosurgery*. 1995;37:273–279.
170. Karaça M, Kiliç E, Yazici B, Demir S, de la Torre JC. Ischemic stroke in elderly patients treated with a free radical scavenger-glycolytic intermediate solution: A preliminary pilot trial. *Neurol Res*. 2002;24(1):73–80.
171. Tazaki Y, Sakai F, Otomo E. Treatment of acute cerebral infarction with a choline precursor in a multi-center, double blind placebo controlled study. *Stroke*. 1988;19:211–216.

Index